THE CULTURE OF PRINT

THE CULTURE OF PRINT

Power and the Uses of Print in Early Modern Europe

Alain Boureau
Roger Chartier
Marie-Elisabeth Ducreux
Christian Jouhaud
Paul Saenger
Catherine Velay-Vallantin

Edited by Roger Chartier

Translated by Lydia G. Cochrane

Princeton University Press
Princeton, New Jersey

© 1987 by Librairie Artheme Fayard
This English translation © 1989 by Polity Press

Published by Princeton University Press
41 William Street, Princeton, New Jersey 08540

Library of Congress Cataloging-in-Publication Data
Usages de l'imprimé. English
   The Culture of print: power and the uses of print in early modern
Europe / edited by Roger Chartier: translated by Lydia G.
Cochrane.
      p.   cm.
   Translation of: Les usages de l'imprimé.
   ISBN 0–691–05580–7
   1. Printing – History.   2. Books and reading – History.   3. Europe –
Civilization.   I. Chartier, Roger.   II. Title.
Z124.U83   1989
686.2'09–dc20

Printed in Great Britain

Contents

PART III   POLITICAL REPRESENTATION
AND PERSUASION   231

Acknowledgements

We wish to thank the following for kindly providing illustrations for this book:

Plate I	Biblioteca Apostolica Vaticana, Rome.
Plate II	Bibliothèque de l'Arsenal, Paris. Photo: Fayard.
Plate III	Bibliothèque Municipale, Douai. Photo: André Parisis.
Plate IV	Bibliothèque Municipale de Troyes.
Plate V	Bibliothèque Municipale de Troyes.
Plate VI	Bibliothèque Nationale, Paris. Photo: Fayard.
Plate VII	Bibliothèque Nationale, Paris. Photo: Fayard.
Plate VIII	Bibliothèque Nationale, Paris. Photo: Fayard.
Plate IX	The Newberry Library, Ms 43, f. 108r, Chicago.
Plate X	Service Photographique Archives des Hospices Civils de Lyon.
Plate XI	Service Photographique Archives des Hospices Civils de Lyon.
Plate XII	Service Photographique Archives des Hospices Civils de Lyon.
Plate XIII	Service Photographique Archives des Hospices Civils de Lyon.
Plate XIV	Service Photographique Archives des Hospices Civils de Lyon.
Plate XV	State Central Archives, Bequest of the Archdiocese, Archives centrales d'État, Fonds of Prague, H 5/2-3 4314. Photo: Fayard.
Plate XVI	State Central Archives, Bequest of the Archdiocese of Prague, H 5/2-3 4314. Photo: Fayard.
Plate XVII	Bibliothèque Nationale, Paris.
Plate XVIII	Bibliothèque Nationale, Paris.

Plate XIX   Bibliothèque Nationale, Paris.
Plate XX   Bibliothèque Nationale, Paris.
Plate XXI   Bibliothèque Nationale, Paris.
Plate XXII   Bibliothèque Nationale, Paris. Photo: Fayard.
Plate XXIII   Bibliothèque Nationale, Paris. Photo: Fayard.
Plate XXIV   Bibliothèque Nationale, Paris. Photo: Fayard.

Figures 7.1, 7.2 and 7.3 Bibliothèque Nationale, Paris. Photo: Fayard.

General Introduction: Print Culture

ROGER CHARTIER

This book proposes a dual definition of print culture. The first and the classic definition refers to the profound transformations that the discovery and then the extended use of the new technique for the reproduction of texts brought to all domains of life, public and private, spiritual and material. This new technique quite evidently encouraged the circulation of the written word on an unheard-of scale, not only because book-making costs were lower when they were spread over the thousand or fifteen hundred copies of one press-run rather than being borne by one copy alone but also because production time, which could be long indeed when books were copied by hand, was shortened. Printing also distributed large quantities of new objects that were easily manipulated, carried on one's person, or posted, and that gave images and texts a more substantial presence and a more familiar reality.

This meant, at least in cities and towns (or first in cities and towns), that where such new means of communication appeared in massive number they modified practices of devotion, of entertainment, of information, and of knowledge, and they redefined men's and women's relations with the sacred, with power, and with their community. After Gutenberg, all culture in western societies can be held to be a culture of the printed word, since what movable type and the printing press produced was not reserved (as in China and Korea) for the administrative use of the ruler but penetrated the entire web of social relations, bore thoughts and brought pleasures and lodged in people's deepest self as well as claiming its place in the public scene.

Print culture can also be understood in a narrower sense, however, as the set of new acts arising out of the production of writing and pictures in a new form. Personalized reading in private by no means exhausted the possible uses of print objects. Their festive, ritual, cultic, civic, and pedagogic uses were by definition collective and postulated decipherment in common, those who knew how to read leading those who did not. Such uses invested the handling of the chapbook, the tract, the broadsheet, or the image with values and intentions that had little to do with those of solitary book reading. With printing, the range of the uses of writing broadened and, as a corollary, an interconnected network of specific practices that defined an original culture came to be formed. All too long this culture has been reduced to reading alone, and to a form of reading that is common today or was practiced by the scholars in medieval and early modern culture. Our joint project has another perspective; it emphasizes the many uses and the plural appropriations of widely distributed works printed from the age of Gutenberg to the nineteenth century, the age of the second revolution of the book.

Recognizing the particularity of some of the products that came out of the invention of type is not to imply breaking the strong continuity between the age of the manuscript and the print era. There was continuity, first, in the physical form the object took. It was during the last centuries of the hand-copied book that a lasting hierarchy was established between the great folio volume, the 'shelf-book' of the universities and of serious study, which had to be propped up to be read, the humanist book, more manageable in its mid-sized format, which served for classical texts and new works of literature, and the portable book, the pocket book or the bedside book of many uses, religious and secular, and of a wider and less selective readership.[1] The printed book was the direct heir of this strict separation of formats, genres and uses, as Lord Chesterfield recalled in the eighteenth century:

Solid *folios* are the people of business, with whom I converse in the morning. *Quartos* are the easier mixed company, with whom I sit after dinner; and I pass my evenings in the light, and often frivolous, *chit-chat* of the smaller *octavos* and *duodecimos*.[2]

The printed book also perpetuated, in the same forms but with greatly increased circulation, the success of genres established by the manuscript *libellus*, first among which were books of hours and lives of the saints, studied here.

A second and powerful element rooted print culture in the long time-span. Well before the Gutenberg invention, a new manner of reading – silently, using the eyes alone – had broken with the oral reading that had long been universally obligatory (or nearly so). In the age of the manuscript this new skill gradually won over monastic copyists, then scholastic and university circles, and finally the lay aristocracies in an advance of decisive importance.[3] With silent reading a new relation with writing was instituted, more private, freer, and totally internalized. The same text could, at that point, be used in different ways. It could be read silently for oneself in the privacy of the study or library or read aloud for others. Or the same book – a book of hours, for example – might contain some parts designed for ritual use, hence for declamation in common, and others designed for personal devotions based on reading with the mouth closed.

The revolution in reading thus preceded the revolution in the book, even though oralized, murmured, 'ruminated' reading long remained the rule among humbler readers. For this reason, even though the present work is devoted to the new possibilities and initiatives arising from a technical innovation of revolutionary consequences – the written word set in movable type and printed by a press – it reserves a generous place for the era of the manuscript, which witnessed the rise of a way of reading that later became widespread, even obligatory, and created a functional and social hierarchy among written objects that was visible in their very form.

The access to print culture we propose is not through a synthesizing, global approach but, quite to the contrary, by means of case studies – more accurately, object studies. We have been guided by three preferences. The first has favoured printed matter other than books or even tracts. Such was the *placard*, a print piece in the form of a broadsheet designed to be read, handled, or posted. Long neglected by a history interested only in more noble objects and conserved by chance thanks to one individual's passion for collecting (as with Pierre de L'Estoile) or an administration's filing practices, such printed sheets, bearing text and image, were the most elementary products of the printing presses. They were also perhaps the most widely distributed, or at least the most visible, since they were pasted up, either on walls in the public streets or in the home. Political broadsides and religious images are the major types of the *placard*, so it is natural that we examine royal propaganda posters in the age of Henry IV and Louis XIII and the *chartes* used for both ritualistic and individual purposes in marriage ceremonies in certain dioceses in France (Lyons in particular).

The second preference we have shared is for particularity over pre-conceived generalization in the consideration of textual and typographical genres. Thus we have opted to study the hagiographic pamphlets concerning *one* particular saint, for example, *one occasionnel* known in one copy alone, *one* body of materials within the lists of the *Bibliothèque bleue* (editions of the tales of Perrault, in this case) or the printed matter concerning *one* event (here, the taking of La Rochelle and the royal entry celebrating the victory of 1628).

Thirdly, we have shared a desire to understand the use of the materials we are investigating within the precise, local, specific context that alone gave them meaning. This context might be ritual (in the case of the books of hours or the marriage *chartes*), political (as with the broadsheets produced following the assassination of Concini or the various texts printed after the surrender of La Rochelle to the King), or at once religious and national (as in the study of a radically different situation in Czech lands in the seventeenth and eighteenth centuries, where a risky but tenacious use of forbidden books preserved group identity in a culture cut off from its roots).

There are several arguments for favouring case studies and object studies.

The first lies in a mistrust of generalizations, which often mask the complexity of materials or practices, when they do not simply miss their mark. Rather than accept the oversimplified contrasts and murky concepts that claim to distinguish 'popular' printed matter from what was not 'popular', for example, we preferred to recall the essential importance of plural uses and interpretations of texts used in common (such as fairy tales) and shared objects (such as books of hours, saints' lives and the current events publications known as *occasionnels*). This book attempts to avoid the too convenient category of the 'popular' – or at least not to rely on it as an a priori criterion for classifying printed matter. To a greater extent than has been thought, widely distributed texts and books crossed social boundaries and drew readers from very different social and economic levels. Hence the need for the precaution of not predetermining their sociological level by dubbing them 'popular' from the outset.

Similarly, the classification of the various sorts of printed materials under previously constituted, fixed, and mutually exclusive rubrics (useful for a first sorting out), must not make us forget that what is even more important is to be aware when a particular form is invested with values usually foreign to it or interests generally expressed elsewhere. This is the case of the hagiographic text or the news item related in the *occasionnel* when they are 'politicized' and used for polemical or propaganda purposes. This is also the case in the 'folklorization' of political pieces when they use motifs from Carnival culture to bolster their persuasiveness. One might also say that the differences between fictional tales in print and printed relations of real events are tenuous indeed, since invented narrative attempts to gain credibility by reiterating the 'proofs' of its authenticity and the event in print exists only as it has been put into the text and picture that represent it to the reader or the viewer.

To look at the question from another viewpoint, the choice of a restricted corpus of materials is a first condition for a study of textual and editorial changes in a given work, hence for an inventory of variants that introduce new meanings and new uses into a work by modifications of text or layout. The long-term analysis of the various publishing formulas that used the tales of Perrault to reach different publics – changing their title or the order and contents of the collection, on occasion transforming the original text, more often changing the vignettes that illustrated it – is a good example of the work that needs to be done on the variations, major or minor, influencing the audience and the status of works that never did have a single fixed meaning expressed in a supposedly stable text. What we need to grasp, then, is the process of the construction of meaning by which readers diversely appropriated the object of their reading. The essays assembled here attempt to move in this direction by combining, for each sort of material considered, a description of formal elements (print format, page layout, the nature and placement of pictorial material) with an identification of the uses, implicit or explicit, that relied on those formal elements. It seems evident that an approach of this sort is possible only on the reduced scale of a limited body of texts, editions, or publications.

The choice of the single object permits 'relocating' objects too hastily considered the common legacy of an immobile, generalized, and supposedly popular culture. Indeed, many widely distributed printed pieces were used, at least at first, in the service of a party or a power, of a religious order or a particular shrine, of a community or an institution. This means that underlying the letter of the tale, be it hagiographic or extraordinary, and behind the evident purpose of glorification of the monarch, lies a hidden or explicit polemical intent to justify, persuade, and rally support. Re-establishing the role of printed matter in conflicts great or small, which affected the fate of the state or settled some very local quarrel, requires reconstruction of the immediate context in which a piece designed to make people act or believe was produced and was first circulated. A broadsheet, the relation of a princely entry – but also a saint's life or an *occasionnel* – is not neutral raw material comprehensible outside the event or the purpose that lay behind its publication and circulation. It is always wound in with a thick skein of tensions and conflicts whose particularity needs to be ascertained.

Working on the level of the particular, however, does not mean that one must forgo making some general judgements, to be revised or ratified by further case studies. The first of these regards print culture as a culture of the image. Pictures are present in all the materials studied here, on the title page or in the body of the text, as a frontispiece or an inset plate, placed at the head of a broadsheet or at the centre of a marriage *charte*, throwing light on the text or being explained by it. A printed image has characteristics that distinguish it from all others. It is to be viewed close to, not at a distance; it is manipulable, easily cut out, pasted up, or carried on the person; it is always connected with the written word, sharing space on the printed page or located in the same book or tract. The essays that follow attempt to describe a number of the functions taken on by print images, and some of the relations that linked them to the written word.

The image was often a proposal or a protocol for reading, suggesting to the reader a correct comprehension and a proper meaning for a text. It could fulfil this function even if it was a re-used plate and not cut for the text it accompanied (as was usually the case in the *occasionnels*, the *canards*, and the volumes of the *Bibliothèque bleue*). It could be used as a place in a mnemonic system, crystallizing in one representation a story, a propaganda message, or a lesson; or it could serve as a moral, symbolic, and analogical figure that gave the overall sense of the text, which the reader might fail to grasp in an intermittent and distracted reading.

Within a book or not, the printed image could also be used autonomously, giving it a function of its own and transmuting it into a ritual object (as with the *chartes de mariage*) or a devotional image (as with confraternity posters, pilgrimage certificates, or the woodcuts or engravings in books of piety). It could also be used in sign of recognition (as with the flysheets of all sorts torn in two, one half to be left with an abandoned child by parents who kept the other half in the hope of an eventual reunion).[4] Linked to the essential acts of life and to important decisions and engagements, the image was invested with an affective charge and an existential value that made these objects, printed in vast numbers, unique to their possessors.

The printed image, like other images, was thought of and used as an important aid to knowledge and as if it were capable of adequate representation of reality. It was considered to engage the unfailing adherence of the beholder and, even more than or better than the text that it accompanied, to induce persuasion or belief. Thus the present book gives ample space to representation by emblem, the image that most perfectly expressed this theory of intellection by the imagination and furnished a model to all pictorial production in the early modern age.

Treated as heraldic devices or demoted to allegories, collected in thick volumes or turned to practical use on broadsides or the programmes for princely entries, employed for the comprehension of the order of the natural world or in the service of the glory of the monarch, emblems belonged fully (though not exclusively) to print culture. They furnished print culture with both a repertory of motifs and figures and a theory of the image that refused to reduce the image to illustration but granted it an efficacy of its own. Of course, not all the pictures printed in mass-distributed books, broadsides and the like obeyed the subtle rhetoric of emblems. Some were clearer and more directly narrative. Nevertheless, for a long time the rhetoric of the emblem provided the means of expression for political and religious representations that used visual persuasion to win over people's minds.

The printed image was subject to two sorts of constraint as it related to the text. First, there were technical limitations. For example, there was the choice between the woodcut, which could be inserted in the same print form as the type characters and be printed by the same press, and the copperplate engraving, which was considered superior as early as the end of the sixteenth century but which required a special printing press, cost more, and was the monopoly of the engravers. When the printer-bookseller used woodcuts, he kept total contol of the production process and could pick his own pictures and text; copperplates obliged him to treat text and illustration as distinct entities and to print them separately, which led to placing pictures outside the text as a frontispiece, on the title page, or on inset plates. Broadsheets and flysheets solved the problem the other way around by engraving written text onto the copper plates along with the picture.

Technical requirements led to a second sort of constraint. It was clear that the significance and the role of printed images were not the same when they were part and parcel of the text, appearing in a topographical proximity that made it easy to pass from picture to writing, and when they were dissociated from the text and the reader had to establish an abstract relation between image and text. In the first instance, the contiguity that the woodcut helped establish between the two media guaranteed control of meaning and assured the unskilled reader an entry into written matter. In the second, the weaker connection between text and image introduced by the copperplate engraving was less constraining to comprehension and permitted a broader range of readings, which could vary according to the skill of the readers. Printed matter is never merely a text, and the authors of these essays have agreed to refuse to separate more learned images from those supposedly good enough for the people, to consider the different functions invested in the image (in printed objects or not), and to distinguish between the various mechanisms that gave order to images, localized them and linked them to the written word.

Not only was early print culture a culture of the image; it was also closely linked to speech. There were, of course, differences between recitation and reading (even reading aloud), or between the spoken tale and the written text. Cervantes describes these differences well in *Don Quixote*, part 1, chapter 20.[5] To pass the time the night before a battle, Sancho Panza offers to tell stories to his master. The way he tells his tale, interrupting the narration by commentaries and digressions, repeating himself and pursuing related thoughts – all of which serve to place the narrator in the thick of his tale and to tie it to the situation at hand[6] – throws his listener into a fit of impatience. 'If that is the way you tell your tale, Sancho,' Don Quixote says, interrupting him, 'repeating everything you are going to say twice, you will not finish it in two days. Go straight on with it, and tell it like a reasonable man, or else say nothing.' A bookish man *par excellence* and to mad excess, Don Quixote is irritated by a tale that lacks the form of his usual readings, and what he really demands is that Sancho Panza's story obey the rules of written style: clear expression, linear development and objectivity. There is an insurmountable distance between the reader's and the listener's expectations and the spoken practice that Sancho Panza is familiar with. Sancho replies, 'Tales are always told in my part of the country in the very way I am telling this, and I cannot tell it in any other, nor is it right of your worship to ask me to adopt new customs.' Resigned but disgruntled, Don Quixote agrees to listen to a text so different from the ones presented in his precious books. 'Tell it as you will,' he exclaims, 'and since fate ordains that I cannot help listening, go on with your tale.'

The gap between the printed written word and the spoken word should lead to a clear differentiation between an oral culture existing without books, as was long true of the culture of the night-time work and socializing

sessions among peasants known as the *veillée*, and a culture in which people
handled books and reading aloud could communicate their full contents
even to those who could not read unaided. Once this difference is
acknowledged (and this is another way of recognizing the specific effects of
the circulation of printed matter), we need to see how print and forms of oral
culture were connected in the early modern age.

One such connection lies in embedding the formulas from oral culture
into the texts destined for a broad public. The oral tradition surfaces in the
imitations of speech patterns in the *occasionnels* (or in some of them, at least),
in reiterated assurances of the truth of the story and the teller repeatedly
giving his word for its veracity, and in variants borrowed from folk tradition
and introduced into more literary tales.

This relation could be reversed, however, and the book or its substitutes
could be an established part of oral social customs. In the public space of the
street, in the festive association or the devotional confraternity, in the bosom
of the family, among friends in the village or the town, the reader's spoken
words could assure the circulation of the written word among the unlettered.
As some scholars have pointed out, the deepest-rooted and most stable
forms of peasant culture (such as the *veillée*) long resisted the penetration of
print culture. Still, as we shall see below, readers in Czech lands attest that
when the written word was bolstered by hymns and songs, familiar texts, or
community reading, it could be practised, memorized, and incorporated into
people's lives.

The aim of the studies assembled in this volume – to reconstruct the
multiple uses of the many forms of print – is not a simple task, as a basic
tension runs through it. To move out from the objects that have been
conserved is in reality to take cognizance of the many constraints imposed
upon the reader and on his or her reading, be they explicit protocols that tell
how to read the book or textual and formal mechanisms that aim at obliging
the reader to an unconscious acceptance of the meaning desired by the
author or the publisher. On the other hand, the minute we go beyond
classification, counts, and descriptions in an investigation of the uses and the
appropriations of typographical materials, we necessarily hold reading to be
an inventive and creative practice that seizes commonly shared objects in
differing ways and endows them with meanings that cannot be reduced to
the authors' and the publishers' intentions alone.[7]

Taking full advantage of this tension between the liberty of the reader
(even a reader of limited skills and unsure inclinations) and the efficacy of
the object presupposes two sorts of investigation. First, we need to describe
the historical modes of reading practices, differentiating them according to
time and place, habitus and circumstance, and the reader's social status,
religious affiliation, sex, and age, and noting their presence as they are
represented in individual testimony (spontaneous or elicited), in fiction, in

painted portraits[8] and in guides to proper reading. Second, we need to take specific pieces of printed matter and consider them as objects – in all their singularity – in order to reconstruct the limits that typographical procedures imposed on free appropriation of texts, directing the way they were read, and to describe transformations in meaning and use by analysing the differences in successive printed versions of a given text.

Evaluations of this sort are always risky, given that the book, which was made to be read, was not in fact always read, and that thaumaturgical, divinatory or propiatiatory utilizations of both manuscript and printed texts are well attested. It is also risky because one cannot deduce behaviour automatically from a description of constraints. Nevertheless, when the study of representations of reading practices is crossed with the study of printed pieces as material objects (the route followed here), we can perhaps reach a new and better comprehension of a major phenomenon in western cultural history, the distribution on a large scale, and for a host of uses, of the written word made possible by the printing press.

Notes

1   Armando Petrucci. 'Alle origini del libro moderno: libri di banco, libri da bisaccia, libretti da mano,' *Italia medioevale e umanistica*, 12(1969), pp. 295-313 (reprinted in Petrucci (ed.), *Libri, scrittura e pubblico nel Rinascimento. Guida storica e critica*, (Laterza, Rome and Bari, 1979), pp. 137-56); Petrucci, 'Il libro manoscritto', *Letteratura italiana* (6 vols, Einaudi, Turin, 1983), vol. 2, *Produzione e consumo*, pp. 499-524.

2   Cited by Roger E. Stoddard, 'Morphology and the book from an American perspective', *Printing History*, 17 (1987), pp. 2-14. [Quoted from *Lord Chesterfield's Worldly Wisdom*, ed. George Birkbeck Hill, (Clarendon Press, Oxford, 1891), p. 11].

3   Paul Saenger, 'Silent reading: Its impact on late medieval script and society', *Viator. Medieval and Renaissance Studies*, 13(1982), pp. 367-414; Saenger, 'Manières de lire médiévales', in Henri-Jean Martin and Roger Chartier (gen. eds) *Histoire de l'édition française* (3 vols, Promodis, Paris, 1982-4), vol. 1, *Le Livre conquérant. Du Moyen Age au milieu du XVIIe siècle*, pp. 131-41.

4   See Jean-Paul Maillet's forthcoming study on confraternity images in Paris in the seventeenth and the eighteenth centuries and, on the images torn in two, Giovanna Cappelleto, 'Infanzia abbandonata e ruoli di mediazione sociale nella Verona del Settecento', *Quaderni Storici*, 53(1983), pp. 421-41.

5   Miguel de Cervantes Saavedra, *L'Ingénieux Hidalgo don Quichotte de la Manche*, tr. Louis Viardot (2 vols, Garnier-Flammarion, Paris, 1969), vol. 1, pp. 184-5. [Quoted from *Don Quixote*, tr. Ormsby, rev. and ed. Joseph R. Jones and Kenneth Douglas (W. W. Norton, NY/London, 1981).]

6   This is the style of telling stories noted by ethnologists in their collections from oral tradition. See, for example, Linda Dégh, *Folktales and Society: Story-Telling in a Hungarian Peasant Community*, tr. Emily M. Schossberger (Indiana University Press, Bloomington, 1969).

7   Michel de Certeau, 'Lire: un braconnage', in his *L'Invention du quotidien*, (2 vols,
    Union Générale d'Editions, 10/18, Paris, 1980), vol. 1, *Arts de faire*, pp. 279–96
    ['Reading as Poaching', in *The Practice of Everyday Life*, tr. Steven F. Rendall
    (University of California Press, Berkeley, 1984), pp. 165–76].
8   See Roger Chartier, *The Cultural Uses of Print in Early Modern France*, tr. Lydia G.
    Cochrane (Princeton University Press, Princeton NJ, 1987); Chartier, 'Les
    Pratiques de l'écrit', in Philippe Ariès and Georges Duby (eds), *Histoire de la vie
    privée* (Le Seuil, Paris, 1986), vol. 3, *De la Renaissance aux Lumières*, pp. 113–61.

PART I

Print to Capture the Imagination

INTRODUCTION

From the Middle Ages to the mid-nineteenth century, many books and tracts, first manuscript then in print, fed the imaginations of a broad range of readers, offering marvellous events, lifelike fictions and stories for their belief or delectation. Out of this mass of texts and forms we have chosen three genres – the saint's life, the *occasionnel* and the tale – because they lasted through the centuries, they accounted for a large part of print production, and they provide obvious points of comparison. Following the principles stated in the foreword, each genre is approached through the study of specific and localized cases exemplary by their very singularity.

Reading has been guided by an attempt to discern, behind the apparent objectivity of narratives that play upon a well-established repertory of plots and motifs, the bitter conflicts, polemical motivations and political designs that led to their writing and publication. Saints' lives, for example, were manipulated to justify the power of one group or the claims of another; they were printed to promote a cult, a sanctuary, or a particular pilgrimage; they were used to reinforce the power of a religious congregation, a political party, or a family. In sixteenth-century Abruzzo, Louis of Anjou was the saint who served to rally Franciscans of the Observance, the barons of the *contado* and the new men of the city of L'Aquila in their struggle against that city's municipal government, its 'notables' and its devotional practices. In seventeenth-century Burgundy, a battle raged around the life and the relics of St Reine of Alise, pitting clergy against laity, the secular against the regular clergy, Franciscan Cordeliers against Benedictines. In both instances, printed matter accompanied, nourished and illustrated these clashes.

In like fashion, the *occasionnels*, although they seem to offer often-repeated stories drawn from ancient stores of folklore or hagiography (such as the story of the hanged woman miraculously saved studied here), could also be suffused with contemporary conflicts – in this case, those involving the League, whose interests were served by the miraculous tale. Politics lurks in unexpected places, and genres that seem totally detached from the political scene are in reality profoundly marked by political aims and political upheavals.

Saints' lives and fairy tales have in common a lasting and important share in the body of texts published in the format of the *Bibliothèque bleue* by printers in Troyes and their competitors. In 1789, for example, Etienne Garnier's inventory of his stock shows that saints' lives accounted for 8.3 per cent of the books, bound and in sheets, in his Troyes warehouse, and fairy tales accounted for another 6.5 per cent. The lasting success of these 'popular' publishers was perhaps due to their skill in making their presses available to the religious authorities (and to their quarrels) and in refashioning texts taken from literary culture for the use of a broader public, as with the fairy tales and other stories printed in the late seventeenth century. It also rested on the flexibility of literate genres but rooted in a culture of action and speech and in narratives that gave fixed form to the motifs of a collective imagination long accustomed by the saints to uncommon lives and by the heroes to prodigious adventures. All the various texts that related edifying or terrifying stories to the humbler sort of readers aimed at belief of the incredible and the extraordinary.

1

Franciscan Piety and Voracity: Uses and Strategems in the Hagiographic Pamphlet

ALAIN BOUREAU

A noble lady of the castle of Galeta suffered from a cyst between her breasts and had found no remedy against the foul odour and the pain that overwhelmed her. One day she entered a church of the Friars to pray. She saw there a tract that contained the life and the miracles of St Francis. She leafed through it with lively interest to know what it contained, and when she had been instructed in the truth, bathed in her own tears, she took the book and placed it on the affected place and said, 'By the truth that is inscribed on this page, O St Francis, deliver me now from this wound by your holy merits.' And for a moment she gave herself, in tears, to intense prayer; then, after having removed her bandages, she found that she was totally cured, to the point that afterwards no trace of a scar could be found.[1]

The Thaumaturgic Book

A hagiographic pamphlet, the *Life and Miracles of St Francis*, effected the cure. The book of miracles had no end. When the booklet was read and handled, it produced once more the supernatural efficacy that was its contents. One clear use of the book here is as an object to transmit the thaumaturgic powers described in its text. This propagation of the sacred, moving from the contents to the container, is not surprising in a religion in a phase of the formation of ritual by cumulative annexation. Still, nothing in this anecdote is simple, neither the nature of the book, nor its status, nor the practices that flourished around it.

The scene is reported in a *Treatise on the Miracles of St Francis* written around 1252 (some twenty-five years after the saint's death) by Thomas of Celano. Thomas, a close companion of St Francis, composed the first biography of the *poverello* of Assisi in two successive versions (*Vita Prima*, *Vita Secunda*) that contributed much to the diffusion of the saint's cult until they were replaced as the official biography when St Bonaventure wrote his life of Francis.² The anecdote in the pamphlet thus belongs to Franciscan literature in an epoch of full harmony between the Friars Minor and the papacy, whose faithful servants they considered themselves. Thus it is far from illustrating autonomous 'popular religiosity'.

What exactly was this pamphlet? And what was it doing in this Friars' church? It may have been a very short text for liturgical use, since the lessons for the canonical hours or a choral 'legend' (saint's life) could dwell briefly on the life and miracles of a saint. If that were the case, it would have been a small work of few pages, and its presence in the church could easily be explained. The verb *perquaesivit* (rendered here as 'leafed through') should then be understood in its classical sense of 'read in its entirety'. The noble lady could easily have absorbed all of a brief antiphonal reading and have been 'instructed in the truth' (*de veritate instructa*). The question of the nature of the book implies a way of reading. If, on the other hand, we take the book to have been a full-length saint's life (Thomas of Celano's, for example), *perquaesivit* should indeed be understood in the sense of 'leaf through' and implies no need to read it or even to know how to read. 'Reading' in this case becomes a magical practice and the noble lady was more 'touched' or 'struck' by the truth than 'instructed' by it. The term used to designate the book – *libellus* – indicated at the time a genre rather than a specific object. In the Middle Ages, the *libellus*, whatever its size, told the life of a saint for devotional purposes but outside liturgical use. A painting of this miracle in the church of St Francis in Pisa seems to confirm this hypothesis, since the book depicted seems substantial. Size proves little, however, given the chronological gap (the painting cannot possibly predate 1260), and, even more, since iconographic codes dictated the depiction of any book as an object of a certain volume.

The painting does show the *libellus* placed on the altar, however, like a Bible or a lectionary. Of course, the troublesome ambiguity of all representations of action (was the book taken up from the altar or placed on it?) makes it impossible to know whether it had sacred status before the noble lady arrived or was promoted to that high place by the miracle. Be that as it may, the anecdote illustrates the sanctification of the hagiographic book after the Biblical model. Sacredness arose out of the text itself. Our translation is powerless to judge the balance between the truth and the efficacy of the text (*sicut vera... quae sunt conscripta... ita liberer*: just as what is written of you is true ... so let me be liberated). The term *pagina* ('what is written on this page') referred not to the physical pamphlet but to the *sacra pagina* that (in the singular) designated the Bible or true doctrine. The sacred status of this booklet, on a par with Holy Writ, explains the thaumaturgic effects of the hagiographic text. Holy Writ had been used for its curative powers from apostolic times: Barnabus, according to the apocryphal Acts that bear his name,³ treated the sick by the laying on of the

Gospels. In a neighbouring domain, the use of a Bible for prophecy and prediction is well known. Opened at random, the Bible predicted the future in the first verse that caught the reader's eye.[4] This practice, far from making a late appearance in the bibliolatry of the Reformation, seems to have been systematic and official from earliest times. A good example of it can be found in Guibert of Nogent in the beginning of the twelfth century, who reports that each of the bishops of Laon practiced *sortes* upon his accession to the episcopal throne.[5]

The thaumaturgic use of saints' lives noted by Thomas of Celano in the mid-thirteenth century was based on three new phenomena: the widespread diffusion of hagiographic texts (on which more later), a new respect for the book, and the near adoration of St Francis, in which written devotions played a large part.

Franciscans and the Cult of the Written Word

Veneration for the book can be understood, first, as the end point of the twelfth-century renewal in theology and philosophy that affirmed (through Aristotle) that truth coincided with being and that established, as Jacques Verger says, 'the quasi-synonymy, frequent at that time, of *pagina* and *doctrina*, *liber* and *scientia*'.[6] The tendency was greatly reinforced by the expansion of the mendicant orders in the early thirteenth century, and it is not by chance that we find extraordinarily high praise of the book, read and *handled*, in a sermon written by a Franciscan, John de la Rochelle (fl. 1225–50). The sermon clarifies the meaning and scope of the word *libellus*. It celebrates the memory of the Franciscan 'evangelical doctor', St Anthony of Padua (d. 1231), guarantor and authority for the 'scholarly' tendencies within the order of the Friars Minor. John applies to Anthony a verse of Revelations (10:8): 'Go; and take the book that is open, from the hand of the Angel who standeth upon the sea, and upon the earth.' John de la Rochelle says,

In this passage, Anthony receives many sorts of praise . . .
Seventh, for his privileged knowledge and practice of Holy Writ. He kept a booklet open in his hand. It is said: an open booklet, a booklet in the hand. The booklet is open because it is understood; it is in the hand, not in a coffer or a purse or on a table, but in the hand, which is to say, put into practice. It is a booklet, not a book, because in this detail his privileged knowledge is emphasized, for he used his memory like a book.[7]

The *libellus* was the site of a continuing incarnation and a special mediation between the hand and the memory, between God and Man. The verse from Revelations pointedly described an angel before it presented the evangelical doctor. John de la Rochelle preferred the 'booklet' to the book because it

manifested being and not having, use (the open booklet) and not possession
(the coffer, the purse, the table). It quite literally functioned as a
memorandum, an external aid to memory. It applies not to a library or to a
thick volume but to the knowing mind. It signalled, it represented,
somewhat as did liturgy. It originated a cultural tradition of the bedside book
– a domestic and religious work that was both intimate and universal, small
and exhaustive, a work to return to again and again, always held, always
open, a hand book and a 'soul book'. The immensely successful career of the
manuscript, and later the printed, hagiographic pamphlet must be
understood as starting from this praise of the hand-held book that, at the
heart of the Middle Ages, removed the book from the learned world and
oriented it towards new uses. After all, in the early thirteenth century the
Franciscans had invented the breviary for universal use, and a general
minister of the order, Haymo of Faversham, gave it nearly definitive form
towards the middle of that century.

Until the early modern period, Franciscans worked to give an even more
concrete and immediate content to this cult of the written word, as Thomas
of Celano's anecdote attests. This cult seems to have been encouraged by
Francis himself, and the oldest Franciscan hagiography, repeated in Jacobus
de Voragine's *Golden Legend*,[8] speaks of a friar who kept a note written in St
Francis's hand on his person to protect himself from temptation. This
tradition was long-lived, as can be seen in François D. Boespflug's study of
Crescentius of Kaufbeuren, an eighteenth- century Franciscan mystic.[9] The
theologian Eusebius Amort speaks of 'Franciscan amulets' in the context of
an affair centring on illicit trade in pious images. According to Boespflug,

The Amort/Bassi report also speaks of *cédules de Luc*, adding that they owed their
name to a Franciscan named Brother Lucas. This object might be compared to the
various 'formulas to be eaten' – that is, paper notes bearing a handwritten or printed
phrase taken from Holy Writ and chosen for their efficacy in conjuration. The paper
was eaten as a precaution against serious danger, but in 'calm times' it could simply be
worn around one's neck.[10]

The dual nature, theological and magical, of the hagiographic book made
it a sacred object that one could manipulate. Like a cult object, it could be
possessed in common and be endowed with sacred power, but like
devotional materials, it was an individual continuation of cultic activities and
the mark of a religious practice. It took its place among medals, pious
images, and pilgrimage tokens. It signalled, recalled, evoked a vow or a past
or ongoing practice. When it was read, leafed through, or put on display it
became a spiritual guide, along with breviaries, missals, and books of hours.

It should be noted that there were two types of *libellus* among the hand-
sized works praised by John de la Rochelle. First, there was the compact
libellus (the breviary and other forms derived from it), which was conceived

to abridge, concentrate, and miniaturize all of liturgy and sacred narrative. Regular use and the techniques of reproduction and binding employed place this sort of work among those books that bore the sheen of constant use. Second, there were the hagiographic pamphlets, which offered the faithful only a small fragment of ritual life (one feast day, one patron saint, one pilgrimage), and belong to the category of books hewn from larger works, fragments of chronicles, readings from saints' lives and lectionaries. The first were books to be used; the second were books to be shelved beside the thick, stubby missal or placed in one's small book cabinet along with a few images and broadsheets.

As in many other domains, the actual practice of reading, venerating, or displaying works such as these escapes us. We would need ten, a hundred, or a thousand Thomas of Celanos, and of varying sorts. We must rely, then, on a closer analysis of the byways of book production.

The Saint's Life: A Definition

The enormous and polymorphous mass of documents is a daunting prospect, however. To risk a tautological definition, the hagiographic *libellus* was a brief work written to celebrate the memory of a Christian saint. The tautology seems lexical only: the diminutive in *livret* or *libellus* designated a contents and a likely use as much as it did a format. The criterion of brevity will be retained, however, in the constitution of our corpus, on the supposition that any longer saint's life implied conditions of use and costs that excluded its 'manual' use and that made learned reading practices obligatory. Celebratory purpose only vaguely circumscribes this body of materials, since such texts stood at varying distances from worship. They could be part of liturgy if the pamphlet contained or elaborated on the readings of the office or included prayers or thanksgiving. They could also fall within individual devotions and meditations, or within community festivity if we include in the genre the innumerable 'dramatic representations' (published in formats closely resembling those of the devotional pamphlets) offering the text of hagiographic dramas performed by confraternal organizations on the occasion of a saint's day.[11] The final element of the definition – sainthood – is an important one. In Christianity, the institution of sainthood acted as a juncture between the various strata of religiosity and culture. Its ambivalence could be directed towards a sort of implicit polytheism contained in and constricted by theologic monotheism. The 'natural' and unmediated status of sainthood compensated for the unbending theoretical violence of trinitarian dogma, so rarely understood by the faithful.

The chronological limits of the potential corpus seem difficult to establish, since hagiographic narrative continued for centuries with little essential variation. If we examine one example among hundreds, a brief anonymous work sold in our own time in the church of Chasteloy-Hérisson in the Bourbonnais (the *Histoire de saint Principin, martyr de Chasteloy*, Montluçon, reprinted in 1961), it is clear that its plan follows the canonical model and differs little from that of its thirteenth-century counterpart. It contains sections on the 'History of St Principin' (St Principius), 'Authorities in support of the history and the cult of St Principin', 'Reports concerning several Cures', 'Hymn', 'Consecration' and 'Prayer'.

But is it feasible to scan the immense period of the development of the hagiographic pamphlet from the third century to our own times? A history of the narrative and doctrinal contents of these works would present few surprises and undoubtedly show slow but continuous shifts in religious sentiment and modes of belief. It is more important to our concerns, however, to focus on the moments of change in the circulation of such pamphlets.

The Three Ages of Hagiography

We can distinguish three periods, three ages of hagiography. In its prehistory (up to the eleventh century), the *libellus* became a form of devotional literature. Its prototypes were many, from the *Passion of Perpetua and Felicitas* (*Passio Sanctarum Perpetuae et Felicitatis*, early third century), the first western martyrdom account, to the *Life of St Martin of Tours* (*Vita Martini Turonensis*, late fourth century) by Sulpicius Severus,[12] often imitated during the first millennium. These were learned works, however, of limited circulation, still far from liturgy and ritual and in competition with the anthologies inspired by the lives of the Desert Fathers, such as the *Dialogues* of Gregory the Great. These works were not produced in any great number, nor did the specific content become fixed into a brief life followed by a list of posthumous miracles until the eighth and ninth centuries, when the Carolingian reforms stabilized the liturgy. Although the Bible was used increasingly in liturgy, revitalized hagiographic narrative was also granted a place. Thus saints' lives were compiled for liturgical use and their biographies were cut up into lessons, brief chapters to be read at matins and occasionally during the Mass. This expanded liturgical use was further increased by a strong demand from monastic communities and parishes, spurred on by lively competition for saintly patrons and a growing cult of relics.[13] Hagiographic texts offered valuable help for proving that an abbey or a monastery possessed authentic relics and for bolstering its prestige. Until the twelfth century, the high point of monastic culture, saints' lives, single or in collections, increased dramatically, but they circulated largely within the learned ecclesiastical milieu, which drew fragments from them for liturgical purposes. Francis Wormald has noted in his study of the illustrated manuscripts of this period that up to the twelfth century hagiographic booklets and anthologies were kept in the monastery's treasury, not in the library.[14] The *libellus* thus appears to have been a source of authority and an act of authentification, but not a work of edification.

Broader diffusion of the *libellus* did not occur until the twelfth and thirteenth centuries, a moment of change that opened the second period in the life of the genre. It then moved outside the monasteries and churches, thanks to the efforts of the mendicant orders, the systematic constitution of hagiographic collections,[15] and more rapid techniques for the reproduction of manuscripts.[16] The new veneration of the book and the tract, which John de la Rochelle's sermon, discussed above, exemplifies, enters into the picture as well. The twelfth century also saw the development of vernacular saints' lives, which on occasion moved them away from the world of the Church towards lay romance. Expression in the vernacular, however, may also have responded to a need for popularization among monks and lay brothers, as Jean-Pierre Perrot has noted in connection with saints' lives in French from northern France.[17] Thomas of Celano's anecdote gives a good picture of the *libellus* brought out from among the objects and charters in the treasury during the course of the thirteenth century.

The third age of the hagiographic pamphlet begins with printing. Of course, affirmations concerning the revolutionary role of typography in the domain of the religious book should be made with caution. Studies following the lead of Father Destrez have shown that manuscripts were produced rapidly and in ample number in the thirteenth century.[18] Nevertheless, mechanical reproduction assured the distribution of the pamphlet everywhere. The universal imposition of the Roman breviary and the Roman calendar after the Council of Trent had the paradoxical effect of favouring the printing of local hagiographic pamphlets, to the detriment of diocesan, regional or national forms of devotion working against the fragmentation of worship. The universal and the liturgical (in this case, the Roman breviary) were superimposed on the local and the narrative (the *libellus*).

Clear proof of this can be found in 1639 in an ordinance of the Bishop of Clermont, which shows how romanization converged with the facilities provided by typography to favour the *libellus*. This ordinance figures in a brief work giving the Latin text of the office for Auvergne-born St Genesius (*Officium sancti Genesii comitis confessoris*), printed at the same time and by the same printer as a hagiographic booklet in French, *La Vie de Sainct Genez comte d'Auvergne confesseur*.[19] The text of the ordinance declares:

Ordinance of Mgr the Most Reverend and Most Illustrious Bishop of Clermont. Joachim Destaing Bishop of Clermont. Given the request presented to us by the Prior, the Parish Priest, and the Chief Vestryman of the Community of Combronde, by which it is remonstrated to us, making our visit to the said parish of Combronde in 1625. And being requested by the said supplicants to change the celebration of the divine Office that they have held to up to now according to the usage of Clermont, that their Breviaries, Missals, Canons and Rubrics were not printed by a press, remaining to them only those that they have put in our hands, which we would have found torn, in bad condition and unserviceable [these were manuscript books giving the text of services according to the diocesan rite]. And not being able so promptly to renew these, they would [like to] have obtained from us permission to celebrate the divine Offices according to the Roman usage because of the facility of printed Books and this by provision. But since we would not have fully provided for the service and offices of the local feasts that must be observed according to ancient usage, we have ordained that this be done. For this purpose now having appeared before us the said Prior, Parish Priest and Chief Vestryman of the said community and having put in our hands an Extract from a Chronicle and Legend of the chapter of Our Lady of Chamalière, held and represented to be ancient and usual in Our Diocese. On which having reduced and divided the Lessons, added the Responses, Antiphons, Versicles and all that is appropriate to the celebration of the canonical Hours and the Mass, in what concerns only the local feast under the title of St Genez, from time immemorial solemnized in the said place ...

This is a prodigious text. It presents a blend of the attraction of the printed book, a disgust for the old manuscript texts, the romanization of the liturgy and of making a pamphlet destined to be printed. This is how the modern ecclesiastical system was built. It was conceived as a centralized institution with multiple outlets in which the independence of churches and rites had to bow to a homogenization that 'by provision' tolerated local adaptation of the universal models of liturgy and narration. This system was not even partially modified until the nineteenth century, when many local cults disappeared and simple specific occurrence of the universal type was substituted for the microscopically small variant. A quantitative study of the production of hagiographic pamphlets thus ought to show a rising curve from the latter sixteenth century, after the Council of Trent, to the end of the seventeenth century.

These differing rhythms of diffusion pose the essential problem of the public for which the *libelli* were destined. Categorization by brevity and celebrative purpose is of little help here. If we trace the career in print of seventeenth-century lives of St Eustachius, for example, we see that they range from the four sheets of poor-quality paper of a *Historia di Santo Eustachio* bearing a crude woodcut and printed in central Italy in the early seventeenth century with no indication of date, place, or printer,[20] to a work at the other end of the scale of social consumption, a small volume printed on a magnificent, cream-coloured silk, *Il martirio di San Eustachio*, an oratorio dedicated to the niece of Pope Alexander VIII published in Rome in 1690.[21] Even without this somewhat facile contrast, it is clear that the problem resides less in too great a diversity in the category than in an apparent lack of differentiation.

To return to the work mentioned above on the life of St Genesius: was this work, printed on handsome paper in Paris by a well-known printer, Sébastien Cramoisy, destined for a public of local notables? Doubtless it was. If we consider, however, that a work of four small-sized sheets promoting the local saint demanded neither considerable expense (or, at least, the symbolic value of a unique object of the sort would permit a certain sacrifice) nor a particularly high reading competence, thanks to the canonical formulation of the text and its function as a quasi-ex-voto, to be contemplated as much as read, we might maintain (unless a humbler version of the same work were to be found) that the work was destined to the whole of the community of Combronde. The purchase of a hagiographic tract had an important symbolic investment that distorted the economic, social, and cultural constraints that played a determining role in the secular book. In other words, the practical (not purely conceptual) notion of the 'community of the faithful' clouded the distinctions between publics.

The multiplication of production sites and centres of diffusion, in combination with the highly local occurrence of cults, disturbs the coherence of the field of observation as well. It brought too many variants into play, and it heightens our uncertainty: how representative are documents surviving from a mass of fragile, rapidly outdated texts that changed as cultic usages shifted? We must thus resign ourselves to a certain indeterminacy, both essential and accidental, concerning the public for which these pamphlets were produced. We must also keep in mind, however, that by the indirect route of the cultic, commemorative, and thaumaturgic power of such humble works, printed matter penetrated into the humblest homes profoundly and in original ways. In what exact proportions and extent and for precisely what purposes will remain a mystery, however, just as great as the notion of popular religion.

Thus I shall take the humbler route of the case study to attempt to reconstruct the publishing career of such works, in the hope that this will sketch out a more general view of expectations and usages. Without prejudice regarding real categories, I have chosen to study tracts that appear 'popular': first, an isolated pamphlet devoted to St Louis of Anjou printed in Italy in the late sixteenth century; second, a varied series of short works on St Reine, in and out of the *Bibliothèque bleue* series, printed in Burgundy and Champagne during the seventeenth century.

A Life of St Louis of Anjou

The *Oratione devotissime del glorioso santo Alvise*, a life of St Louis of Anjou (1274–97) in Italian and in verse, was published anonymously in the late sixteenth century by Giuseppe Cacchio, a printer in the city of L'Aquila in the Abruzzo. I choose this work because it lends itself to a precise reconstitution of its genealogy. We know a good deal about its production, its place in developing ritual, and changes in its text:

We know something of the printer, Cacchio, and the milieu in which he worked.

The choice of a later and historical saint enables us to follow the stages of his veneration and the motivations behind it, in particular in the Kingdom of Naples, of which the Abruzzo was a part. There is hope of finding all the narrative, homiletic, and liturgical texts written subsequent to his canonization. Furthermore, the ideological implications in the life of a saint who was at once heir to the throne of Naples, a fervent Franciscan, and a bishop are exceptionally rich.

The place of publication of this work, L'Aquila, a city founded comparatively late (mid-thirteenth century), favours analysis of the connections between a place, a saint, and a cult.

In more general fashion, late sixteenth-century Italy offers elements uniquely interesting for our purposes. The production of 'popular' hagiographic pamphlets appeared there earlier than in France, and it developed abundantly after the 1560s all over the Italian peninsula, but in particular in central Italy (Lazio, Umbria) in Tuscany (Florence, for the most part), and, to a lesser degree, in Venice. One gets the impression that printers in Italy continued to enjoy a measure of autonomy from the Church.

Lorenzo Baldacchini has located an interesting instance of this liberty (but also of its suppression) in the inquisitorial archives in Modena.[22] In 1594, the Inquisition in Modena pursued a certain Francesco Gadaldini, a printer, accused of having published an *orazione* to St Martha (by metonymy, the word designated not only 'prayer' but all sorts of brief hagiographic tracts). The accused defended himself in terms that throw light on one of the possible sources of a true popular hagiography:

I did not print it. . . . It is true that my father, as he was coming from Bologna with an *orazione* of St Martha written by hand, begged me to find him another, and in order to satisfy him, I set myself to inquiring and in particular of a certain Margarita Chiaponna, who I asked whether she knew the *orazione* of St Martha. She told me that she knew it by heart; I noted it down in writing with my own hand and gave it to my father, who in a few days printed it. I took only one copy and I think, without being able to swear to it, that I gave it to the said Margarita.

This is a highly interesting and frustrating text. Obviously, we do not have the text of this *orazione*. We are left reiterating a sad truth: we lack the 'hottest' texts in this ephemeral production. It is imaginable that a text concerning Martha (the image of salvation through labour in the great ideological diptych of Mary and Martha) might reflect the refractory and anticlerical aspects of popular Christianity in northern Italy. Still, we cannot exclude the hypothesis that Gadaldini was arrested for the simple misdemeanour of printing a religious text without authorization. This text at least proves the existence of a clandestine or semi-clandestine production of hagiographic tracts, a characteristic not found in seventeenth-century France, and one that will have a bearing on interpretation of the work on Louis of Anjou. I might also note the extreme rapidity and simplicity of the cycle from oral memory to printed text: a walk around Modena and a few days of work were all it took. This may have been where the popular pamphlet's subversion lay, in reproducing quickly the ancient and oral forms of religiosity.

The L'Aquila pamphlet (plate I) is known in only one copy that owes its survival to the passion for book collecting of a great prelate, Cardinal Orazio Capponi, who gathered together several dozen hagiographic pamphlets in the portfolios donated to the Vatican Library and known as the Fondo Capponi.[23]

The pamphlet is an octavo of one signature of four sheets measuring 11 cm × 17 cm. The eight pages bear no page or folio numbers and only one signature mark (A2, page 3). The paper is extremely thin and of poor quality, letting the ink show through from the reverse side of the page. The irregular pages are untrimmed. The first page bears the title, ORATIONE / DEVOTISSIMA / DEL GLORIOSO / SANTO ALVISE in capital letters (somewhat larger for the second line of the title). Underneath there is a woodcut (8.2 cm × 6.1 cm) inside a guilloche border and trimmed at the top showing a bearded bishop with a halo, wearing his mitre and holding his crosier in his left hand. He is kneeling before a representation of Christ that occupies the upper left corner of the woodcut. Before the bishop and underneath the figure of Christ lies an open book. The decor, highly simplified, is reduced to two broken columns and two courses of brick to indicate the ground level. The extremely summary rendering of the central figure resembles that of a good many contemporary tracts; since no specific detail clearly identifies the bishop, the use of an all-purpose woodcut seems probable. Moreover, in contemporary iconography the beard systematically signified age, whereas Louis died at the age of twenty-three and was always portrayed in learned ecclesiastical paintings of the later Middle Ages as beardless and almost childlike. The place of publication and the name of the printer appear under the woodcut: 'NELL'AQUILA / Appresso Gioseppe Cacchio'.

The seven pages of the text were printed in small-sized Roman type. Each page bears three and a half eight-line stanzas of verse, except for the eighth page, which has one stanza followed by an indication of ending ('IL FINE') and the text, in Latin, of the prayer announced by the title (*Oratione devotissima*). There is a catchword at the foot of every page but page 2; each stanza begins with a display initial; subsequent lines begin with lower case letters. The stanzas follow the same model (with a good many exceptions or metrical and verse 'errors'): eight dodecasyllabic lines with a rhyme scheme of *abababcc*. This form, known as the *ottava*, was the current form for Italian popular poetry after Ariosto. Two pamphlet titles hint at this popular and religious adoption of Ariosto. First, in 1589 a certain Goro da Colcellalto printed a religious adaptation of *Orlando furioso* entitled *Primo Canto del Furioso traslato in spirituale* for the famous Florentine bookshop, Alla Scala di Badia. Second, Giulio Cesare Croce, an active and versatile writer, proposed a volume of *Rime compassionevoli cioè il primo canto dell'Ariosto tradotto in spirituale* (Viterbo, 1676).[24] The form of a series of *ottave* was thus of learned origin, but its appearance in popular works points to a probable use for these poetical pamphlets in which the liturgical was reduced to little or nothing. Until recent times saints' lives were recited – half sung, half chanted – in the marketplaces of Italy, in Tuscany in particular.

The prayer that ends the pamphlet seems in no way surprising. Publishers of hagiographic tracts habitually offered a short Latin text of the sort. Central to the commemoration of the saint, it was repeated at each reading of the saint's office, or else it was the only text referring to the saint in particular when the office or the Mass was said according to the Common of the Saints. For Louis of Anjou, this was frequently the case, as he seldom found a permanent place in liturgy and most of the breviaries and missals that celebrate him do so according to the Common of confessors and bishops. The Franciscan order, however, gave Louis the right to an office of his own. The prayer, a short and highly conventional (hence easily memorized) text, was the indispensable and minimal liturgical baggage for the devotee of a particular saint. This is why a great many pamphlets, though essentially narrative, are entitled *Oratione* or *Devotissima oratione*. We can thus accept the likely fiction that these booklets were offered for sale as a token or a souvenir by street and marketplace singers who recited or 'lined out' the Latin prayer after singing or chanting the narrative stanzas that encouraged veneration of the the saint.

The only surprising element in this *oratione* supports the hypothesis of a lay publication for popular use. The printed prayer came from the office for another St Louis, Louis IX, king of France, the great-uncle of Louis of Anjou. After the fourteenth century, almost all the breviaries that mention the feast of Louis of Anjou repeated the same version of his prayer. The few variants, which differ greatly from the prayer to Louis IX, are in older liturgical texts, whereas the form of the prayer to Louis IX was fixed in the early fourteenth century. Moreover, the simplest country parson could recognize the error, since the prayer to this Louis the Confessor did not bear the indispensable mention of his episcopal title, even though he was clearly indicated as a bishop in the preceding narrative. Typographical errors, which were few in the Italian text, abounded here, even in the most stereotyped formulas (*de tarreno* for *de terreno*; *Regni regum* for *Regi regum*). Finally, Louis was designated by the name 'Alvise', a Venetian dialectal form of 'Lodovico', which was the form that appeared in all Italian breviaries and missals and by which Louis was distinguished from his great-uncle, 'Luigi'. This form was not meant to be dialectal, however, since the Italian text, in pure Tuscan, bears no other Venetian or Abruzzese forms.[25] For the moment, the most we can say is that Louis's name does not come from the Church here. Without any doubt, the text was written by a lay person not overly conversant with liturgy who had consulted a breviary containing only the prayer to Louis IX, whose saint's day was close to Louis of Anjou's (25 August for 'Luigi', 20 August for 'Lodovico'/ 'Alvise').

The Two Careers of Giuseppe Cacchio

Fortunately, we know a good deal about the printer-publisher, Giuseppe Cacchio (or Cacchi).[26] The mystery surrounding the pamphlet only thickens, however, for Cacchio hardly corresponded to the common idea of the popular publisher. Born around 1533 in L'Aquila, he was schooled in Naples, where he learned typography with the great Orazio Salviani. Around 1565 he returned to his native city to open a print shop there. Printing had had a brilliant beginning in L'Aquila, thanks to the presence of Adam of Rottwill, a cleric from the diocese of Metz who was active for sixteen years in various parts of Italy.[27] In Rome (1471-4) Adam published both juridical texts and the *Mirabilia Romae* and the *Soty of Two Lovers* of Eneas Silvius Piccolomini; in Venice (1476-81) he printed Latin texts, breviaries and a German-Italian dictionary. He introduced printing in L'Aquila when he settled there with a privilege of 1481, publishing a handsome edition of the first Plutarch in Italian (1482). After that date the printer's art disappeared from L'Aquila until Cacchio's return eighty years later. The town honoured Cacchio and offered him a generous welcome. A communal statute of 1566 gave him a subsidy to bolster the further establishment of his craft. In 1569, however, he began to publish in Naples, though he kept his bookshop and his print shop in L'Aquila. Naples was the true theatre of his activities, and of the 165 titles that he published from 1566 until his death in 1592, 136 come from his print shop in Naples and only twelve from L'Aquila, while seventeen others were printed in Vico Equense, a small town near Naples where he worked after

1581, for the most part for the local bishop, Paolo Regio, of whom more later.

This career is quite comprehensible if one omits the pamphlet on Louis of Anjou. Cacchio, a talented printer, became dissatisfied with the narrow range of opportunities in L'Aquila and gradually shifted his work to the metropolis towards which L'Aquila had always looked. The episode of Vico Equense, which in no way interrupted Cacchio's activities in Naples, can be explained by his attachment to a devout humanist prelate, a friend to books and himself a prolific writer.

The quality of Cacchio's work is evident from a first glance at his publications. The typography is clear and varied (using twenty-four Roman and twenty-three cursive characters), attention has been paid to page layout; the paper is of good quality. The nature of the texts he published confirms this impression: more than two-thirds were by contemporary authors, nobles or churchmen in the kingdom of Naples and members of its cultural elite. More than half of these texts (84 out of 164) are in Latin. A necessarily summary classification by genres shows 24 per cent works of literature (poetry for the most part), 35 per cent works of jurisprudence or local erudition (descriptions of the towns and cities of the kingdom, statutes), 28 per cent religious texts (theology, devotions, hagiography), 13 per cent various titles (grammar, medicine, physics, etc.). There are no popular works on Cacchio's lists, and hagiography enters only in the learned form of anthologies concerning the patron saints and protectors of Naples and Sorrento. Even when he published a pocked-sized work (octavo, 40 sheets, 165 mm × 110 mm) devoted to a saint, *Officium Sanctae Fortunatae* (Naples, 1568), a liturgical office in Latin of clerical origin, it is in the tradition of carefully produced devotional books. The twelve works printed in L'Aquila (in 1566–7 and 1578–81) show the same general characteristics, even though their authors belonged to the socio-cultural world of the Abruzzo.

How, then, are we to understand the incongruity between our pamphlet and its publisher's customary practice? Four hypotheses are possible:

1 The pamphlet might be a counterfeit using the prestigious name of Cacchio – something not unprecedented in the history of publishing. But could one get away with a fraud of the sort in a small city like L'Aquila without immediately being found out and sued? Outside the Abruzzo region the name of a publisher in L'Aquila would confer no prestige.

2 The booklet might have been published at the very beginning of Cacchio's career, when his print shop had neither the personnel nor the physical equipment that would later make his reputation, perhaps even earlier than his official installation, the granting of his privilege, and the receipt of a subsidy from the city government. Cacchio, in this case, would have tried his wings with a genre that was produced inexpensively and

executed rapidly. Examination of the first works that Cacchio published in L'Aquila seem to support this thesis. The five works, although nicely printed, are small in format (octavo or duodecimo) and short (the longest has 66 sheets). It was in Naples that Cacchio's production gradually became more ambitious. The first work printed in his new print shop in 1567, a handsome folio, presented four ordinances of Pedro de Ribera, viceroy of Naples. Furthermore, two of the five works of the first period in L'Aquila mention the same address as the pamphlet ('Nell'Aquila / Appresso Giuseppe Cacchio'). Later, an unfortunate consonance between the printer's name and an obscenity made him change to the more anodyne Cacchi (Cacchius for the works printed in Latin).[28] Unfortunately for this hypothesis, the more scabrous spelling reappears when Cacchio returned to L'Aquila in 1578–81.

3 A third hypothesis seems better founded. Up to 1582, Cacchio kept up his activities in both L'Aquila and Naples, which we know because in 1583 he ceded his property in L'Aquila to two printers, Bernardino Cacchi, a relative, and Marino d'Alessandro. Between 1567 and 1578, Giuseppe Cacchio published nothing in L'Aquila, so we might conjecture that his workers and future successors used the scanty equipment that he had not bothered to take to Naples to continue to print (with lower standards and at a slower pace) under the name of their proprietor, who held the printing monopoly in the city until 1582, when Giorgio Dagano also obtained a privilege.[29] That Cacchio ceded his property in 1583 seems a good indication that a market existed in L'Aquila and that Cacchio's absence had harmed his local business. Unfortunately, we have no other trace of any 'franchised' activity (to use a modern term) that took place between the visits of the master printer, when he brought his skill and his equipment.

4 A fourth hypothesis gives a totally different dimension to Cacchio by attributing to him a dual, though not clandestine, activity. This interpretation rests on a document that is unfortunately unique. In 1576–8, Giuseppe Cacchio was called before the diocesan court of Naples for having printed without license 'false indulgences in connection with the grains' (with recitation of the Rosary).[30] We can guess what this meant. The accusation must have been aimed not so much at the printed text as at the laity's having taken devotional practices into their own hands. Indulgences were still an important pontifical prerogative in the late sixteenth century. The devotion of the Rosary, introduced in the late Middle Ages by the Dominicans, was not in itself suspect, but it facilitated the development of an individual piety that eluded Church control.

Cacchio's pamphlet (no trace of which remains) evinced lay devotional autonomy. Cacchio failed to appear before the court and was excommunicated *in absentia*; the sentence was proclaimed publicly 'to the sound of bells and by the posting of bills in all the usual public places' in Naples.

We know that this work did not bear Cacchio's name, since at the second trial the Inquisitor, Sigillardus, performed a task of veritable expertise on it, comparing the type characters in the pamphlet under attack to the works of Bishop Paolo Regio that Cacchio had printed in Naples. The examination must have been conclusive since the excommunication was raised, but the printer was forced to abjure and do penance, and he was sentenced to two years of surveillance and prohibition from publishing. This prohibition, which may have been limited to the diocese of Naples, may explain Cacchio's second sojourn in L'Aquila from 1578 to 1581 and his later return to Vico Equense near a bishop with whom he had close connections. I might note that the pamphlet on Louis bears no permission to print, unlike all other texts printed by Cacchio. It also bears no date: Cacchio may have tried to get away with publishing it during his two-year prohibition.

Cacchio's return to L'Aquila is easy to imagine. Prohibited from publishing in Naples and under suspicion – he who had worked for the archdiocese and for the court – he felt his situation was compromised, so he attempted to build up business in the Abruzzo once again. Perhaps he thought of diversifying his output by attempting a popular book. Such publications did well in neighboring Lazio but were less frequent in the kingdom of Naples. He thus chose a saint honoured in the kingdom and, for fear of being refused, he decided to do without support and without ecclesiastical permission. No other work confirms this new, 'popular' orientation, but the rest of his output in this second period in the Abruzzo shows a desire for novelty and for local roots. Among the seven titles remaining from this period is a slim, small-sized pamphlet (octavo, 15 cm × 10 cm), a book of songs printed as verse in 19 *ottava* stanzas and a sonnet, *Canzone alla Siciliana sentenziose e belle, non mai piu poste in luce e aggiuntovi un lamento d'uno giovane sopra della morte. Cose molte piacevole da inteneere* (1580). The inflated style of the title suggests popular works that were sold by public cry of their title. In the same year, Cacchio put out another small work in *ottave* on the model of Ariosto. The other works printed during this time show greater interest in local affairs than those published in 1566-7: the first work published after Cacchio's return was a description of Sulmona, a town near L'Aquila (1578); in a similar vein he put out a patriotic account of L'Aquila's war against the *condottiere*, Braccio da Montone, accompanied by pieces to honour the Camponeschi family, who were powerful in the area (on whom more later).

Paradoxically, Cacchio seemed to have found a way out of his disgrace through the same religious literature that had led him into trouble. This time, however, and until the end of his life, he took cover behind pious prelates. In 1581, his last and most prestigious publication in L'Aquila presented the decrees (*Decreta diocesana*) of Mgr Racciaccari, L'Aquila's new bishop (1579-92) and a Franciscan of the Observance. Three years later, in

1584, Cacchio returned to Naples and Vico Equense and to his association with Bishop Paolo Regio, who was to write (among other hagiographic works) a life of St Louis of Anjou.

The mystery of the dual career of Giuseppe Cacchio may perhaps never be cleared up directly, but we can attempt a closer look at the interests defended in our pamphlet by examining the nature and the functions of veneration of Louis of Anjou in the Abruzzo in the late sixteenth century.

Louis as Angevin and Franciscan

Louis was born in 1274 in Brignoles, in Provence, the son of Marie of Hungary and Charles II of Anjou, himself the son of Charles I, King of Sicily and brother of Louis IX of France. The throne of Sicily included what was to become the Kingdom of Naples, and it fell to the Anjou family as the result of the popes' manoeuvres against the Hohenstaufens.[31] When Louis was ten years old, his father was captured by his rival, Alfonso III of Aragon. When he was fourteen, he and his two brothers went to Catalonia as hostages (in 1288, after the treaty of Canfranc) to guarantee the peace and in exchange for his father. He remained in prison there for seven years. During that time he was influenced by two Franciscan friars, François Brun, his confessor, and Piero Scarerii, a Catalan, both of whom belonged to the Spiritual branch of the order who, unlike the Conventuals, insisted on a return to the poverty and asceticism of the first followers of Francis. In 1294, when Louis was still in Catalonia, pope Celestine V, the former Abruzzo hermit, permitted Louis to be tonsured and to take the first four minor orders. He also charged Louis with the administration of the archdiocese of Lyons. When Louis's older brother, Charles Martel, died in 1295, Louis became heir to the kingdom of Sicily. Freed from prison in October 1295, he was made a subdeacon by the pope in December of that year. In January 1296 he renounced the throne in favour of his younger brother Robert; in May he was ordained a priest. In December of that year Boniface VIII, Celestine's successor, named Louis bishop of Toulouse, a charge that he accepted on the condition that he be allowed to enter the Franciscan order. He pronounced secret vows in December 1296, renewed publicly in February 1297. He then went to Paris, to Toulouse, and next to Catalonia for a peace mission, returning to Italy through Provence, where he died at Brignoles on 20 August 1297. He was buried, as he wished, in the Franciscan church in Marseilles, and was canonized by Pope John XXII in 1317.

This brief life is extraordinary in that it is a perfect digest of sainthood in the central Middle Ages. A prince renounces his crown for Christ; a bishop named against his will asks nothing better than to return to pious solitude and refuses the honours of his charge to serve the poor, to whom he distributes food daily. For the Church his career had something of the perfect. The Church had never really promoted royal sainthood. A good king was a dead king, possibly a martyr. This explains the cult during the high Middle Ages of assassinated kings, even when their death had little to do with fighting for the faith.[32] A prince who renounced the throne thus represented the pinnacle of royal virtue.

Canonization, although rapid, aroused little enthusiasm and the saint's cult remained modest. Edith Pásztor has clearly shown the reasons for this reticence,[33] which was connected both with conflicts within the Franciscan order and with the misfortunes of the Angevin dynasty. As early as 1300 Charles II launched a movement for canonization, but it received no support from the Franciscans, even though the order had at the time only two saints (Francis himself and Anthony of Padua). The Franciscan cardinal Matthaeus of Aquasparta strongly opposed the move. In 1307, when Pope Clement V decided to open an investigation with a view to canonization, no Franciscans appeared among the procurators favouring the cause, contrary to usage. The procurator served as a lawyer, taking responsibility for composing the candidate's biography according to established modes and models, so it would seem normal and desirable that the members of a congregation support one of their own. Louis, however, belonged to the spiritual tendency of the order at a time when rivalry between the two branches had become ferocious, a state of affairs that continued until 1323, when John XXII condemned the *fraticelli* – the 'fundamentalist' Franciscans – in a move that was intended to reach beyond those extremists to strike at the Spirituals. The personal orientation of Louis was all too clearly written in his life. As Edith Pásztor has shown, Louis's practices as a bishop were clearly based on the theory of absolute poverty, even in cases of episcopal responsibility, as expressed by Peter John Olivi, the great Spiritual Franciscan from Narbonne, in his *Quaestiones de perfectione evangelica*.[34] The Franciscan bishop, he wrote, must abandon nothing of Christ-like poverty, contrary to the affirmations of the 'Common Brothers', the future Conventuals, who argued for adapting to the times.

The canonization went forward, however, thanks to pressure from the royal house. The Anjou family was strongly established in Italy and still had close ties with its French origins and the Capetian dynasty, and it spared no effort. The canonization was accomplished, though, at the cost of a compromise that rewrote the saint's life during the process of canonization and provided the basis for hagiographic tradition. All tensions disappeared from Louis's life and his memory paled.

The cult of St Louis of Anjou followed the fortunes of the Angevin dynasty. Charles II, Robert, and Joanna I, celebrating Louis as both Angevin prince and Spiritual Franciscan, contributed largely to the establishment of devotions in Naples and in Provence.[35] The royal house was in fact constant in its support of the Spirituals. Robert of Anjou gave energetic support to Michael of Cesena, who was elected minister general of the Friars Minor in Naples in 1316 and was soon deposed for his Spiritual leanings. Robert himself composed more than two hundred sermons (three on his brother Louis) and a treatise, *On poverty*, and was buried in the habit of a friar minor. His queen, Sancia of Majorca, whose brother Philip had renounced the Majorcan throne to take the Franciscan habit, protected fifty disciples of Michael of Cesena near Castellammare.

The royal couple's deep devotion to the Spirituals assured the public veneration of Louis,[36] as attested by the handsome portrait of him by Simone Martini now in the National Museum of Capodimonte in Naples,[37] but it also thrust the cult into a polemical, political, and religious context. During the fourteenth century, the Angevin dynasty made itself unpopular and provoked an anti-French reaction that peaked under the reign of Joanna I (1343–82), Louis's great-niece, dethroned by her cousin of the Durazzo branch, Charles III, who strongly resented the French entourage of the court of Naples. They continued royal support of the cult of the saint, however, as Charles III and his successor Ladislas had their eyes on the throne of Hungary, and Louis's mother was Marie of Hungary. Among the various traces of this cult, there is a mention of him and an illustration in a missal,[38] written in Church Slavonic in Glagolitic script written for Hroje Vukčić Hrvatinić, governor of Croatia and Dalmatia, and King Ladislas's regent in Croatia, Dalmatia and Bosnia in 1403–4, at the time of the King's expedition to Zara, before his definitive renunciation of the throne.

In 1378, the Great Schism, which at its start opposed the Italian (and Neapolitan) Urban VI to the Avignon pope, Clement VII, destroyed the religious influence of the Angevin dynasty in Naples and in the kingdom, since the Angevin party was obedient to Avignon, against the religious and nationalistic desires of the Neapolitans. Furthermore, after the disappearance of the first House of Anjou, the Angevin party, opposed to the Durazzos, pledged itself to the second House of Anjou and was still closer to Provence and to the Avignon papacy. In the fifteenth century, then, after Ladislas's setback in Hungary and after brief conquests of power by an abhorred Angevin dynasty (Louis II, the nephew of Charles V of France, reigned from 1387 to 1399 and René of Anjou from 1435 to 1442), the fortunes of the saint tottered. The Durazzo branch of the family ended with the death of Joanna II, and the Aragon dynasty took over. Louis, linked to a suspect religious faction and to an unpopular dynasty, could not have been the object of much devotion. His memory faded, and even in the Angevin party in Naples, Louis of France was more honoured than Louis of Anjou: three kings in the second House of Anjou were called Luigi and not Lodovico. In a symbolic gesture, Alfonso of Aragon, before chasing the Angevins and the Durazzos out of Naples, sent his buccaneers in 1423 to remove Louis's body from the Franciscan monastery in Marseilles. After the saint's remains were returned to Catalonia, Louis's former prison, they inspired no veneration, and by the mid-fifteenth century, Louis has sunk into obscurity. And yet the unfortunate Franciscan prince and his pathetic fate were remembered in L'Aquila a century later. Why?

The Saint's Party: Observant Franciscans, Barons, and New Men

At the end of the Middle Ages, the Observants, the solidly institutionalized Franciscans of reforming tendency, heirs to the Spiritual Franciscans, found the Abruzzo region much to their liking. From the twelfth century, the mountains of the Abruzzo had nurtured a strong hermit tradition. In the late thirteenth century, Peter of Morrone, who later became Pope Celestine V (1294) and who gave the archdiocese of Lyons into Louis's charge, established a hermitage and an order (known as the Celestines) near Sulmona. Celestine was buried in L'Aquila and was particularly venerated there. As it happens, the Celestines welcomed into their midst the *fraticelli*, the extremist branch of the Spirituals. In the early fifteenth century, the Observance developed vigorously in the Abruzzo, not far from the Franciscans' place of origin in Umbria, thanks to the efforts of such great figures as Giovanni da Capistrano, Jacopo delle Marche, and, above all, Bernardino da Siena, who came to L'Aquila in 1438 to settle the question of the status of the Observants. Bernardino preached extensively in the city, and died there on 20 May 1444. Rapidly canonized (in 1450), he was much honoured in L'Aquila, even after his body was translated elsewhere in 1472. In 1456, a privilege of Alfonso of Aragon ordered that his feast be celebrated in the city, and the pilgrimage to Bernardino's tomb was so well attended that Giovanni da Capistrano exclaimed, 'Your city has grown fat thanks to St Bernardino.' Observant monasteries multiplied in the Abruzzo, which was entitled, after 1457, 'province of St Bernardino'. There were four Observant monasteries in 1420, fourteen in 1450, nineteen in 1495, and in 1452 the general chapter of the Observance was held in L'Aquila.[39] This tradition continued until the early modern period, and the bishop of L'Aquila for whom Cacchio worked in 1581 was an Observant. Without doubt, Louis of Anjou benefited from the dynamism of the Observants in the Abruzzo and he is constantly associated with Francis of Assisi, Bernardino da Siena, and Giovanni da Capistrano in frescos and paintings in churches throughout the Abruzzo and Umbria.[40]

The structure of sociopolitical life in L'Aquila from the fourteenth to the sixteenth century assured the saint's cult a particular role that may explain how the pamphlet Cacchio published came to be written. L'Aquila was founded late, in 1254, probably by Conrad IV,[41] successor to Frederick II

Hohenstaufen, who was interested in bolstering the imperial hold on central Italy by establishing a stronghold to resist the rebellions of the powerful counts of Celano. Pope Alexander IV soon granted communal status to the town, and in 1257 transferred the episcopal seat from Forcone to L'Aquila. In 1259, Manfred, the illegitimate son of Frederick II, destroyed the city; Charles I, King of Sicily, authorized and encouraged its reconstruction, which perhaps explains the city's at least partial loyalty to the House of Anjou. The Abruzzo lay at the borders of the kingdom of Sicily, however, and this encouraged the city government (the *Comune*) and the barons of the surrounding territory to make a play for autonomy, after the assassination of Charles III of Durazzo, by supporting the second House of Anjou against the Naples-based Durazzos. At the end of the fifteenth century, L'Aquila backed René of Anjou against Alfonso of Aragon, and it fell to the lot of the pro-Angevin *condottiere*, Jacopo Caldora. Later, in 1495, the city welcomed the French king, Charles VIII, who was fighting the Aragonese. In 1527-8, the *Comune* of L'Aquila again supported the French against the emperor Charles V before it lost its autonomy.

On closer examination, political attitudes were even more complex, since three groups were active in city life: the barons of the *contado* surrounding the city, the people, and the merchants and burghers who controlled the city government until the early sixteenth century. The *Comune*, which coined money from 1385 to 1556, was organized both topographically and corporatively, and it favoured the merchants, who traded in wool, skins, and saffron and looked towards Umbria and the north rather than towards Naples. The barons attempted to impose their power against both Naples and the *Comune*. The people, or rather the 'new men' like Lodovico (the name is noteworthy!) Franchi, the head of a powerful faction in the early sixteenth century, forged tactical alliances with one baron or another in an attempt to break the merchants' hold on the city government.

Religious cults in the city must be seen with these politico-social oppositions in mind. We know from the statutes of the city, extant up to the Renaissance,[42] that the *Comune* placed strict regulations on devotional practices in the city and its wards. For example, a statute of 1371 stipulated that 'no inhabitant of L'Aquila may leave his ward to make his offering during a feast day or an indulgence unless it is in groups of ten men or more, separated by spaces, exception made for the feast of the Annunciation of the Virgin.' A 1315 statute lists the official feast days of the *Comune* for the first time: Christmas, Easter, Ascension Day, Pentecost, the Feast of the Virgin, the feast of the Eucharist (Corpus Christi), All Saints' Day, Passion Sunday, the feast of the Apostles, and the feast days of saints Maximinus, George, Nicholas, Mary Magdalene, Michael Archangel, John the Baptist, Lawrence, Benedict, Peter the Confessor (Peter Celestine), Blaise, Agnes, and Catherine, plus the feast of the Exaltation of the Holy Cross. In 1357, the

Comune added the feasts of St Sebastian and the Visitation; in 1400, SS Leonard and Flavianus; in 1408, St Salvius; in 1423, St Claire, credited with a miraculous victory over the *condottiere*, Braccio da Montone. In 1460, St Nicholas of Tolentino, recently canonized (1446), was added to the city's list. Nicholas's appearance should be understood in connection with Louis's absence: Nicholas of Tolentino, a contemporary of Louis of Anjou who died in 1305, was an Augustinian hermit born to a merchant family in the Marches near Ancona (a market for Aquilan products). He embodied the communal spirit far better than did the Angevin prince. Finally, in 1507, the city government added to its list St Roch, a saint whose popularity was on the rise throughout Europe at the end of the Middle Ages.

Communal hagiography was thus limited to universal saints and to three local or regional saints. St Maximinus was a martyr from the times of the Roman persecultion, who played the common role of founder of local Christianity. St Peter the Confessor (Pope Celestine V) protected the city by the real presence of his body (a statute of 1434 speaks of him as 'our protector and defender'). St Nicholas of Tolentino, a 'bourgeois' saint, epitomized the daily activity of the Aquilanians. The *Comune* resisted imposed cults such as that of Louis IX, which Ladislas attempted to found by an ordinance of 1407, and that of Bernardino, ordered, as we have seen, by Alfonso of Aragon.

The barons of the *contado*, though, seem to have venerated Louis of Anjou. In 1393 Count Roger II of Celano was buried, dressed in the Franciscan habit, in Castelvecchio Subequo, near L'Aquila, in a church that contained two statues of Louis.[43] Several members of the Camponeschi family, who dominated the region in the fifteenth and sixteenth centuries, were named Loyso, a dialect form of Lodovico close to the Venetian Aloyso or Alvise. As we have seen, this was the form of Louis used in our pamphlet, though exactly how it made its way from Venice to the Abruzzo is not clear. Their Angevin connections were important to the Camponeschis, however, as in 1460 one of the family proclaimed René of Anjou King of Naples, unseemly as it was.

We can imagine three groups, then, allied by their piety and their hostility to the *Comune* towards the end of the sixteenth century, when L'Aquila had lost its autonomy under Spanish rule. There were the Observant Franciscans, who benefited from the memory of Bernardino of Siena's popular and populist preaching; there were the people, or the 'new men', defeated in 1521–7; there were the nobles, who, closely hemmed in by the new Spanish administration, dreamed nostalgically of the Angevin era.[44] This ideological convergence perhaps occurred within the mystical and evangelical current centring on Juan de Valdés in Naples towards the middle of the century. If this were the case it would explain Cacchio's difficulties with the Inquisition in Naples and his relationship with Paolo Regio, bishop and biographer of Jacopo delle Marche and Louis of Anjou. Furthermore, if we return to the Aquilanian authors that Cacchio published, we find the melancholy world of a provincial elite, half devout, half simply futile and defeated by history, ready to be taken in hand by the friars or the court.

The first author published by Cacchio in L'Aquila in 1566 was Giovanni Cantelmo, the author of a pastoral comedy entitled *La Psiche*. The Cantelmos had originally come from Provence, where their family name was Gentiaume, and they had accompanied Charles I to the kingdom of Sicily. Their sovereign rewarded them with the title of counts of Popoli. The first Cantelmo, Jacques, who died in 1288, was succeeded by Béranger, Seneschal of Provence, after whom came Restaino, the father-in-law of Bernard of Les Baux. As they became integrated into the society of the Abruzzo, they allied themselves and intermarried with the venerable family of the counts of Celano. Giovanni Cantelmo, our author, served the Aragonese regime as Captain General of the Abruzzo, but he inclined towards the pontifical party. A nephew by marriage of Paul IV, he married a Colonna, and in 1555 he commanded the pontifical army. He quarrelled with the pope and died in L'Aquila in 1560. *La Psiche* is the only known work of this representative of what later would be the ultra-Catholic, 'black' noble class in the kingdom of Naples. Another work published by Cacchio in 1580 recounting the city's struggle against the *condottiere*, Braccio da Montone, gives a quite different impression of the nostalgia of these nobles. Written by a local scholar of the late fifteenth century in celebration of the Camponeschi family, it ends with a eulogy of Pietro Lalle Camponeschi, last of the counts of Montorio, who had married a countess of Popoli. On the religious side, in 1566 Cacchio published the works of St Bonaventure, the great Franciscan doctor, in the translation into Italian by Vincenzo Belprato, an author known for his sympathy with Juan de Valdés and his circle. In 1581, as we have seen, Cacchio published the diocesan decrees of the Franciscan bishop of L'Aquila.

To situate the publication of the *oratione* in honour of St Louis of Anjou: it was part of the old traditionalist dream of the people and the nobility joining together, beyond municipal hagiography and outside the official Church, to celebrate a man who had been both prince and poor, bishop and Franciscan, in memory of the Angevin era and as an appeal for its return. The enigma remains, however: was Cacchio reflecting a general state of mind, or was he undertaking a specific commercial, religious, or civic operation? Perhaps we will never know, since the work was not followed by others known to us. It seems probable, as we have seen, that the booklet was written outside ecclesiastical circles. That the attempt to promote the cult failed would thus indicate that a broad-based devotional culture could not exist outside of the clerical framework and the ecclesiastical sanctification of the book. When his attempt to bend social and religious rules proved insufficient, Cacchio abandoned L'Aquila and returned to Vico Equense. Speaking to the people was not that easy a task, and in Christian circles the most effective way to reach them was not through nostalgia but through liturgy.

Roman liturgy had literally swallowed up all memory of St Louis of Anjou at the end of the sixteenth century, unbeknownst to the writers of our pamphlet. Up to this point, the genealogy of the text itself has not been discussed since its hagiographic content is so meagre and vague that both identifying autonomous elements and comparing versions remain hazardous. We need to superimpose the living genealogy of the uses and cultic appropriations of this tract on the archaeological model of a small erratic block, eroded but unique, deposited in L'Aquila by the sociocultural current that brought it there.

A Plural Biography

The first biographical sketches of Louis of Anjou portray vividly the internal tensions within both the Church and the Franciscan order as they emphasized or played down the voluntary poverty of the bishop of Toulouse, and hence Louis's participation in or distance from the Spiritual tradition. Hagiographic differences were soon neutralized, however, rendering the narrative sterile. The papal bull of Clement V, *Ineffabilis providentia* (1 August 1307) proposed a Spiritual reading of the life of Louis, but six months later the procurators who drew up the fifty-five articles of the preliminary questionnaire in the canonization proceedings followed the stereotyped model of the traditional saint's biography. The narrow precision of the questionnaire definitively shaped later biographical tradition, which then no longer drew on any other source. The bull of canonization, *Sol oriens*, promulgated by John XXII in 1317, made this standardization final.[45]

The Spirituals did not give up writing their version of Louis's biography. A long life by an anonymous author, later and arbitrarily identified as a certain Johannes de Orta,[46] was probably the work of a Spiritual close to the Angevin court written around 1320. It was none the less cast in the rigid form imposed by the Curia. Johannes de Orta made use of the canonization procedures, and his biography is organized by listing the saint's virtues, following both the outline of the 1317 bull and the prestigious model of St Bonaventure's life of St Francis that had replaced the more chronologically organized life by Thomas of Celano. The uniqueness of Louis's life was dissipated in these rewritings. Johannes de Orta's life was read, of course. There is a summary of it in Catalan written as early as 1320, and another life, a manuscript of which is in the Bibliothèque Nationale in Paris,[47] reiterates Louis's penchant for primitive Franciscanism. Still, when the Spirituals were condemned in 1323 in the bull *Cum inter nonnullos*, Johannes de Orta's life was ranged among the polemical texts, and between 1319 and 1330 Paolino of Venice, Inquisitor, trusted advisor to John XXII and a Conventual Franciscan, confirmed the sense of the bull *Sol oriens*. The institutionalization of the Observance and its attempt to include Louis among its own without presenting him as an extremist later eclipsed the Spiritualist versions of his life.

It is noteworthy that nearly all the breviaries that celebrate Louis, even in Observant circles, make use of the bull *Sol oriens*, even though they do not copy it mechanically, dividing the text into lessons in quite different ways. Literal citation of a papal bull was rare but not unprecedented, and it usually did not preclude the elaboration of new liturgical and narrative versions. In this case, however, the bull remained in printed breviaries up to the early modern age. Moreover, editions of the *Golden Legend* in both Latin and Italian added to Jacobus de Voragine's text a chapter on Louis that reproduced *Sol oriens* in its entirety.[48] This was the probable source of our pamphlet. The twenty-two stanzas of the text follow the outline of the bull exactly; certain parallel expressions are derivative; no specific detail appears in the pontifical version alone. Only an insistence in the pamphlet on the theme of Louis following the way of Christ (suggestive of Franciscan origin) signals a vaguely Observant tendency. Cacchio's text could not have come from a breviary, however, for in this case the author would have reproduced the right prayer. Moreover, the existence of the *Golden Legend* among the incunabula of libraries in the Abruzzo is documented. By this paradoxical and circuitous route, the Curia diffused its orthodox text in a lay, 'popular' and contestatory version. The cult did not escape the hold of the Church.

Another confirmation of the Church's hold and of the capital role of the Franciscans can be seen, changing epoch and place, by studying the many tracts devoted to St Reine of Alise throughout the seventeenth century, both in the *Bibliothèque bleue* and elsewhere. In this second case study we will examine not a single object but a series of works.

St Reine in Alise and Flavigny

The *Bibliothèque bleue* of Troyes, which has so often attracted the interest of historians in the last twenty years,[49] presented a body of literature defined as 'popular' or 'of popular destination', since the slim volumes grouped under that name, made cheaply using poor-quality paper, were for the most part sold in great numbers by pedlars. Religious literature accounted for much of this output (approximately one-quarter of all titles), and a number of lives of universal (more frequently, regional) saints were among them. Such books were most probably sold on feast days and at pilgrimage sites.

One of the pamphlets most often reprinted and rewritten recounts the life of St Reine of Alise (also spelled Regina). Alfred Morin's catalogue lists eleven different Troyes editions.[50] Several other titles from Troyes in the Bibliothèque Nationale and the Musée des Arts et Traditions Populaires in Paris, along with other 'popular' or semi-popular works published in Paris, Dijon, and Autun, should be added to the list.

A cursory reading of these pamphlets gives a disappointing impression of banality. Reine's biography was imitated from the life of a well-known universal saint, St Margaret. Reine, the daughter of a pagan lord, instructed in the Christian faith by her nurse, arouses the desires of the Roman prefect, Olibrius (the same name appears in the life of Margaret), who was smitten with love for her when he saw her guarding sheep. In the name of her faith, she refuses to marry him, proclaims herself a Christian, and undergoes martyrdom after miraculously surviving the cruellest conceivable tortures.[51] The site of her martyrdom, sanctified by a miraculous spring, became a pilgrimage site. What we see is the banal creation of a local cult based on a universal model that had withstood the test of time. The more interesting question is for what reasons there was the intense cultic activity and the proliferation of texts indicated by the number and variety of surviving tracts.

The little town of Alise-Sainte-Reine is located in the northern part of the modern département of the Côte-d'Or and it lies in the diocese of Autun. Geographically, then, it was within the zone of diffusion of the *Bibliothèque bleue*, which corresponded, roughly speaking, to the dioceses of Troyes, Rheims, Langres, Besançon and Autun. Alise stands on a hill facing another hill on which the great Benedictine abbey of Flavigny was built in the early eighth century. The cult of Reine, saint and martyr, is attested as early as the sixth century by a martyrology and by mentions in the liturgical calendars of Usuard and Ado (ninth century). The Flavigny breviary includes Latin litanies of the ninth century. A church in Alise, perhaps with Reine as patron saint, was first transformed into a monastery and then became a parish church. Benedictine tradition explains the foundation of this church in the fifth century by a miraculous revelation, celebrated on 13 July, of where the saint's relics lay. Still according to this tradition, Abbot Egilo of Flavigny had Reine's relics moved from Alise to Flavigny on 21 March (St Benedict's feast day) 864, an occasion on which, according to the Flavigny breviary, six miracles took place.[52] Up to this point, this historico-legendary schema is easily comprehensible, as the Norman invasions indeed produced a movement for the withdrawal of ritual activities into monasteries. Furthermore, from the ninth century on, the growth of the cult of relics and rivalry among monasteries and between monasteries and parishes provoked a frantic search for proofs of sanctity that went as far as the 'holy thefts' (*furta sacra*), described by Patrick J. Geary.[53]

We know nothing whatsoever of the fortunes of St Reine during the

Middle Ages, but it does not seem that any active pilgrimage developed. Reine's relics were placed in the rich collection of holy remains in the abbey in Flavigny, and Benedictine practice after the twelfth century dictated an extremely moderate and selective use of relics, exhibited only to important visitors. They functioned as parts of a treasury, not as means for edification or the propagation of the faith. The parish church of Alise, on the facing hillside, had given up Reine as patron saint and taken the title of St Léger, bishop of Autun.

Still, the memory of Reine did not completely die out in Alise and seems even to have revived towards the end of the fifteenth century, a period of renewal of saints' cults.[54] We have a record of this still modest resurgence in a conflict in the 1490s that pitted the parish priest of Alise, Michel Gueneau, against the *curé* of a neighbouring village, Julien Clerget, who claimed the right to set up an open-air altar near Reine's spring, at the site of her martyrdom outside the village. The bishop of Autun, Antoine de Châlon, ruled in favour of Julien Clerget and in 1498 gave him permission to build an autonomous chapel. The bishop acted on dual authority, both as bishop and as the local lord, since the territory of Alise was part of the mensal lands of the diocese of Autun. The account of this quarrel (by a Benedictine, Dom Viole, to whom we shall return) adds that the altar was in the diocesan vineyard, which indicates that the pilgrimage to Alise was little frequented. In 1501 the new bishop, Jean Rolin III, modified the agreement and placed the chapel under the parish authority of the *curé* of Alise.

The only traces remaining of the cult of St Reine in the sixteenth century are a stone statue in the church and a first hagiographic pamphlet, unfortunately not extant but mentioned in 1854 by Charles Nisard in his *Histoire des livres populaires ou de la littérature de colportage*. According to Nisard, '*La Vie et légende de madame sainte Reine* was printed by Jehan Lecoq in Troyes without date (*c.* 1510) in octavo gothic, in sixteen sheets, and reprinted by several other printers of the same city.'[55] It is reasonable to think that Julien Clerget's efforts launched both the pilgrimage and the publication of guides and souvenir texts.

The veneration of St Reine reached its height in the seventeenth century. In 1590 the chapel had been enlarged. The pilgrimage prospered: in 1598, the *bailli* of Auxois prohibited the sale of candles and medals in the chapel and ordered that the offerings box be locked with a triple lock, the three keys to be kept by the parish priest, himself, and the citizens of the town. His decision was overturned by the Parlement in 1600, after which the offerings box was locked with two locks and the keys were held by the priest and the head vestryman. The chapel of St Reine was enlarged again in 1613. Troyes continued to furnish tracts to accompany this development, and Alfred Morin notes a *Vie et martyre de madame sainte Reine* that he dates 1606.[56] The pamphlets were probably sold by pedlars in the nearby country areas, but they were also available at the pilgrimage site. The bishop's vineyard must have been covered with shops, for in 1611 a court decree forbade the construction of shops less than fifteen feet from the chapel. This little trade prospered throughout the century, and in 1674, the *bailli* of Touillon and Alise put out an ordinance against the public disturbances 'which are made by various persons, women, girls or others, through the commerce that they carry on in the chapel of St Reine in the sale and distribution of candles and tapers . . . and against those who work and sell [there], merchants, taverners, haberdashers, and bakers.'[57] The role of the haberdashers in the distribution of the *Bibliothèque bleue* volumes is well known.

The Undertakings of Curé Cadiou: Holy Waters and Hagiography

Reine's glory, however, was to set more important interests simmering and to launch a veritable village and regional war that had a profound influence on the production of hagiographic tracts. The pilgrimage to the miraculous waters of Alise gained nearly pan-European status: one pilgrims' guide to Alise notes the access routes from Italy and Flanders, and in 1670 the bishop of Autun estimated that 70,000 pilgrims visited Alise each year. The holy waters were bottled and sold in Paris, and wagons making the trip to the capital and back every week carried 40,000 bottles a year. The Queen, Anne of Austria, used the water, and Alise continued to profit from the partly balné, partly devotional sojourns of illustrious visitors throughout the century. A hospital opened in 1661 that could serve 20,000 indigent or infirm pilgrims a year.

Profits from the pilgrimage, both material and symbolic, set off a violent rivalry after 1628, when Jean-Baptiste Cadiou, priest and doctor of theology, was named parish priest. An energetic man, he undertook the systematic exploitation of the site, delegating a vicar to the parish church and taking over the management of the chapel himself, with another vicar. He persuaded the bishop to grant him a monopoly of religious services at the chapel, he had baptismal fonts built, and he had a large basin for devotional ablutions dug downhill from the chapel. This remarkable man understood the importance of popular hagiographic publications, and around 1630 he wrote a pamphlet of which a somewhat later version (before 1638) remains (Troyes, chez Nicolas Oudot rüe Nostre Dame), the *Vie de Saincte Reine. Avec la Messe et Miracles nouvellement faits*. We can postulate the existence of an earlier edition because the approval of the doctors of the Church, dated 6 March 1630 and signed (an important detail, as we shall see) by the Dominicans of Autun, gives the title as *La vie de Sainte Reine avec les admirables effets de l'eau de sa fontaine*. The permission of the bishop, Claude de Ragny (D[4] verso of the extant copy) gives another title and cites a second pamphlet by Cadiou, discussed below. It says, 'We permit the printing, sale, and distribution of the *Vie, les Miracles, l'Instruction du Pèlerin* and the *Cantiques Spirituels composez à la louange de Ste Reine.*'

This small volume (14.5 cm × 8 cm, 18 unpaginated sheets with signatures A to D) presents the usual characteristics of the books in the *Bibliothèque bleue* in spite of its panoply of unctious and saccharine forewords and messages to the reader.[58] In his dedication to the bishop of Autun, Claude de Ragny, Cadiou borrows the ecclesiastical rhetoric of the 'unworthy author', and in his foreword this energetic entrepreneur presents himself as in search of a subject. He wrote, he says,

after surveying the subjects that might serve as a theme for more prolific pens and at the urging of my friends, whom I honestly dared not refuse, and I would none the less not have given them this satisfaction were it not for the particular duty that I owe to this saint for [my] being, though unworthy, chaplain of the chapel in which she is venerated and where so many miracles take place daily.

Cadiou's foreword is interesting for its viewpoint on an earlier source judged to be 'popular', which could only be the *Vie* published in Troyes in 1606 (perhaps as early as 1510). He says,

Light was shed some time ago on the story of this life by some person who was mistaken in some points for not having researched it at its source and for having taken it from the talk of common folk, with whom the most certain truths change with the passing of time just as surely as waters lose their purity as they flow away from their place of origin. But this treatise, the errors of which the author has corrected, will suffice to deduce this life in all its original clarity and truth as drawn from the Venerable Bede, Usuard, Mombrizio and other authors.

Nonetheless, the extreme meagreness of biographical data on Reine hardly lent itself to radically divergent possibilities. Cadiou's version differed most from previous tracts in its ample Church authentification. What the *curé* was attempting to do was to legitimate reappropriation of a lay narrative by borrowing its form. His own chief contribution was most probably his urgent encouragement of pilgrimages. Two 'devout prayers' in French that follow his narrative invited the faithful with paraliturgical formulas such as 'How great is my joy to have arrived in your holy Chapel.' The pilgrim's souvenir booklet also included the Latin text of the Mass of St Reine with the Gospel reading for the day (the only work in the corpus of materials on St Reine to do so). The pamphlet ended with a calendar of feasts, which had a certain importance, as we shall see. The two main feasts of St Reine mentioned were 7 September, the 'Vigil of the Nativity of Our Lady', and the 'Revelation of 13 July'. The list continues: 'The attendant feast days of St Reine are on the feasts of the Annunciation of Our Lady, Pentecost, St Claude, St Reine, and All Saints' Day.' About the same time (between 1630 and 1638, according to the indication furnished by the publishing data), the *curé* of Alise wrote *Les Cantiques spirituels. Nouvellement composez à la louange de Dieu et de Madame Ste Reyne Vierge et Martyre*, 'by J.B. Cadiou, Chaplain of the Chapel in which the saint is venerated'. The imprint was given as 'A Troyes, chez Nicolas Oudot demeurant en la ruë Nostre Dame au Chappon d'or couronné'. By 1638, the year of Nicolas Oudot's death, there had already been several editions of Cadiou's text, as indicated by the mention, 'Revised, corrected and newly augmented from the previous printings'. Of the same format as the *Vie* (14.5 cm × 8 cm), the *Cantiques* (12 unpaginated sheets, signatures A to B) contained seven hymns in French, indicating the melody by reference to secular songs, 'On the air of "Enfin celle que j'aymois tant" ', 'of "Cessez, cessez vos pleurs" ', 'of "Léandre" ', 'of "Capucin, rendre je me veux"', 'of "Du fond de" ', 'of "Léandre" ', 'of "Destin qui séparez" ', and so forth.

If we can trust the catalogues of the *Bibliothèque bleue*, this was one of the first occurrences of a collection of hymns to be sung to secular airs. It is interesting that this popular touch came from a churchman, who had no compunctions about using an off-colour and anticlerical song ('Capucin, rendre je me veux') for his hymns. This may be the first trace of the fight unto death (literally) that some years later was to pit father Cadiou against the Cordeliers of Alise, as we shall see. Popular culture, brought back under the control of the Church, had little autonomy in the *Bibliothèque bleue*. We have paused over this pamphlet both because of the striking personality of Cadiou and because its text served as a model for all later *Bibliothèque bleue* editions, which merged the narration and the hymns, somewhat abridged and adapted, into one volume.

The publication of pamphlets was an important issue in subsequent developments, and each camp added its nuances, more noticeable in the accompanying liturgical portions than in the narrative. It was as if rapidly formed ecclesiastical struggles roiled and seethed underneath an immutable layer of closely related texts, unbeknownst to the simple worshipper who bought a pamphlet from a pedlar or at the haberdasher's shop in Alise. The people, dispossessed of their own forms of expression (the pamphlet, the song), unconsciously aroused the appetites of warring powers.

The Badier Family and the Cordeliers

The fight was becoming fierce. In 1631, the Maréchale of Saint-Luc attempted to establish a true congregation at the chapel of St Reine, with five canons and four chaplains. The project failed, but so did Cadiou's one-man campaign, which collapsed before the hostility of the inhabitants of Alise, who had watched profits from the pilgrimage elude them after Cadiou gained control of all the chapel revenues. Heading the villagers was Philibert

Badier, the innkeeper of *A l'écu de France* in Alise. His resentment of Cadiou went back to 1628, when Cadiou had become parish priest, a position Badier had wanted for his son Jacques. The Badier family offered a remarkable example of social promotion, certainly thanks to the temporal benefits of St Reine. The eldest of the sons, Thomas, became *bailli* of Flavigny; Jacques was a parish priest; Etienne was a monk (in Flavigny); Philibert the younger was the official *procureur* of Flavigny. This up-and-coming family sued Cadiou on the issue of his monopoly of the income from the chapel, arguing that the diocese owned the holy site. A decree of Parlement dated 26 December 1635 ruled in favour of the inhabitants of Alise, stipulating that, in conformity with the earlier 1600 decree, the *habitants* of the village and the *curé* were to share the revenues equally. On 28 July 1640, however, the Parlement clarified its decision and charged the head vestryman, as representative of the community of the faithful, with accounting for all moneys received in this fashion between 1600 and 1628, at the time when Cadiou was appointed parish priest, which amounted to a considerable sum. Simultaneously, Jacques Badier attempted to attack the original source of the conflict, the joining of the chapel to the parish, by bringing suit before the pontifical courts and the Parlement, which rejected it on 22 March 1640.

In 1640, the threat to examine the church's financial records added fear to jealousy. The Badiers and the other villagers did not feel they could take on the expenses of a trial; above all, they did not want to risk having to reimburse moneys that the village had received above and beyond the legal norms. They appealed to the Cordeliers, and succeeding developments proved the accuracy of La Fontaine's famous fable of the cat, the weasel and the little rabbit. The Badiers, who already had ties with the Cordeliers in Dijon and Autun in 1640, chose Philibert Badier the younger as 'temporal father' of their enterprise.

As is known, the branch of the Franciscan order that had separated off from Conventual orthodoxy in the fourteenth century under the name of the Observance were commonly called Cordeliers. The Observant friars, whom we have already seen in the Abruzzo in the context of Cacchio's pamphlet, claimed to return to the spirit of St Francis, rejecting the institutional compromises of the Conventuals, a move that was repeated in the sixteenth century by a third branch of the Franciscans, the Capuchins, when they in turn split off from the Friars Minor of the Observance. The Cordeliers in Burgundy could assume the heavy financial risks involved in the Alise operation. Their economic power, founded on systematic collections of alms and appeals for bequests, gave them a solid financial base that was enhanced by the brothers' scrupulous respect of the principle of personal poverty and by a rapid circulation of investments that, after the Middle Ages, made the order a symbolic model of dynamic capitalism. Furthermore, in the seventeenth century the Cordeliers played a politico-religious role of capital importance as confessors to the mighty (a function that they shared with Jesuits and with Capuchins such as the famous Father Joseph of Paris). The Cordeliers were firmly entrenched in Burgundy, the province of St Bonaventure, in Dijon and Autun in particular.

If the Cordeliers took charge of the chapel, it would enable the Badiers to get rid of Cadiou, to take on the risk of having to reimburse the offerings collected to assure their own political (and hence judiciary) power. The man chosen to be guardian of the Cordeliers' monastery in Alise, François Marmesse, confessor to the duke of Longueville, had the ear of the prince of Condé (the governor of Burgundy), the duke of Enghien, and the prince of Conti, princes of the blood and close kin to the duke of Longueville. Through the prince of Condé and as allies of the Cordeliers, the Badier clan could hope for the juridical support of the Parlement of Dijon. Moreover, the affection that the Queen, Anne of Austria, bore the Cordeliers was well known, as was her personal devotion to St Reine. The Badiers hoped, then, that even if the Cordeliers demanded the lion's share of the spoils Father Cadiou would be removed from the parish, leaving the post to Jacques Badier.

On 29 July 1644, then, a contract was drawn up between the Cordeliers and the citizens of Alise. The Franciscans were to have the right to establish a monastery with a donation of seven *journaux* of land. They were to have all offerings, in exchange for paying an annuity to the parish priest, for agreeing to provide aid and alms for poor pilgrims, and, above all, for promising to take care of any eventual reimbursements when the parish account books had been gone over. The bishop of Autun, Claude de la Magdeleine de Ragny, whose family had ties with the Retz and the Lesdiguières families, delayed giving his approval to a contract that diminished his episcopal and seigneurial rights and that installed the Cordeliers, close allies of the Condé clan. Anne of Austria wrote twice to the bishop, who was finally obliged to cede. On 24 August 1644, *lettres patentes* from the King sealed the agreement. Although the bishop and the parish priest kept nominal direction of the pilgrimage, the Cordeliers' authority increased with each passing year and even each passing month. In October 1644 they obtained permission to use the chapel for their conventual offices; later they obtained the right of patronage over the parish and of nomination of the parish priest.

The parish priest, Cadiou, still had not been ousted, and for another twenty-six years he continued to resist with all his might. Brawls took place in the chapel in 1645. In 1646, Cadiou appealed the decisions of the Parlement of Dijon, which removed him from his cure. In 1647, the Parlement of Grenoble re-established him in all his rights. The suits continued until 1670, renewed each time the Cordeliers named a parish priest. In spite of several decisions of the king's Council in favour of Cadiou, demanding a stay to his expulsion, he was obliged to resign his charge twice, in

1649 and in 1650, under direct pressure from Anne of Austria. Each time, Cadiou took up his judicial battle again, arguing that he had been subject to constraint when he resigned. He was defeated in the long run, and the Cordeliers ended up with total control of the pilgrimage. The Cordeliers and the Badiers, impatient with judicial delays, tried tougher tactics by charging Cadiou with 'spiritual incest' with a parishioner, Anne Trivelet, who later admitted that she had been paid to testify. On 6 August 1649, Cadiou was sentenced to be hanged and was indeed executed in effigy, which did not prevent him from becoming canon of Autun in 1651 or from continuing his vain judicial campaign.

Cordeliers versus Benedictines: The War of the Texts

In the meantime, however, the Cordeliers had pursued their conquest of Alise on another front. It brought a new actor onto the scene, who was also destined to be crushed, but after a more equal fight.

In 1648, François Marmesse, Cordelier and guardian of the Alise monastery, accompanied the duke of Longueville in his embassy to Münster for the signing of the peace of Westphalia. Marmesse took advantage of the opportunity to ask the bishop of Osnabrück in Saxony for relics of St Reine, who was venerated there because (according to a local legend that the Cordeliers soon propagated in Alise) Charlemagne had had the saint's body sent from Alise to Osnabrück when he conquered Saxony and converted the Saxons. Thanks to his powerful connections, Marmesse obtained a radius from the saint's forearm, which he took in triumph to Paris, then to Alise, taking the precaution of obtaining letters of authenticity and of approval from the duke of Longueville, the bishop and the cathedral chapter of Osnabrück, the bishop of Paris, and the bishop of Autun.

Marmesse wasted no time – perhaps cleverly, perhaps cynically – in having republished the life of Reine written by his arch enemy, Cadiou, adding a word on the miraculous novelty. The text, unfortunately lost, is known only through Dom Viole, who says,

The Reverend Father François [Marmesse] had it [news of the arrival of Reine's forearm bone] preached publicly in Alise; [he] obtained approvals from Our Lords the Archbishops of Paris and Autun and had several little tracts written, in particular, several additions to the life of St Reine written by Jean-Baptiste Cadiou, parish priest of Alise, and [had them] printed in the year 1648.[59]

Marmesse's move constituted a real declaration of war with the nearby abbey of Flavigny, which claimed to possess all of Reine's mortal remains. As we have seen, the monks of Flavigny had done little to promote the cult of St Reine, although Nisard affirmed in 1854 (without offering any proof or evidence) that mysteries centering on Reine were performed at the abbey.[60] The best Dom Viole could say in defence of the abbey was that these relics were 'quite often shown to Pilgrims, according to the persons' merit'. His reticence is worth noting: a Benedictine scholar as impassioned and as well-prepared as Dom Viole would not have passed up a chance to brandish proofs of the antiquity and importance of the pilgrimage to Flavigny.

Until 1648, the abbey stayed out of the quarrels on the neighbouring hillside of Alise, and, like many Benedictine monasteries in the early modern era, it was undergoing a slow decline. In 1620, it was forced to sell the priory of Couches to the Society of Jesus, and in 1642 it had to cede its precious treasury of medieval manuscripts, again to the Jesuits. We can imagine that the Benedictines were not displeased as they watched the Badiers' first attacks on Father Cadiou, and we have already seen the Badier clan's many ties with Flavigny, which were reinforced when Badier, the weasel, was cheated of his prey by the Cordelier cat. The situation changed around 1640, however, when the Benedictine reform of Saint-Maur was imposed on the abbey in Flavigny.

This revival enabled the abbey to react more forcefully to counter the Cordeliers' attacks, and Dom Georges Viole (1598–1669), a Benedictine scholar who came from a prominent family of Parisian magistrates, was delegated to mount a counterattack. As was the rule with the Benedictines, the attack took a learned (or semi-learned) form in the publication of a slim volume (18 cm × 12 cm; 85 pages) published in Paris in 1649 by Claude Huot, *La Vie de Sainte Reine, vierge et martyre, avec une apologie pour prouver que l'Abbaye de Flavigny, ordre de Saint-Benoît, au Diocèse d'Autun, est en possession du sacré corps de cette sainte*. This narrative, neatly broken up into chapters, differs little from the canonic version of Reine's life, which was beginning to be distributed and known even outside the region of Alise. This broader distribution was aided by the publication of the adaptation in French of the *Flos sanctorum*, the *Fleurs des vies des saints et fêtes de toute l'année*, written at the end of the sixteenth century by the Spanish Jesuit, Pedro de Ribadeneyra. This was the most widely distributed work of the sort in France in the seventeenth century: Orgon, in Molière's *Le Tartuffe*, had a copy of it in his house. The 1646 edition of the French adaptation of the work contained a chapter on St Reine. Dom Viole brought to the argument an extremely detailed refutation of the Germanic and Cordelier legend of the translation of Reine's relics to Osnabrück. We need to summarize the various stages of the polemic briefly before returning to some of the arguments contained in it.

The Cordeliers responded with ardour and on two fronts. First, they respected the polemical and scholarly mode by publishing (Paris, 1651, printed by Edme Martin) an *Eclaircissement sur la véritable relique de Ste Reine d'Alyse donnée à Monseigneur de Longueville par l'Evesque et Chapitre d'Osnabrück*.

Pour servir de Reponse à une libelle intitulé Apologie pour les véritables reliques de Flavigny... by 'les religieux de saincte Reyne d'Alyse'. Second, in the same year of 1651 and fully aware of the importance of the popular pamphlet, they abandoned Cadiou's narrative and wrote their own pamphlet, published by the same printer and in the same format (13.5 cm × 8 cm; 52 pages), *Histoire de Ste Reyne vierge et martyre comprenant sa naissance, sa vie et sa mort; l'élévation et translation de ses saintes reliques; une authentique approbation de celle qui est de présent en sa Chapelle à Alyse; ensuitte un petit Office et des litanies; le tout consacré à sa gloire,* 'by an Observant religious of the province of St Bonaventure'. The Cordeliers abandoned their Trojan – or Troyesian – horse (Cadiou's *Life* published first by the bookseller-printer Oudot and then by his widow) and doubtless took over the distribution of their pamphlet themselves.

The Franciscans returned to the time-honoured tradition of the pamphlet on a saint's life and miracles and adapted it to the new forms of private devotion and pilgrimage. They substituted the 'minor office' in Latin for the text of the Mass. This office was modelled on conventual hours and it included salutation, invocation, hymn, psalm, antiphon, versicle, response, and prayer, followed by a prayer in French. It was aimed at individual devotions as well as parish and public celebrations on 7 September, the principal feast day of St Reine. Pilgrimages must have taken place at other times as well, to judge by the list of days when a visit to St Reine brought an indulgence (published at the end of the volume): 'Plenary indulgence conceded by our holy Father Pope Paul V exclusively for the brothers of St Reine ... 7 September, All Saints' Day, 25 March, the three days of Pentecost, 6 June'. Immediately following the narrative (pp. 1–22) came words of praise for the water of St Reine (mentioned very discreetly in the earlier texts). The virtues of the waters were described, along with 'its antipathy towards all illnesses, towards fever, kidney stones, gout, paralysis, scabies, dropsy ... [and] the deadly and shameful illnesses caused by impurity'. The Cordeliers, who exploited the holy waters, were quick to praise their efficacy in curing venereal diseases, a virtue that assured a good part of their celebrity in the seventeenth century, as we can see in a malicious anecdote of Tallemant des Réaux. 'The archbishop of Paris', he writes (Cardinal Henri de Retz, the famous author's uncle and a man known for the freedom of his mores), 'had had a chapel built which he had dedicated to I know not what saint. "I cannot think", Bautru said, "that it could be dedicated to any but St Reine".'[61]

After praising the waters of Alise, the Cordelier author went on to establish the legitimacy of the new relic by recounting briefly the translation of the saint's remains in the ninth century and the return of the sacred bone 'to Alyze, otherwise called Saincte Reine, into the hands of the Fathers of the Observance of St Francis established by their Majesties'. A miracle attested to the reality of the returned relic: one Catherine Le Blanc from Beaune had been cured, as proved by a court report, which,

signed by two royal notaries, by several doctors and by many irreprochable persons, approved by Mgr the diocesan Bishop of Autun and printed in Paris, will be able to satisfy in full measure the curiosity of those who believe only great things and who only accept the truth when they are amply instructed in all the circumstances. (p. 27)

We should note in passing the Franciscans' tendency to print everything – apology, pamphlet, court proceedings. The truth got printed.

The Saint with Three Radii

Dom Viole did not give up. He attacked on the same two fronts, although with a strong preference for the learned and polemical. He brought in experts (in 1649, 1651, and 1659) to testify that both of Reine's radii were indeed in Flavigny, and that the Alise radius, which was bigger, was out of proportion to the small stature of the young martyr. The learned Benedictine gives the impression, however, of distaste at having to stoop to the spectacular and somewhat vulgar methods of the Cordeliers, whom he calls 'petty merchants who traffic in everything'. Later, and with misgivings, he resolved to exhibit the true radii:

In the year 1655, on the occasion of a certain arm supposedly of St Reine that the Reverend Father François Marmesse, Cordelier, brought from Germany and exposed publicly in Alize as the arm of St Reine, we were obliged to draw from their ancient reliquary the two arms of our saint.[62]

The battle continued, for the Cordeliers intended to oust the Benedictines completely. In 1671, the procession from Flavigny to Alise, an ancient and discrete tradition among the monks of Flavigny, was confined to the abbey, and in the following year, 1672, the Franciscans created their own procession on the same days (7 September and Trinity Sunday). Father Bazin, the new guardian of the Cordeliers, went so far as to demand the translation of all of Reine's relics from Flavigny to Alise, but thanks to the arbitration of the bishop of Autun in 1693, the rival orders were permitted to exhibit their relics concurrently, and Reine officially had three forearms.

Earlier, Dom Viole had responded swiftly to the Franciscans' allegations by reprinting his 1649 *Apologie* (Paris, 1653, printed by Jean Piot), adding to the title, 'Second edition augmented with several specific reflections as a Response to a booklet entitled *Esclaircissement sur la véritable relique de saincte Reine d'Alize*. With proofs taken from the foundation and other charters and ancient manuscripts of Flavigny'. The strictly historical argument concerning the ninth-century translation of the relics interests us less here than noting the difference in attitude toward printed matter between the Benedictines and the Franciscans. Dom Viole places his trust uniquely in ancient texts and manuscripts, whereas the Franciscans give up historical arguments, always a possibility in an age so rich in charters and contradictory chronicles, and surround themselves with contemporary proofs and acts of authentication, no sooner in hand than printed. Typography worked wonders for veracity.

Marmesse brought back from Saxony a printed calendar that proved that Reine was venerated in Osnabrück, and in 1651 the Cordeliers complained of Dom Viole's scorn of this handsome object:

You say that the Reverend Father François thought to have found the truth of his relic by showing small yellow images in which the image of St Reine can be seen. . . . What you call yellow images are none other than an extremely handsome calendar that he brought back from Osnabrück on which the image of St Reyne is engraved and her feast marked as for a Patron [saint] of that country.[63]

Dom Viole was untouched by the offended sighs of the Cordeliers and with a sardonic smile he went on to sharper mockery in 1653:

As for the little images, I confess in truth that I thought that he [Marmesse] had brought them back for no other purpose than to serve as prizes at the Catechism of the small children of Alize; and since good housekeepers find a use for everything, I am grateful to him for proving in images a translation that by this means he will render doubly imaginary. If it were true that a saint is everywhere that his image is distributed, the Reverend Father would soon have the price of his, since he would have not only the arm but the entire body of St Reine.[64]

The episode clearly shows the Franciscans' attitudes towards the written and the printed word. In the tradition of Francis, Thomas of Celano, John de la Rochelle, and Brother Lucas, they learned how to use techniques made available by print to surround the faithful with sacred or magical writings – pamphlets, talismanic formulas, images – that functioned as variants on representations, relics, or the Eucharist. Whether paper was swallowed, chewed, or ruminated upon, the written word went to nourish the pious.

Dom Viole also continued to give battle on the front of popular literature by giving a more 'popular' cast to his 1649 life of St Reine, which went through a large number of printings in the provinces (fifteen at least before the eighteenth century). Many of these editions are lost to us, but we do have a seventh edition printed in Autun in 1669 by Blaise Simonnot, 'Printer of Mgr the Most Illustrious and Reverend Bishop'. Small in format (15 cm × 10 cm), it was more roughly produced than the first edition, and the earlier frontispiece is presented as a woodcut rather than a copperplate engraving. Dom Viole swept aside his original argument to present the work as an official guide to worship: *La Vie de Sainte Reyne, vierge et martyre. Avec son petit office, et un catalogue des principalles reliques de l'Abbaye de Saint-Pierre de Flavigny.*

A new chapter division shows an attempt to cater to a humbler public, but it is even more evident in the translation of the office into French. The polemical thrust of the pamphlet is reduced to a few rapid notes at the end of the narrative, a complete list of the relics held in Flavigny and a calendar of the pilgrimage days in which Viole plays down the feast of 13 July (the elevation of the relics 'celebrated with particular ceremonies in Alize') in favour of the Benedictine feast of the procession from Flavigny to Alize on Trinity Sunday, called 'one of the most beautiful ceremonies to be seen'.

Dom Viole stresses the legitimacy of the Benedictines' claims more visibly elsewhere. The arms of the bishop of Autun adorn the cover papers (of mediocre quality), and the pamphlet is well supplied with ecclesiastical authentications: a letter to Mgr Gabriel de la Roquette (pp. 3–10), an approval of the doctors of the Church dated 1649, and a permission from the bishop. A foreword written by the Bishop of Autun for the fifth edition of 1659 (p. 68) limited the scope of the feast day, which had become overextended:

But because the too great multitude of feasts that have been introduced gradually and with some sort of abuse in several places has obliged several of our predecessors and ourselves as well to cut back a portion of them for very good reasons, we have limited the command to abstain from work on this day to the district of Alise and the district of Auxois alone, and to all the archpriestly territories of Toüillon and Duesme.

This ordinance is difficult to interpret. It might reflect attempts to establish new working hours to correct medieval practice: at the end of the Middle Ages *pigritia* (laziness) took the place of *acedia-tristitia* (depressive melancholy) as one of the seven capital sins. Or it might mean that the bishop was attempting to control the flood of worshippers, increased through the efforts of the Cordeliers, who by now were solidly entrenched in Alise.

The attitude of the various bishops of Autun would seem to confirm the latter hypothesis. In 1652, Mgr d'Attichy succeeded Claude de Ragny and attempted to oppose the inroads of the Cordeliers with greater vigour than his predecessor. His successor after 1667, Mgr de La Roquette, returned the parish of Alise to the secular priesthood in 1669, the year of the seventh edition of the *Vie de Sainte Reyne* discussed above. This decision was overturned by the Parlement of Dijon, as always loyal to the Franciscans, but was upheld in 1673 by the King's Council, and for several years the Cordeliers lost control over the saint's spring, only to be recalled by the bishop, who soon repented and missed their talents as managers and propagandists. The second installation of the Franciscans was to last until the Revolution. The Cordeliers even obtained the right to join their own chapel to the chapel of St Reine, with its immersion pool, by roofing over the street between the two. The bishop's decision to authorize the dual exhibition of relics sealed the definitive victory of the Franciscans.

Our 1669 pamphlet gives evidence of the bitterness of the battles: Mgr de La Roquette's permission to print attempts to impose a monopoly for Dom Viole's pamphlet that is certainly aimed at the Cordeliers' tract:

We permit Blaise Simonnot, our printer, to print, sell and distribute the Life of St Reine Virgin and Martyr composed by Dom Georges Viole, Benedictine religious of the congregation of Saint-Maur, with prohibition to all Haberdashers, Pedlars and others on selling other [such lives] in our diocese, and in particular in the territory of our Seigneurie of Alise, on pain of confiscation of the said copies and of all expenses, damages and interests, enjoining furthermore our Officers to enforce this.

Dual episcopal and seigneurial power gave Mgr de La Roqette powers to execute his decree, and, in point of fact, we can find no trace of the rival pamphlet up to the Troyes editions of the 1680s, whereas Dom Viole's book circulated widely. There was a ninth edition in Châtillon, printed by Claude Bourut in 1684, and unnumbered editions in Semur, printed by Michard in 1715, and in Dijon, printed by Joseph Sirot in 1724.

The Trojan Horse

The Cordeliers' silence did not signify total success for Dom Viole's literary efforts. When they found it difficult to distribute their own pamphlet because of the 1669 ordinance, they seem to have passed Reine's torch to the publishers in Troyes, whose books could be sold in the neighbouring dioceses of Langres and Dijon (a diocese created in 1731), and even in Champagne in the dioceses of Troyes and Rheims, cities from which the pilgrimage started. The first anonymous pamphlet printed in Troyes by Pierre Garnier carried an approval by 'Bernard Gonthier, priest, doctor of theology, episcopal curate of Langres', dated 1678. Its title is a close imitation of Dom Viole's: *La Vie de Ste Reine, vierge-martire. Avec son petit Office en François, ses Litanies, Cantiques et Oraisons. En faveur des dévots Pèlerins qui visitent son sanctuaire*. The volume (duodecimo, 24 unnumbered sheets, 14 cm × 7.5 cm) fitted perfectly into the tradition of the Troyes *Bibliothèque bleue*, and it contained nine woodcuts, each printed on a left-hand page facing a right-hand page of narration. The text was a simplification of the Cadiou-Cordelier version, and the six hymns given were Cadiou's melodies and words with only slight abbreviations and some modernization in language.

The work contains no polemics to show its rivalry with the Cordeliers' text, since that would have made it inaccessible to the people. Rather it accentuates the sacredness of Alise, the 'sanctuary' designated in its title. The narrative prologue opens with a celebration of the holy waters; prayers in French follow the stages of the pilgrimage to Alise as the Cordeliers had planned it: 'Prayer to be recited in the chapel of St Reine of Alise . . . Prayer before the Crucifix . . . Prayer going to the spring . . . Prayer entering into the bath . . . Prayer at the three crosses'. The only text in Latin, that of the Litanies, refers to Franciscan mediation:

They [the litanies] are sung every evening in four parts by the Religious of the Convent in the Chapel of Alise, before the altar on which there is the miraculous image of this Saint, to satisfy the devotions of the Pilgrims and the poor sick people and the persons of piety who are in attendance from all places. . . . This Antiphon is then sung to the honour of the good Saint to satisfy the Novenas of Prayer that the Pilgrims ask the religious to undertake.

Finally, the liturgical calendar registered the Cordeliers' take-over of the procession of Trinity Sunday in 1673: 'Trinity Sunday: On this day the Procession to Flavigny is made, where relics are carried.'

The tactical flexibility of the Cordeliers was an updating of the Trojan (or Troyesan) horse. Although the printers of Troyes were by no means autonomous or out of the range of Church quarrels, a growing number of such pamphlets was put out in the eighteenth century by Pierre Garnier, by his widow, by Jean-Antoine Garnier, and by A. P. F. André. Parallelling the great battle taking place at Alise, another work was published, first by those same printers (a dozen editions from the end of the seventeenth century to the end of the eighteenth), and then in Châtillon and Paris. It was *Le Martyre de la glorieuse Sainte Reine d'Alize*, a tragedy in five acts by Claude Ternet, 'professor of mathematics and surveyor of the King in the Châlonnais', dedicated to the bishop and first published in Autun in 1682. This half-devotional, half-burlesque play follows in the tradition of a series of semi-scholarly provincial works such as *Le Chariot de triomphe tiré par deux aigles de la glorieuse, noble et illustre bergère, sainte Reine d'Alise, vierge et martyre* by Hugues Millotet, a tragedy in five acts (Autun, 1664); *Le Triomphe de l'amour divin de Sainte Reine, vierge et martyre*, a *tragédie en machines* in five acts dedicated to the queen by Alexandre Le Grand, sieur d'Argicour (Paris, 1671); and *La Victoire spirituelle de la glorieuse saincte Reyne, remportée sur le tiran Olibre*, a tragedy in five acts 'newly composed' by Pierre-Corneille Blessebois (Autun, 1686).

The battle thus ended in songs and theatricals. The decline in the production of hagiographic works, which lacked invention after Pierre Garnier's pamphlet (1680?), corresponded to the Cordeliers' total domination of Alise and to the related decline of the pilgrimage, which turned secular during the eighteenth century. The waters of St Reine joined in the nascent vogue for water therapy, and although the Cordeliers refused in 1713 to have the waters of Reine's spring analysed by medical experts, they specialized increasingly in bottling the water.

The 'popular' pamphlets on Reine had totally escaped the people, however that entity is defined. From Julien Clerget at the end of the fifteenth century to Fathers Marmesse and Bazin at the end of the seventeenth, both cult and propaganda proceeded from the Church and from its rival agents (the secular clergy, the Cordeliers and the Benedictines). Apparently popular forms – the use of the French language, of popular melodies, of woodcuts; distribution by pedlars, etc. – all derived from precise tactics that were directed not so much towards a particular public as against ecclesiastical rivals. This does not exclude (indeed, it implies) a cultural, commemorative, and almost magical use of the pamphlet, which the Franciscans fitted into their more general campaign of bombarding the faithful with works to touch the imagination, a popular counterpart of the great Jesuit enterprise of seduction by the image and by drama that grew out of their schools in the seventeenth century. This is the way religion was peddled.

The Franciscans assured a mediation of capital importance between the mute adoration of sacred books and the rival orders' long-winded devotions. They nonetheless ended up where they had started, with an unequal division of knowledge and beliefs. The history of our hagiographic pamphlets superimposes three overlapping but autonomous uses:

the popular consumption of a printed object, sold on the saint's feast day and carefully preserved as a small monument to a bookish worship that was both magical and Christian;

the manipulation of memory according to the circumstances in opposing and unstable configurations of social and religious references;

an absolute domination of the texts and of worship by the Church.

Flow between these uses seems to have been assured here by the Franciscan order. Their role in the diffusion of a Christianity that remained strictly Catholic and Roman but that welcomed popular practices still awaits full evaluation. The Franciscans, with their admirably ramified yet centralized structure, acted as coherent agents for Christian acculturation and for the alienation of devotional practices from the people. The hagiographic pamphlet, even when it was conceived as an instrument for reappropriation, in reality led to the Romanization of Christian Europe, in Combronde as in L'Aquila and in Alise. It assured a new and strong intimacy with the world of writing, and it contributed to the foundation of one of the original characteristics of European popular culture, its affinity with print.

Notes

1   Thomas of Celano, *Tractatus de miraculis*, miracle no. 193, published in *Analecta Francescana* (1926–41), vol. 10, p. 328. My thanks to Chiara Frugoni of the University of Pisa for bringing this text to my attention.
2   See the texts collected and commented on in Théophile Desbonnets, OFM, and Damien Vorrieux, OFM, *Saint François d'Assise, documents écrits et premières biographies*... (Editions franciscaines, Paris, 1968).

3 Jacobus de Voragine reports this miracle in his chapter on Barnabus (*La Légende dorée*, tr. J.-B. M. Roze (3 vols, E. Rouveyre, Paris, 1902; 1967), vol. 1, p. 393).

4 See Joseph-Claude Poulain, 'Entre magie et religion. Recherches sur les utilisations marginales de l'écrit dans la culture populaire du haut Moyen Age', in Pierre Boglioni (ed.), *La Culture populaire au Moyen Age* (L'Aurore, Montreal, 1979), pp. 187–243.

5 Guibert of Nogent, *Autobiographie (De Vita sua)*, ed. and tr. Edmond-René Labande (Belles-lettres, Paris, 1981), pp. 396–7.

6 Jacques Verger, 'Condition de l'intellectuel aux XIIIe et XIVe siècles', in Ruedi Imbach and Maryse-Hélène Méléard (eds), *Philosophes médiévaux des XIIIe et XIVe siècle: Anthologie de textes philosophiques* (Union Générale d'Editions, 10/18, Paris, 1986), p. 31.

7 John de la Rochelle, 'Sermon pour la fête de saint Antoine de Padoue', presented and tr. Louis Jacques Bataillon in Imbach and Méléard, *Philosophes médiévaux*, p. 60.

8 Jacobus de Voragine, *Légende dorée*, vol. 2, p. 263.

9 François D. Boespflug, *Dieu dans l'art. Sollicitudini Nostrae de Benoît XIV (1745) et l'affaire Crescence de Kaufbeuren* (Editions du Cerf, Paris, 1984).

10 ibid., pp. 133–4.

11 See Vincenzo de Bartholomaeis, *Laude drammatiche e rappresentazioni sacre* (F. Le Monnier, Florence, 1943).

12 Sulpicius Severus, *Vie de saint Martin*, intro. and tr. Jacques Fontaine, (3 vols, Editions du Cerf, Paris, 1967–9).

13 See Patrick J. Geary, *Furta sacra: Theft of Relics in the Central Middle Ages* (Princeton University Press, Princeton, NJ, 1978).

14 Francis Wormald, 'Some Illustrated Manuscripts of the Lives of the Saints', *Bulletin of the John Rylands Library of Manchester*, 35, 1(1952), pp. 248–66.

15 See Alain Boureau, *La Légende dorée. Le système narratif de Jacques de Voragine (1298)* (Editions du Cerf, Paris, 1984); Boureau, 'Barthélemy de Trente et l'invention de la legenda nova', in S. Boesch Gajano (ed.), *Le Raccolte de vite di santi dal XIII al XVIII secolo*, forthcoming.

16 See Jean Destrez, *La Pecia dans les manuscrits universitaires au XIIIe et au XIVe siècle* (J. Vautrain, Paris, 1935); Carla Bozzolo and Ezio Ornato, *Pour une histoire du livre manuscrit au Moyen Age. Trois essais de codicologie quantitative* (Editions du Centre National de la Recherche Scientifique, Paris, 1980; 1983); Graham Pollard, 'The pecia system in the medieval universities', in M. B. Parkes and Andrew G. Watson (eds), *Medieval Scribes, Manuscripts and Libraries: Essays Presented to N. R. Ker* (Scolar Press, London, 1978).

17 Jean-Pierre Perrot, *Le Passionnaire français*, unpublished dissertation for the Doctorat d'Etat (Paris, 1980).

18 See note 16.

19 Printed by Sébastien Cramoisy in Paris.

20 Vatican Library, Capponi archive IV, 907, 37.

21 Printed by Giovanni Giacomo Komarek in Rome.

22 Lorenzo Baldacchini, *Bibliografia delle stampe popolari religiose del XVI-XVII secolo: Biblioteche Vaticana, Alessandrina, Estense* (L. S. Olschki, Florence, 1980), p. 8.

23 See the list of works in the Fondo Capponi in Baldacchini, *Bibliografia*.

24   I should note that Ariosto passed rapidly into popular culture. *Orlando furioso*, published in 1516, appeared in a popular edition in Venice as early as 1525.

25   My thanks to professor Alfredo Stussi of the University of Pisa for his dialectological expertise concerning this text.

26   See Pietro Manzi's extremely useful study, *La Tipografia napoletana nel'500. Annali di Giuseppe Cacchi, Giovanni Battista Cappelli e tipografi minori, 1566-1600* (6 vols, L. S. Olschki, Florence, 1974), vol. 5. Manzi does not mention the booklet on St Louis of Anjou, however.

27   See Anita Mondolfo, 'Adamo di Rotwill', in *Dizionario biografico degli Italiani* (Istituto della Enciclopedia italiana, Rome, 1960-), vol. 1, pp. 243-4.

28   The word *cacchio* is a euphemistic version of a vulgar designation for the male sex organ.

29   See the information on Cacchio in Manzi, *La Tipografia napoletana*.

30   See Manzi, *La Tipografia napoletana*, p. 16; Carlo de Frede, 'Tipografi editori librai italiani del Cinquecento coinvolti in processi di eresia', *Rivista di Storia della Chiesa in Italia*, 23(1969), pp. 21-53. On the diocesan courts and the Inquisition in Naples, see the remarkable study by Jean-Michel Sallman, *Chercheurs de trésors et jeteuses de sorts. La quête du surnaturel à Naples au XVIe siècle* (Aubier Montaigne, Paris, 1986).

31   See Emile G. Léonard, *Les Angevins de Naples* (Presses Universitaires de France, Paris, 1954).

32   See Robert Folz, *Les Saints Rois du Moyen Age en occident (VIe-XIIIe siècles)* (Société des Bollandistes, Brussels, 1984).

33   Edith Pásztor, *Per la storia di san Ludovico d'Angiò (1274-1297)* (Istituto storico italiano per il Medio Evo, Rome, 1955).

34   See Mercedes van Heuckelum, *Spiritualische Strömungen an den Höfen von Aragon und Anjou während der Höhe des Armutsreifes* (Rotschild, Leipzig/Berlin, 1912).

35   See Marie-Hyacinthe Laurent, *Le Culte de S. Louis d'Anjou à Marseille au XIVe siècle* (Edizioni di storia e letteratura, Rome, 1954).

36   See Ronald G. Musto, 'Queen Sancia of Naples (1286-1345) and the Spiritual Franciscans', in Julius Kirshner and Suzanne F. Wemple (eds), *Women of the Medieval World: Essays in Honor of John H. Mundy* (Basil Blackwell, Oxford and New York, 1985).

37   See Fernando Bologna, *I Pittori alla corte angioina di Napoli. 1266-1414, e un riesame dell'arte nell'età fridericiana* (U. Bozzi, Rome, 1969).

38   See the facsimile edition of Biserka Grabar, Anica Nazor, Marija Pantelić and Vickoslav Stefanić (eds), *Missale Hervoiae ducis spalatensis Croatico-glagoliticum* (2 vols, facsimile edition Akademische Druck- u. Verlagsanstelt, Graz, 1973).

39   See Giacinto D'Agostino, *San Francesco e i francescani negli Abruzzi* (3 vols, Tip. G. Carabbal, Lanciano, 1913-29).

40   See Giacinto Marinangeli, *Bernardino da Siena all'Aquila* (L'Aquila, 1979).

41   See Antonio de Stefano, 'Le Origini di Aquila', *Bullettino Abruzzese*, 3, 14(1923-37).

42   *Statuta Civitatis Aquile*, Alessandro Clementi (ed.) (Istituto Storico del Medio Evo, Rome, 1977).

43   Anonymous, 'Gli Affreschi del secolo XIV nella chiesa di San Francesco d'Assisi in Castelvecchio Subequo', *Rivista abruzzese* 7, 2(1954), pp. 58-62.

44 See Raffaele Colapietra, *Gli ultimi anni delle libertà communali aquilane (1521-1529)* (ESI, Naples, 1963).

45 Text published in *Analecta Franciscana*, 7(1951), pp. 395-9.

46 ibid., pp. 335-80.

47 See Geneviève Brunel, 'Les Saints franciscains dans les versions en langue d'oc et en catalan de la *Legenda aurea*', in Brenda Dunn-Lardeau (ed.), *Legenda aurea. Sept siècles de diffusion* (Bellarmin, Montreal, and J. Vrin, Paris, 1986), pp. 103-15.

48 The text of *Sol oriens*, used as a reading on St Louis of Anjou can be found in all the older Latin and Italian editions. I have used an Italian edition published in Venice in 1521 (fo. 133).

49 For a bibliography of such works up to 1980, see Giovanni Dotoli and Paolo Carile, 'Appendice bibliographica', in *La Bibliothèque bleue nel Seicento o della letteratura per il popolo* (Adriatica, Bari and Nizet, Paris, 1981), pp. 183-98. For an overview, see Roger Chartier, 'Livres bleues et lectures populaires', in Henri-Jean Martin and Roger Chartier (gen. eds), *Histoire de l'édition française* (3 vols, Promodis, Paris, 1982-4), vol. 2, pp. 498-511 ['The Bibliothèque bleue and Popular Reading', in Roger Chartier, *The Cultural Uses of Print in Early Modern France*, tr. Lydia G. Cochrane (Princeton University Press, Princeton NJ, 1988), pp. 240-64].

50 Alfred Morin, *Catalogue descriptif de la Bibliothèque bleue de Troyes* (Droz, Geneva, 1974).

51 The dossier of older hagiographic texts can be found in the *Acta Sanctorum*, September, vol. 3, *Dies VII-XI* (Antwerp, 1750), pp. 24-43.

52 See Dom Georges Viole, *La Vie de Saincte Reyne, vierge et martyre* (Autun, 1669), ch. 14.

53 See note 13.

54 See Jean Delumeau, *Histoire vécue du peuple chrétien* (2 vols, Toulouse, Privat, 1979).

55 Charles Nisard, *Histoire des livres populaires ou de la littérature de colportage* (2 vols, 2nd edition, rev. and corr., E. Dentu, Paris, 1864).

56 Morin, *Catalogue descriptif*, p. 468.

57 For the history of Alise in the seventeenth century I have used the remarkable monograph of Dr. Marcel Bolotte, *Alise Sainte Reine aux XVIIe et XVIIIe siècles. Les pèlerinages, la station thermale. Histoire de l'hôpital* (Dijon, 1970).

58 This highly important text published by a 'standard' Troyes publisher does not appear in the catalogues of the *Bibliothèque bleue* (Geneviève Bollème, Alfred Morin), perhaps because it is not anonymous. The copy I have used is in the Bibliothèque Nationale in Paris (Rés. Ll[27] 33378).

59 Dom Georges Viole, *Apologie pour la véritable présence du corps de Saincte Reyne…* (2nd edition, Paris, 1653), p. 6.

60 Delumeau, *Histoire vécue*, passim.

61 Gédéon Tallemant des Réaux, *Historiettes*, ed. A. Adam (2 vols, Gallimard, Bibliothèque de la Pléiade, Paris, 1960-1), vol. 1, p. 368.

62 *La Vie de Saincte Reyne, vierge et martyre*, p. 49.

63 *Esclaircissement sur la véritable relique* (Paris, 1651), p. 40.

64 Viole, *Apologie*, p. 42.

2

The Hanged Woman
Miraculously Saved:
An *occasionnel*

ROGER CHARTIER

The page is headed by a long title in the style of the age: *Discours miraculeux et véritable advenu nouvellement, en la personne d'une fille nommée Anne Belthumier, servante en l'Hostellerie du Pot d'Estain, en la Ville de Mont-fort entre Nantes et Rennes en Bretaigne, laquelle a esté pendue III jours & 3. nuits sans mourir. Avec Confession de plusieurs dudit Mont-fort, comme l'on pourra voir par ce présent discours*. At the foot of the page there is an imprint and a date: 'A Douay, Chez la Vefve Boscart, selon la copie imprimée à Paris M. D. LXXXIX. Avec permission.'

Between the title and the imprint, a woodcut shows a hanged woman and a group of men gathered near the gallows who are pointing at her. Their gestures explain the miracle – the hanged woman is still alive. On the verso of the title page another woodcut shows the crucifixion. It is followed by the promised account, which closes at the fifteenth page with a picture of the Virgin and Child. The one known extant copy of this tract printed by the Widow Boscard of Douai is now in the municipal library of that city.[1]

The slim volume belongs to a clearly defined print genre, the *occasionnel* or *canard*. Similar tracts, produced in large quantities in the sixteenth and seventeenth centuries, told of natural calamities (floods, earthquakes, storms, lightning strikes), of abominable crimes and the capital punishment that usually ensued, and of extraordinary phenomena that transgressed the laws of nature. Among the latter, along with ghostly apparitions and enchantments, were miracles. There is nothing odd, then, about the account published in Douai. A modest and inexpensive chapbook, it was directed, like dozens of others, at the most 'popular' readers – which does not mean, however, that its buyers were all artisans or merchants.[2]

Labelling a text or classifying it as belonging in a certain category of printed matter tells us nothing about the author's or the publisher's intentions, however, nor about how its various sorts of readers may have interpreted it. Still, if we are to understand the function of the *occasionnels*, the reason for their publication, or the various uses to which they were put, this is precisely what is essential. With the almanacs, *occasionnels* were doubtless the most widely distributed non-religious books and the first works of 'popular' print literature. The present case study attempts to analyse a common object and an unusual tale, to link the particularity of a story and the banality of a form, and to focus on the question of the effects, desired or actual, produced by the printing of a tale such as this. Hence I shall attempt, on the only scale on which this is truly possible, to analyse a text and its editions, and to reconstruct the various meanings inherent in the narrational and typographical mechanisms of wide-scale publishing.

One Text, Two Publishers

The last statement needs explanation if it is to be applied to the *occasionnels*. The press run of each of their editions is impossible to ascertain exactly, but it was perhaps on the order of 2,000 or 2,500 copies, a figure often encountered for other printed works of widespread use such as breviaries.[3] This figure should be at least doubled for the *Discours miraculeux et véritable* that occupies us here, since Jacques Boscard's widow in Douai, publishing 'A l'Ecu de Bourgogne', was not the work's only publisher. Another edition, with the same title (a few typographical variants apart) and the same woodcut, came from the printshop of Jean Bogart, 'A la Bible d'Or', in the same year, 1589. The imprint reads, 'A Douay, chez Jean Bogart, selon la copie imprimée à Paris par Geoffroy du Pont M. D. LXXXIX. Avec permission.'[4] This supposes the existence of a previous edition of the same text in Paris attributed to a certain Geoffroy du Pont.

It was common usage in the sixteenth and seventeenth centuries for publishers in the provinces to take over titles that had already been published by Parisians. This was the case in 1594, for example, when the same Jacques Bogart printed in Douai a *Sermon de la simulée conversion et nullité de la prétendue absolution de Henry de Bourbon*, a pro-League piece published towards the end of the preceding year for the Paris booksellers Guillaume Chaudière, Robert Nivelle, and Rolin Thierry.[5] Bogart could thus have done the same thing five years earlier with the account of the miraculous hanging. The absence of any trace of the supposed Geoffroy du Pont in notarial archives or the records of the Paris printers' guild raises a doubt, however.[6] Was this a false name, masking the name of the real publisher of the book, pirated by Bogart? Or did the first edition never exist, mention of it being put into the Douai edition to make the miraculous event more credible by seeming to repeat another telling taken as true? Either is possible. In any event, with two and perhaps three editions, the *Discours miraculeux et véritable* of Anne Belthumier's adventure had a distribution of at least 4,000–5,000 copies and perhaps more.

The two editions that are known, in both cases thanks to one surviving copy (but only one, which makes one think that other editions may have disappeared with no trace), were printed in Douai. At the end of the sixteenth century, the city was part of the Spanish Netherlands. It had a university, founded in 1562, and was known as a bastion of the Catholic Church, the Society of Jesus, and pro-Spanish sentiment. The presses of Bogart and the Boscards – Jacques, who died in 1580, and his widow, who took up the business after 1585 – turned out religious books (treatises on theology, devotional works, books of spirituality) and works in support of Spain and the League that on occasion were reprinted in Paris after the Douai texts.

In 1589 neither Jean Bogart nor the widow Boscard appear as regular publishers of *occasionnels*.[7] Furthermore, the difference between the two firms was great. Jean Bogart had been in business since 1574 and had put out 196 works in fifteen years, for the most part thick volumes on religious subjects or classics (in particular, Aristotle and Luis of Grenada), nearly half of which were in Latin. In 1589, besides the *Discours miraculeux*, he published seven books, four in Latin (including Aristotle, *De moribus ad Nicomachum*, one *comeodia sacra*, and a history of Belgium from the death of Charles V to the arrival of the Duke of Alba dedicated to Alessandro Farnese), and three in French (a new translation of the *Imitation of Christ*, the *Exposition sur la reigle de Monsieur S. Augustin* by Hugh of Saint Victor, a text written in the twelfth century, and a volume of Jesuit accounts of China and Japan).

At that same period, the widow Boscard's catalogue was still slim, showing only three titles published in 1585 and 1586: a Latin poem celebrating the taking of Antwerp by Alessandro Farnese, a royal ordinance concerning prices and wages, and a political piece. In 1589 the *Discours miraculeux* was accompanied by eight other titles. Six were *libelles* (lampoons) of from eight to ten pages on the assassination of the Guises in Blois the preceding year, and one made use of a standard topic for *occasionnels*, celestial phenomena, to kindle support for the League: *Signes merveilleux apparus sur la ville et chasteau de Blois en la présence du Roy: & l'assistance du peuple. Ensamble les signes et commette apparuz près Paris le douziesme de janvier, 1589 comme voyez par ce présent portraict.*

The publication of the tale of the miraculous escape from hanging supposes collaboration and agreement between its two Douai publishers (since they used the same woodcut), but it fitted into totally different schemes of publishing activity. On the one side, we see the flourishing trade of the biggest publisher in the city, who printed political broadsheets in Flemish as well as texts for the University colleges and major works of the new spirituality. On the other, we see the start in business of a master printer's widow who specialized in pro-League pieces of topical interest. The printing of *occasionnels* was thus not a specialized activity or the province of only some publishers (as the publishing of the *Bibliothèque bleue* works may have been when that publishing formula began in Troyes). It fitted in with quite different business formulas. Still, that an *occasionnel* on a miracle was put out by two publishers whose presses worked for Spain, the League, and the Church raises a question: was there a connection between the publication of this text, which belongs, in the last analysis, to a genre seldom represented in the Douai catalogues,[8] and the usual orientation of the works published in that city – Jesuit, pro-Guise, and Tridentine, like the city itself?[9] I shall return to the question.

An *occasionnel* was first a title made to be cried through the streets by vendors (the *porte-paniers* or *contre-porteux* spoken of by Pierre de L'Estoile), but also made to catch the eye, to be seen and read when displayed at the bookseller's shop. The Douai *Discours* makes use of the usual strategies of the genre. It piles up concrete details about the heroine (designated by her full name), about places (Montfort, situated geographically, the inn, identified by its sign) and about elapsed time (the three days and three nights for the hanging). Set in a real place, made concrete by details that seem true because they could be verified (such as the existence of the 'Pot d'Etain' or a person named Anne Belthumier), the story seems authentic and appears to be a true story. The direct citation of eyewitnesses lends an incontestable guarantee to the truth of the tale. The 'confession of several of the said Montfort' enables the discourse to pose as *véritable* – that is, as a report of what actually took place, and took place *nouvellement* (recently).

As is usually the case with the *occasionnels*, which played on the supposed reality of the event to proclaim the 'novelty' of their story, the Douai *Discours* claims to relate a very recent event and give fresh information on it. In this case, however, the novelty and the truth are of the special order of the miracle. The typographical layout indicates as much. The key word in the title, *miraculeux*, appears in Jean Bogart's version in large, thick capital letters carefully detached from the rest of the title; in that of the widow Boscard it is presented more awkwardly, since the word is cut in two, perhaps by an inexperienced compositor more familiar with political lampoons and ephemeral pieces. *Discours miraculeux et véritable* – the adjective modifies not the past event but the report of it that the reader holds in his or her hands, as if the text itself had something of the miraculous about it and were imbued with the sacred character of the event it relates. The title of the pamphlets thus make use of several techniques to capture the attention of the readers, announcing the novelty of a history never before read, the assured authenticity of an event out of the ordinary, and the Christian thrust of a story that declared the power of God. Each reader could find a good reason (or several) to buy the pamphlet. And to read it. It runs thus:

Miraculous and true discourse [that] happened
in the person of a young woman named
Anne Belthumier, servant in the inn
of the Pewter Pot, in Montfort, between Nantes
and Rennes in Brittany

There is no reasonable creature in Christendom who does not know or should not
know that God naturally is alone All-powerful and how often he has given us
confirmation of His Omnipotence by many fine miracles mentioned in his Holy
Scripture. None the less, it seems that he has wanted to refresh our memories by a
miracle that he permitted four or five months ago to occur in the person of a girl of
noble race and of Brittany named Anne Belthumier, who was in a hostelry of Montfort
in Brittany, at the sign of the Pewter Pot. But for the little means that her father had,
[she] was obliged to put herself to service. While she was in service, the daughter of
the house in which she lived abandoned herself to a lover, with whom she erred and
became large with child; being delivered of which, she sought all possible ways to save
her honour and make it so that her sin not be known by the world. The Devil, who
knows all things past and consequently who knew the fault she had committed,
counselled her to suffocate her child, which she did, and which ill-advised girls such as
she usually do under the subjection of the evil spirit whose only counsel is to do
wrong. Thinking [she had] been well counselled and in order to muffle her deed with
a wet sack, she decided to win over her father and her mother so as to certify to justice
that the child came from their chambermaid and that she had killed it deliberately. At
the eighth hour of the morning she [the servant] was taken prisoner, and within the
hour sentenced to be executed at ten. Before being led to execution, she begged the
judge to let her confess and receive her Creator, which he accorded her, although that
was not ordinarily done. After she had made all the devotions and prayers that she
desired, the hangman arrived to put the rope around her neck, as is the custom with
those who are to be hanged and strangled, but the rope broke two or three times. In
spite of this, the hangman finally found a way to hang her and [he] left her for dead,
having omitted nothing of his customary [acts] when justice makes him hang and
strangle some malefactor. And when she was on the ladder, ready to be executed, she
recommended herself so to all the Our Ladies to which one goes on voyages of
devotion, and principally to Our Lady of Liesse, that she was hanged three days and
three nights without dying. For on the third day an individual passing along the road
where she had been hanged, seeing her move her legs, so suspected that she was not
dead that he gave word of what he had seen to several of the town, who went to see
her. Having found the informer's report true, they cut the cord by which she was
hanged and they brought her into a house to get her to regain her breath and her wits,
in which it would have been difficult that she not be somewhat troubled because of
the pain and the torment she had endured. Several days after she had regained her
senses, she herself took the ropes with which she had been hanged to a Chapel which
is in the Church of the Jacobins in Nantes, founded for Our Lady of Good News. This
miracle can serve as an example and a mirror to the administrators of Justice, to make
it clear to them that they must take care not to condemn a criminal to death lightly
when someone comes before them. On the contrary, that there is great need, if they
wish to discharge their conscience well, to examine the witnesses closely so that they

judge no one on trust, and that they not be cause to make [anyone] despair and thus lose the soul of a poor Christian, which is to be prayed for and respected above all the goods that the inhabitants of this world hold as sovereign, as is easy to know by an infinity of passages mentioned in the Holy Scriptures. This miracle will also serve as instruction to the followers of the new Religion, to demonstrate and prove to them how much a Christian profits from imploring the Saints and the Virgin Mary at his death, from which the servant girl spoken of in this miracle was preserved for having principally appealed to and implored the aid of Our Lady of Liesse when she was to be put to death.

[ornament]

The statements made here above must be believed to be true, for I have learned them from one of the most upright men and the greatest in wealth and in quality that there could be in the city of Chartres, who told me he had been present when the girl in question was interrogated. And although it is not necessary in the present day to give miracles to Christians, because of the assurance that they have or should have in God's Omnipotence, none the less we must firmly believe that the Rector of the universe willed such a miracle to happen anew, either to manifest his admirable secrets and deeds or to lay bare the wickedness and sin of those who attempted to put this poor girl to death on hearsay so as to have them punished for their wrongdoing, as they have been, and deservedly, for the Bailiff of Montfort, [on] knowing that such a case had been committed, had taken prisoner the father, the mother and the daughter who had plotted the death of their servant girl by the false report of two Midwives and of a Surgeon, who deposed [that] the child had come from the Chambermaid in order to save the honour of the daughter of the house, in which they were disappointed and foiled, for instead of keeping their evil deed secret, as they thought, it was all the more published abroad by means of the innocence that was found in the accused girl.

The day that she arrived in the city of Chartres to fulfil the vow that she had made to go and salute Our Lady of Chartres if she could be preserved from death, My Lords of the Chapter assembled to ascertain from her if the miracle that people said had happened on her person was true. Being advised of which, My Lords of Justice sent for her, and after having interrogated her on several points and articles, in the end they asked her if she had brought her sentence, to which she answered that she had taken it to show it in all the places and the towns through which she might pass, but a certain man that had been furnished to lead her to all the places of devotion that she wanted to visit had stolen and taken away her said sentence, with the twenty-five *écus* she had been given to pay the costs and expenses that she might incur along the roads. The said Lords or President of Chartres would not have been content with this response if at the very moment that they were interrogating her a Gentleman from her part of Brittany had not come along, who assured them of the fact for having seen it, because of which they left her in peace by giving her permission to make her way to Paris to perform there the devotions that she intended. My Lords of the Chapter, to help her carry out her intention, had delivered to her a sum of money that served her until she arrived in the city of Paris.

Christians should not find this miracle too strange and marvellous, for others like it have occurred, and to personages who said that they were favourites of God. In spite of this, they did not have as much belief in Him as in the virgin and holy Mary his Mother, as others, steeped in Christian Philosophy, call on St James, in whose favour one day was spared the life of a young man about to be put to death wrongly and without cause, as by devotion he was on his way to visit the Church that has been built in his honour because his body reposes there. The conspiracy of his death was made by a serving girl in the hostelry in which one day he was lodged, for having found him handsome and to her liking at his arrival, [she] became infatuated with him, so that by fine words, joyous and laughing, in short, by all the means ordinarily practised by those who make love, she did her best to make him condescend to her actions and her purpose, which was only to take her carnal pleasures with him. But seeing that he paid no attention to her entreaty, she conceived an enmity against him such that she plotted his death by means of a silver cup that she put, early in the morning before he had awakened, in his bag so that he would be accused of theft and so hanged and strangled. The poor Pilgrim, giving no thought to the drink that had been brewed for him, picked up his bag early in the morning without looking into it, as he was accustomed to do, and set off over the fields to finish his voyage as soon as he could. As soon as he had left the Hostelry, the chambermaid, to execute her plan and her pernicious enterprise, began to cry out throughout the house and to torment herself, giving her master to understand that someone had stolen a silver cup and that she suspected no other than the young Pilgrim who had just left the house.

The master, thinking that the serving girl's statements were true, sent after this young man, who was found in possession of the silver cup as the chambermaid had said, for which he put him into the hands of Justice, to have him punished as he thought that he had merited. The administrators of Justice, lending credence to the deposition of the chambermaid and her master, sentenced this poor innocent youth to be hanged and strangled, without inquiring deeply into the truth. It is to be presumed that when this good Pilgrim saw himself sentenced to death he invoked St James to his aid and prayed to him with such great emotion and devotion that he saved his life, by the grace of God, in which we can see a beautiful miracle and example to put before the eyes of those who will not believe that the Saints have the power to heal all sorts of diseases, from which they have delivered those who were tormented by them, since they have the power to preserve someone from death, which is a much greater miracle, without comparison with protecting a person from a simple illness.

A True Exemplum

The text begins like a sermon, not like a story. The author first states its 'theme' – the demonstration by miracle of the omnipotence of God. Of course, God's power has no need to be proved, since everyone is aware of it 'naturally', but in Scripture or history, miracle manifests it and makes it visible. This immediately gives the status of an *exemplum* to the tale about to be told – that is, of an account given as true, charged with embodying in a specific example the universal truth of the proposition that it illustrates.[10] An essential technique in medieval preaching, the *exemplum* had not been lost in the sixteenth century, and here it provides structure to the account, given as an 'example and mirror' in the service of a lesson that was both worldly and religious.

The first pages of the pamphlet tell the salient points of the story: the plot hatched by the daughter of the innkeeper at the Pot d'Etain, with the complicity of her parents, to impute to the serving girl, Anne, the daughter of an impoverished nobleman, the infanticide she had committed; the hasty trial that resulted in Anne's sentencing to be hanged, carried out after she has received confession and recommended her soul to Our Lady of Liesse, even though the rope broke 'two or three times'; the discovery, first by 'an individual', then by 'several of the town', of the miracle that had taken place to save Anne, 'hanged three days and three nights without dying'; finally, Anne's trip to Nantes to offer the ropes with which she had been hanged as an ex-voto to the chapel of Notre-Dame de Bonnes-Nouvelles in the church of the Jacobins. This first part of the account, set off by the woodcut that occupies two-thirds of page 8, ends with a dual moral message, in part directed at the judiciary to point out the danger of taking any deposition at face value, risking judicial error, in part directed at all Christians, proving the efficacy of veneration of the saints and the Virgin, who can effect saving miracles.

After a pause, which has practical reasons, since it corresponds to the end of the first of the two quires that make up the *occasionnel* and were perhaps composed at different times, the narrative starts again in another key, with the direct intervention of the narrator and of someone else who has told him the story, 'one of the most upright men and the greatest in wealth and in quality that there could be in the City of Chartres'. He opens the way to the epilogue in which everyone receives his deserved fate: in Montfort, the guilty parties and their accomplices are confounded; in Chartres, Anne's story is recognized as true thanks to the declaration of a Breton gentleman who says he had been an eyewitness to it. With the authorization of the judges of the *bailliage* and the canons of the cathedral chapter of Chartres, the miraculous survivor is free to continue her pious Marian voyage, which had already brought her from Nantes to Chartres and was to lead her, at the point where the text leaves her, to Notre-Dame in Paris.

Although the story of Anne Belthumier finishes there, the *occasionnel* continues for several more pages to add a second *exemplum* about a young pilgrim unjustly condemned on the word of false witnesses and saved from hanging by the protection of St James, whom he had invoked. The two stories are similar and show a like sequence of events – a false accusation, an unjust sentence, a final miracle – to state the same lesson: the legitimacy of the cult of the saints, protectors and healers just like the Virgin.

As this story is presented to its 'popular' readers, avid for novelty and tales of the extraordinary, a number of questions remain unclear, even from a formal perspective, and the tale's silences, its short cuts and its open ending are somewhat bewildering. Why and how did Anne arrive at the inn? Who are the *mignon* (lover) who seduces the innkeeper's daughter and the *quidam* (individual) who notices that Anne is still alive, the 'certain man' who accompanied her on her pious voyage or the 'gentleman from her part of Brittany' who confirms her story in Chartres? What is the meaning of the episode of the theft of her *écus* and her written copy of her sentence on the way to Chartres? What happened to Anne after she left for Paris?

There is a strong contrast between the repeated doctrinal statements that frame the text and drive it forward and the imprecision and haziness of the story itself, which is shot through with ellipses and short cuts. This stylistic discrepancy shows clearly the tale's status as an *exemplum*, in which details matter less than the religious truth to be demonstrated – the legitimacy, proved by the miracle, of devotion to the Virgin and to the saints. The insistent reiteration of this verity has the function of constraining interpretation and indicating how the text is to be understood – of pinning down its meaning, as it were. In contrast, the plot line remains loose in the two exemplary tales, as if the story could be or should be completed by the reader, who is called on to lend consistency to barely sketched characters, to link relations and sentiments, and to imagine the unspoken motivations and missing episodes. Another text might possibly be produced by reading, a text written wholly in the reader's imagination out of elements from the printed text, a free text arising out of what is left unsaid.

The same might perhaps be said of all the *occasionnels*. Rather than offering finished, polished narratives, they provided material for invention; they solicited and supposed an effort of the imagination. The *Discours miraculeux et véritable* of the Douai printers followed the time-honoured formulas of the sermon, joining the use of a technique for imposing an explicit and univocal meaning – the 'theme' – with a narrative form making use of stories that served as examples – barely sketched-out stories that readers could complete on their own.

The narrator must persuade the reader that his story is true; he must give proof that it is not a fable but speaks of an event that really occurred. In order to do this, he makes use of a series of reliable witnesses. The narrator's word is backed up by that of the 'upright man' from Chartres (perhaps a *bailliage* judge) who 'tells' the narrator that he was present at Anne's interrogation. His word is in turn guaranteed by that of the Breton gentleman, who 'assures' the judges of the truthfulness of Anne's declarations. The credibility of the text is thus produced by the social authority of those who certify the truth of the story being told and who, by their social status, are judged worthy of trust. This explains the superimposition of time schemes – the time of the writing of the *occasionnel*, that of Anne's interrogation and of the testimony of the Breton gentleman and that of the miracle of Montfort itself.

The author manipulates the discrepancies between the time of his writing, the time of the events reported, and the times of the various statements that tell the tale, first with no narrator present, then, after the break following page 8, with a series of secondary narrators inserted into the narrative itself. In this way, although the story has taken on written form, it still does not lose the force of credibility traditionally accorded to oral witness and the corroborating word of authoritative persons. This enables the author to do two things. First, he lends credit to his tale as a true story; second, he makes Anne the true narrator, since it is her own word, authenticated by the Breton gentleman, repeated by the 'upright man' and set down in writing by the author of the *occasionnel*, that in point of fact tells of the miracle that saved her.

The girl thus proclaims by her own word and demonstrates by her own life the power of Mary, and, by that token, the full legitimacy of the cult of the Virgin. Anne's trip from one shrine of Notre-Dame to another, from Nantes to Chartres and then from Chartres to Paris, transforms her, as fifty years later Jeanne des Anges is transformed, into a veritable 'walking miracle'.[11] In 1589, a demonstration of the sort would necessarily be a spectacular reaffirmation of the truth of the Church's doctrine against the Protestant heresy. The *occasionnel* is thus put to the service of Catholic apologetics as it picks up the major themes of obligatory devotion to Mary, the justness of invoking protecting and healing saints (denounced as papist idolatry by the reformers),[12] and (perhaps above all) the real presence in the Eucharist, recalled by the phrase, 'She begged the judge to let her confess and receive her Creator.' This sets up a connection in the reader's mind between the miracle and communion, as if it were the sacrament that gave force to Anne's prayers; as if the miracle were credited to the omnipotence of the 'Rector of the universe', the living God received with faith. The text operates on two levels, one that reaffirms the demonstrated truth of the dogmas denied by the Reformation, another that aims at linking the miracle and the Eucharist, thus preserving from all taint of superstition the legitimate certitude of the 'grace of God' that alone founds the miraculous powers of the Virgin or the saints.

An Abominable Infanticide

The story charged with enouncing this teaching combines two motifs, infanticide falsely imputed to an innocent girl, and the miracle of a hanged woman preserved from death by her faith in the Virgin. Understanding the narrative that joins these two motifs may help to reconstruct the meaning that readers of the late sixteenth century might have given them. For them, infanticide was incontestably one of the most frequent but one of the most abominable crimes. In 1586, only three years before our story was published,

Henry III ordered parish priests and their curates to read in their sermons the edict of Henry II of February 1556, 'which pronounces the pain of death against young women who, having concealed their pregnancy and their childbirth, let their infants perish without receiving Baptism'. Read in all solemnity from the pulpit and heard several times a year, a message of the sort accustomed people to considering infanticide as a frequent crime but a terrible one.[13] It was frequent, as the preamble to this edict declares:

Being duly advised of a most enormous and execrable crime frequent in our Kingdom, which is that many women, having conceived infants by disreputable means, or otherwise persuaded by ill will and [evil] counsel, disguise, conceal and hide their pregnancies without informing of them and declaring them; and when the time of their parturition and the deliverance of their fruit is come, secretly they are delivered of it, then they suffocate, torture, and in other ways suppress it.

It was terrible because, as the text continues,

Without having the holy Sacrament of Baptism imparted to [the infants], this done, they throw them in hidden and filthy places or bury them in unconsecrated ground, depriving them by this of the customary burial of Christians.

To combat this 'most enormous and execrable' crime, the edict threatens the death penalty to women judged to have concealed their pregnancies and their childbirths without declaring them, and to have failed to gather sufficient testimony on the death of their infants before burial without the sacraments. The story in our *occasionnel* presents elements presuming infanticide (secret pregnancy and childbirth) in order to make it believable that Anne should be accused and sentenced as the infanticide mother of the suffocated child. The text even directly repeats a formula used in the edict of 1556 by imputing to 'ill will and [evil] counsel' – here the Devil's – the homicidal act of the innkeeper's daughter.

The false imputation of infanticide was a motif necessarily familiar to readers of this tale, both because they heard regular condemnations of the crime from their parish priests and, perhaps, because the act of obliterating the unwanted fruits of extramarital relations, as was the case with the innkeeper's daughter and her *mignon*, was itself frequent enough in the sixteenth century.[14] It was a familiar motif, then, and one with a strong emotional charge, evoking horror at a crime that deprived a soul of eternal salvation. When our tale opens with the dual abomination perpetrated by the young mistress of the Pot d'Etain – against her child, lost for all eternity, and against Anne, unjustly sentenced – it promises to be one of the 'horrible', 'frightful', 'cruel', 'tragic' stories that so appealed to the readers of *occasionnels* at the end of the century.

The theme of the mother who murders her child was one of their very favourites. Before the two pamphlets from Douai, it figures, in the somewhat different form of infanticide through desperate poverty, in a publication 'following the copy printed in Toulouse' in 1584 and in a *canard* printed in Rouen. The *occasionnel* from Toulouse was entitled *Histoire sanguinaire, cruelle et émerveillable, d'une femme de Cahors en Quercy près Montaubant, qui désespérée pour le mauvais Gouvernement et ménage de son mary, et pour ne pouvoir apaiser la famine insuportable de sa Famille, massacra inhumainement ses deux petits enffans.* The Rouen publication was called *Un Discours lamentable et pitoyable sur la calamite cherté du temps présent. Ensemble, ce qui est advenu au Pays et Conté de Henaut d'une pauvre femme veufve chargée de trois petits enfans masles qui n'ayant moyen de leur subvenir en pendit et estrangla deux puis après se pendit et estrangla.*[15]

The motif shifts with the story of Anne Belthumier, perhaps under the influence of the new publicity given to the edict of Henry II when it was applied to infanticide committed to hide the fruits of illicit love or to take revenge on the lover. In this guise, the theme inspired several writers of *occasionnels*. In Troyes in 1608 there was a *Histoire prodigieuse d'une jeune damoiselle de Dole, en la Franche Conté, laquelle fit manger le foye de son enfant à un jeune Gentilhomme qui avoit violé sa pudicité sous ombre d'un mariage prétendu.* In Paris in 1618 the same story was shifted to Bresse and the gentleman became a *lansquenet* (mercenary soldier). In Lyons in 1618 the title announced: *Histoire lamentable d'une jeune damoiselle laquelle a eu la teste tranchée dans la ville de Bourdeaux pour avoir enterré son enfant tout vif au profond d'une cave, lequel au bout de six jours fust treuvé miraculeusement tout en vie et ayant reçeu le Baptesme rendit son âme à Dieu.*[16] Infanticide, along with its inverse, parricide, was one of the crimes that most surely attracted buyers fascinated by the monstrous cruelty of the topic.

The Miracle of the Hanged Man or the Miracle of the Virgin

The infanticide recounted in the Douai versions of 1589, unlike those in the later tales, was not committed by the woman accused of it. It was the story of an innocent woman perfidiously accused, unjustly punished, and miraculously saved; the story of a hanged woman who failed to die. Anne was not the first to receive such a boon. The motif of the hanged man saved by miracle runs throughout medieval hagiographic literature, and it is this tradition that we need to examine as a repertory of stories and motifs on which the authors of *occasionnels* in the age of print could draw.

For such writers, the most immediately available anthology was still the *Golden Legend*.[17] Often republished in the age of the incunabula (141 known editions, 91 of them in Latin, and 20 in French), this work was still being published in the first half of the sixteenth century (14 editions between 1502 and 1540), and it often found a place in private libraries. In Amiens, for example, it appears in 45 inventories after death between 1503 and 1576, making it the most frequently found title, second only to books of hours. It is also one of the most 'popular' titles, as twelve merchants and ten artisans in that city are given as owning it.[18] The *Golden Legend*, compiled by the Dominican Jacobus de Voragine around 1260, was thus a work familiar to sixteenth-century people, a title that sold well (the Parisian bookseller, Galliot du Pré, had thirteen copies in stock in 1561)[19] and a gold mine for authors in search of inspiration.

The *Golden Legend*, under the heading of the Birth of the Blessed Virgin Mary, tells the story of a miraculous hanging. It says:

There was a thief that often stole, but he had always great devotion to the Virgin Mary, and saluted her oft. It was so that on a time he was taken and judged to be hanged. And when he was hanged the blessed Virgin sustained and hanged him up with her hands three days that he died not ne had no hurt, and they that hanged him passed by adventure thereby, and found him living and of glad cheer. And then they supposed that the cord had not been well strained, and would have slain him with a sword, and have cut his throat, but our blessed Lady set on her hand tofore the strokes so that they might not slay him ne grieve him, and then knew they by that he told to them that the blessed Mother of God helped him, and then they marvelled, and took him off and let him go, in the honour of the Virgin Mary, and then he went and entered into a monastery, and was in the service of the Mother of God as long as he lived [tr. William Caxton].[20]

There are, of course, obvious differences between this text and that of the *occasionnels*. Here the person who receives the miracle is a man, a true robber justly punished after having been caught in the act, not a young and innocent woman unjustly sentenced. The description of the miracle also differs: here it is effected quite concretely by the Virgin, who holds up the body of the condemned man and parries the blows directed at him, while Anne is saved, in the Douai versions, by grace manifested by no visible act. The same is true, moreover, concerning the young pilgrim making his way to Compostela protected by St James. One detail suggests a possible relation between the two tales, however: the thief faithful to Mary survives for three days, held up by his protectress, and Anne also remains hanging three days and three nights. Finally, the texts have similar intentions: the miracle, merited through faith in the Virgin, expresses Her glory and is solemnized in one case by retirement to a monastery and in the other by the devotional voyage.

The story Jacobus de Voragine told in the late thirteenth century was often used by preachers as an *exemplum*. Frederic C. Tubach's inventory mentions twenty-seven such occurrences.[21] Before and after the *Golden Legend*, anthologies presented a robber faithful to the Virgin and miraculously safeguarded for his loyalty. The *Tractatus de diversis materiis predicabilibus* of the Dominican Etienne de Bourbon, written in the mid-thirteenth century, states:

In the same fashion, one reads that a certain thief judged it good to fast on bread and water on the vigils of feasts of the blessed Mary and when he set off to rob he always said *Ave, Maria*, asking the Virgin not to let him die in a like sin. But having been taken, he was hanged, and remained suspended for three days without dying. Because he asked those who passed by to call a priest for him, one arrived and joined with the others around him. The thief was then taken down from the gallows, saying that the most beautiful Virgin had supported him by the feet for three days, and as he promised to correct his ways, they let him go free.[22]

In the *Liber exemplorum*, written between 1275 and 1279, the story is even closer to that of the *Golden Legend*, which perhaps inspired it. The thief, here named Ebbo, is physically protected by the Virgin, who not only 'during two days supported him with her holy hands' but also, to defend him against those who were trying to cut his throat, 'put her hands on his throat and did not allow it to be slit'. When the miracle was recognized, Ebbo was freed and became a monk, a servant of God and of the Virgin to the end of his life.[23]

Ebbo (or Eppo) the thief appears often in anthologies written to aid preachers because the story is one of the miracles of the Virgin most frequently encountered in hagiographic compilations, not only in Latin[24] but also in the vernacular. From the thirteenth to the fifteenth century he can be found in various forms in verse and prose. In the early thirteenth century, Gautier de Coincy includes him in his miracles in verse. The title of the tale varies with the manuscript, hesitating over the number of days the miracle lasted. In some, it took place over three days, as in the *Golden Legend* or in Etienne de Bourbon (for example, in the illuminated manuscript BN, n.a.f. 24541, *Du larron que Nostre Dame soustint par trois jours as fourches pendant et le délivra de mort*). Other manuscripts, the majority, reduce it to two days (*Dou larron pendu que Nostre Dame soustint par deuz jors*).[25] In Gautier de Coincy's poem, Mary comes to the aid of her thief twice, once at the hanging:

> She whom none of her own forgets
> Came most quickly to his aid;
> Her white hands under his feet [she] held
> And two whole days supported him
> Whom no suffering nor pangs did pain;

and once again to protect him from the sword blows:

> For then put her hands on him
> The mother of the king, who soon cried out.
> At which the thief exclaimed:
> 'Flee, Flee! All is in vain!
> Know all you to your profit
> That my lady holy Mary
> To my help and aid is.
> The sweet lady holds me
> And on my throat her hand holds;
> The sweet debonair lady
> Permits no evil to me.

The poet draws the lesson from this story of his *lerres*, his thief saved by miracle because of his faith in the Virgin:

> The mother of God heals all sins.
> No sooner do we sinners enter into her healing
> Than now we are healed.

Nostre Dame sainte Marie never fails those who love her and serve her, even if they are sinners, and even if they are guilty.

In the mid-sixteenth century, in the *Miracles de Nostre Dame* of Jean Miélot, the story is given as *Autre miracle d'un larron qui fut pendu au gibet, mais la Vierge Marie le préserva de morir lors*. The miracle runs thus:

And when his feet were hanging in the air, then the holy virgin Mary, mother of God, appeared to his aid, who, as it seemed to him, for the length of two days supported him in the air with her saintly hands and never permitted him to suffer the least wound, but surely when those who had hanged him returned to the place where he was hanging, which they had left a short while before, and saw him in life with a hale and hearty visage as if he suffered no harm, they thought they had not well strangled him with the rope. And approaching, as they went to cut his throat, the holy Virgin, mother of God, suddenly put her hands to the throat of the said robber and would not suffer them to cut his throat. These, then, knowing by the report of the said hanged man that the most holy mother of God was aiding him, as it has been said, marvelled much and let him go his way for the love of God.[26]

From Hagiography to the occasionnel

Both Gautier de Coincy's poem and Jean Miélot's narrative inspired pictorial matter in some manuscript versions. In the former's *Miracles de Nostre Dame*, the miniature shows both of the Virgin's acts. With her right hand, as if effortlessly, she supports the hanged man's body by its side; with her left hand, she parries the sword threatening him.[27] Thus the two life-saving gestures are represented simultaneously, to the astonishment of those present, whose hands and fingers point upwards and whose averted faces and exchanged glances express their wonder. The artist who illustrated Jean Miélot's account portrays literally the passage stating that the Virgin 'supported him in the air with her saintly hands' at his hanging.[28] Here the image shows the viewer what those present at the scene could not see. The miracle is represented as a totally physical operation made possible by a concrete act at once supernatural and ordinary that places divine intervention on the plane of human actions.

The woodcut placed on the title page of the widow Boscard's and Jean Bogart's publications is of a totally different sort. It shows not the miraculous operation but its result: the hanged girl, discovered by the three men in the foreground, is still alive. The Virgin no longer shows in the picture, and nothing of her mediation is visible. As in the text, the miracle is affirmed, recognized, but not described as an action. It is as if a realistic representation of it would diminish its mystery; as if showing how it came to pass would destroy its occult operation. From one picture to the other we can discern both a shift in religious sensitivity and the prudence of the producers of *occasionnels*, ever attentive to eliminate all taint of superstition from the reality of the miracle.

The identity of the person receiving the miracle also changes from the medieval texts to the late sixteenth-century pamphlets. In the earlier versions a man and a genuine robber is hanged; in the later versions, it is an entirely innocent woman. The sovereign Virgin who protects her faithful even when they are guilty cedes to the Virgin who remedies the injustice of men. The miracle itself shifts, since it is no longer a sign of grace granted by Mary to the likes of Ebbo, who, although a sinner, ' The sweet mother of the king of glory / Had in so high recall' (Gautier de Coincy), and who honoured with all his heart / The glorious Virgin Mary, mother of God'(Jean Miélot). The miracle becomes a divine means for righting wrongs perpetrated on the innocent. In the Middle Ages, the Virgin saves the robber because he is one of her own, of her house and her clientage; in the late sixteenth century, she assists Anne because she is an unsullied and defenceless victim. With the Catholic Reformation, the miracle becomes related to an emphasis on morality that judged the story of a guilty man protected uniquely for his Marian piety as a poor *exemplum*. The tale told in the *occasionnel* is of another sort. It teaches that the Virgin is a supreme judge whose miraculous intervention punishes the wicked and safeguards the innocent. Faith in her and confidence in her grace thus cannot be separated from an upright and pure life.

In medieval hagiographic literature the Virgin is not alone in saving hanged criminals. A similar miracle is imputed to a great many saints, male and female, who sometimes intervene during their lifetimes, but more often posthumously. They generally act to preserve from death innocent persons falsely accused and unjustly punished, but the figure of Ebbo, the devout and repentant thief, none the less persists under other names.[29] In the various lives of the saints the miracle is effected in one of two ways. Either the hanging rope breaks or, as in the *Golden Legend*, the condemned person is held up by his saintly protector. Among the saintly protectors of the hanged, St James is without contest the most frequently cited, and his miraculous aid can be seen in stained glass windows, bas-reliefs, frescos, and paintings. The account that closes the *occasionnel* published in Douai – the adventure of the young pilgrim victimized by the amorous servant girl but saved in the end by the saint whose shrine he was on his way to visit – is based on a story that can be documented as early as the beginning of the fifteenth century. For example, in 1418 the *Voyaige d'oultremer en Jhérusalem* of the seigneur de

Caumont enjoyed a huge success and brought new life to the repertory of the miracles of St James. The writer of our tract borrows from hagiography directly and literally, even though he euphemizes the miracle by not relating how it was effected. He also omits the ending of the tale, in which the judge declares to the young man's parents that he will not believe that their son has survived his ordeal unless the fowls he is about to eat resuscitate – which is just what happens, since, as the 1418 manuscript puts it, 'suddenly the cock and the hen rose up from the platter and sang.'

But if the miracle of St James, as told in the 1589 *occasionnel*, comes directly out of the written tradition of the lives of that saint, the same cannot be said of the miracle of the Virgin in favour of the hanged girl. Between the late sixteenth-century story and the others from the miracles of the Virgin, there is the important difference of the sex of the victim rescued. In both Marian literature and all other hagiographic accounts, the beneficiary of miraculous pardon is always, without exception, a man, whether he is a sinner or sinless, pilgrim or robber, young or not so young. The figure of the hanged *woman* miraculously saved seems foreign to the medieval repertory. Thus, even though the 1589 *occasionnel* re-uses old religious forms in the service of a new genre and a new apologetics, it does so by changing the identity of the person who attests to the miracle. Now it is an innocent young girl who, by her presence in the supposed reality of the devotional voyage or by her story, outlined for all to read in print, is charged with that task.

How are we to understand this change of sex that cautions us against any hasty judgement that the author of this *canard* has simply borrowed his tale from the legenda of Mary and of the saints? After all, why could we not take the story as true? Why should we not think that Anne Belthumier actually existed, along with the innkeeper and his infanticide daughter, and that she actually was unjustly sentenced as the result of an odious plot, and that when she survived hanging she became certain proof of the miracles of Mary and of the absolute power of God the Creator?

An Identical Story

Before accepting this conclusion, however, we need to read another text and another *occasionnel*. Although the title is laconic, it suggests a possible relation to the story printed in Douai: *Discours d'une Histoire et miracle advenu en la ville de Mont-fort, cinq lieües près Rennes en Bretaigne*.[30] The title page bears a geometric rosette motif. It gives no imprint, only the indication, 'Imprimé à Rennes, M. C.LXXXVIII'. This slim volume of six leaves follows the account of the miracle with an *Oraison à nostre Dame de Lyesse* in four stanzas, set in different characters than the ones used for the *Discours*, paginated apart and printed on its own signature, all of which attests to the independence of two texts joined after the fact.

What then is this *Histoire et miracle* offered to the reading public a year before the Douai *occasionnel?* The text runs:

Discourse on a story and miracle
[that] occurred in the town of Montfort,
five leagues from Rennes, in Brittany

Damoiselle Anne des Grez, daughter of the late Guillaume des Gres, Esquire, and of Damoiselle Perrine de Thimière, her Father and Mother, poor in wealth for [it] having been lost by the said late Father and Mother because of the heresy and new opinion of which they died infected, was given by her Grandfather to Madame de Crapador, to be nourished and instructed by the said Lady. And having been a goodly amount of time with that Lady, the said Lady was begged many times by one named Jehan Sucquet, rich Merchant and Innkeeper, living in the said Montfort where hangs the Sign of the Pewter Platter, to do him the favour and honour of permitting the said Damoiselle to dwell some time in his house, to teach and help to learn many civilities (of which she had goodly store) to his only daughter, already twenty-four years of age and not married. And he would give her for doing this each year the sum of twenty *écus*. And having lived there about two years [the following] occurred two days before My Lord St John the Baptist, 22nd day of June 1588 past. The night before, the Daughter of the said Merchant gave birth to a child. The Father and Mother, seeing what had happened, to their great dishonour, and [thinking] that their neighbours had perhaps heard something of her delivery, led by the Evil Spirit, killed the said Infant, [and] having broken its neck, carried it secretly into the bed in which the said Demoiselle was lying and put it near her feet without touching them, of which the said Demoiselle was unaware, and arose without having any cognizance of it. This done, they made great haste – the Father, the Mother and the Daughter – to call in people and Midwives, to whom they gave a goodly sum of coins to impute the crime to the said Demoiselle. And so quickly did the Bailiff of St Main, who was taken in Gonaisy in the absence of the judge of the said Montfort, proceed to the matter of the trial that in all diligence the said trial was done and finished at an hour after midday. The Surgeon and two Midwives had, according to the said Surgeon, he twenty-five *écus* [and] the said Midwives each ten, and on seeing their report was the said Demoiselle Sentenced to be hanged and strangled that very day. And when the Damoiselle was near the Gallows, she remembered to recommend herself to God, and even to call to her aid (as she was wont to do) Our Lady of Liesse and our lord St Servais. She prayed her Confessor that if God permitted in three times twenty-four hours the truth of the act to be known, it was her pleasure that he go for her to Our Lady of Liesse. Being thrown, her rope broke and [she] fell to the ground, was picked up, and thrown again with two ropes, the which broke, [and] immediately [she] was raised up and thrown again with three ropes at her neck, which broke again, [and she] was raised up and thrown for the fourth time with six ropes at her neck and left for dead. A small child five years old who was her godchild, for the sorrow he felt for his Godmother went every day to look at the gallows. The following Saturday morning, arriving near the gallows, he saw her raise her hands on high. The said child returned [and] told his

Mother that his Godmother was not dead and that he knew it well for having seen her that very day. To which the said Mother replied that if he did not keep quiet she would spank him. The said child, who had begun to go to School, went to his Master's house and said to him, 'I would like to tell you something, but I would be afraid of being whipped.' His Master promised him that he would not be. Then he told him, 'My Godmother is not dead. I just saw her a little while ago raising her hands up high.' The said School-Master, who was a man of the Church and who had heard her last confession, having celebrated the Mass with the best devotion that he possibly could, went to see if the report of the said child was true. And when he came near the said gallows, he saw how she raised her eyes upwards. This done, he went to his house to find the Judge who had sentenced her, where he found him with good company ready to put themselves to table to dine. Drawing him apart in private, he said to him, 'My Lord Bailiff, that poor girl whom you sentenced to death Wednesday last is still in life. Please think what you would like to do about it.' The said Judge began to laugh and joke at these words, and out of Irony to repeat publicly what had been told him in secret. And among other things, he said these words: 'It is just as true that she is still alive as it is that I am galloping across these viands on the table.' And suddenly the said Judge began to gallop across the said table most horribly, in sight of all the company, and all were much moved by this happening and went of a common accord with a great number of people. After having sent for the hangman, who, hearing what had happened, said that he was sure that she was dead and that he would cut her throat if that was not true, or else it was a miraculous thing done by the will of God, and that if this were the case, that it was no affair of his. The said Judge and other officers of Justice and [the] hangman set off on their way to go to the place where the said gallows stood, and they could not come close to it for a most horrifying vision that they had along the way, and the body of the said Damoiselle was taken down by the said man of the Church, her confessor, with no tremor or difficulty. And the said body being taken down, it was covered and wrapped, and then carried to a sure and honourable house to be warmed and helped. At the day's end the said body had regained its natural strength and virtue. So much was this true, that now she has set off on her way to carry out the voyage for our Lady of Liesse and of My Lord St Servais that she had begged [her confessor] to make for her. The truth of the affair being known, and as soon as she had been taken down from the gallows, the innkeeper, his wife, their daughter, [the] surgeon and [the] midwives were taken prisoners, and so did the trial go against them that they were sentenced to death by My Lords of Justice of Rennes. The same against the hangman, who was sentenced to be flogged for having kicked her ribs, which exceeded the judgement and penalty that had been ordered, from which blow she then felt more pain than she did from all the other torments and excesses that she had endured. At the execution of the abovesaid father, mother and daughter, they freely declared the innocence of the abovesaid Damoiselle, saying [that] what they had done to her had been thinking to protect their honour and that of their daughter, who confessed that this was the sixth infant that she had had, all done away with without having given any knowledge of them.

We have an identical story, then, with a concealed infanticide, a false accusation, an unjust execution carried out, and a miracle of Our Lady. An identical story – *the* identical story, one might say. And yet, from Rennes to

Douai the two texts differ immensely in their detail, their manner, and their status. The difference is visible even in the title. In Rennes, the title gives a generic indication of the class of tale – the miracle – in which the reader is to place the story he is about to hear. Nothing particularizes the tale but the town of Montfort, which was near the place of publication.

Should this account be taken to be true? There are reasons for suspicion. First, the Rennes print shop was a small business in 1588 and had produced no other texts comparable to the *Discours d'une histoire et miracle*.[31] Why this hapax? Is it to be taken as proof of the authenticity of the extraordinary event that it reports, and is an unusual publication to be explained by geographical proximity? Or should we take the near coincidence of the supposed site of the miracle and the place of printing of the pamphlet as an effort to add credibility that implies nothing either about the reality of the story or about the true place of publishing? Even more, a tactic of creating authenticity by giving a nearby site of the event and place of publication might suggest that the work was not printed in Rennes at all, since it was in Brittany that people would best have known if the story were true or not.

This argument is possible, but it is not totally persuasive. Nothing says that for sixteenth-century readers the known, verifiable reality of the acts related in the *canards* was of any importance for reading about them. It is in fact possible that readers went along with the realistic effects imbedded in the texts without believing that what they read was true – indeed, that they knew very well that nothing of the sort had actually taken place. In this perspective, we cannot reject the notion that a story was printed in Rennes that did not take place in Montfort or anywhere else.

From Rennes to Douai: The Differences

There are significant differences between the Rennes and the Douai title pages (plates II and III). First, the long title used in the Douai version, which piles up details and summarizes the central episode of the story, seems better designed to be 'cried' by the print vendors and thus to attract eventual readers. Second, by substituting the formula *Discours miraculeux et véritable* for *Discours d'une histoire et miracle*, the Douai printers shifted the status of their text. No longer a simple narration that takes its distance from the event that it relates, the text now has a closer participatory involvement in the miracle that it attests and proclaims. The reader's relation to the narrative is, by that token, singularly changed, implying not just an astonished curiosity for an out-of-the-ordinary happening but reverence for a text that shows traces of divine intervention.

The image showing the rescue of the hanged woman that replaces the geometric motif of the Rennes edition has the same effect, since it focuses reading on what is to follow and on the essential theme of the supernatural pardon accorded to the innocent girl. It also brings the reader into the miracle by inviting him or her to identify with the villagers who discover Anne still alive. When the customer for the Douai *occasionnel* was projected into the image – hence into the story – he or she could not easily refuse the truth stated there or reject the mystery guaranteed by the authority of the Church and portrayed in the image itself. The Rennes text, then, is one *occasionnel*, one relation of an extraordinary event, among many; the Douai version is a text that can also be handled like a relic, since it bears within it something of the sacred quality of the miracle it describes.

The Rennes *occasionnel*, unlike its Douai counterpart, does not begin by stating a moral precept or theme to be illustrated. It begins like a story, the story of a noble young girl, born in a Protestant family and placed, after the confiscation of her parents' wealth, as preceptress to the daughter of a rich innkeeper in Montfort. The first of the details that differ in the Rennes and the Douai texts concerns names. In the Rennes version the girl's name is not Anne Belthumier but Anne des Grez; the inn is not at the sign of the 'Pot d'Etain' but of the 'Plat d'Etain'. It often happened that the *occasionnels* used the same plot but changed the identity of the protagonists and the place in which the story supposedly took place.

Here, however, when the Douai author transformed the name of the heroine, he (perhaps thoughtlessly) eliminated one of the elements that gave Breton credibility to the story. Des Gretz is in fact the name of a Breton family that retained its nobility after the 1669 reforms. It is thus a name that rang true and Breton because it was authentic. It may also have had connotations attached to a legend concerning the town of Montfort, where St Méen was said to have fought the pagans, defeating them in the end by crushing them with the enormous stone they used as their sacrificial altar, known as the *grès de St Méen* or the *grès* of Montfort (and which could still be seen near the town).[32] By eliminating the regional and local connotation of the name of the hanged woman given in the Rennes version, the Douai text also changes her social condition. No longer a preceptress who tutors her charge in 'civilities'(not too much of a loss of caste for a young noblewoman), she becomes a chambermaid forced into domestic service, which makes her unhappy fate even more piteous.

To turn to the story itself, the *occasionnel* given as printed in Rennes gives a linear, sequential, carefully dated narration from which an exact chronology can easily be re-established. On the night of 21 June, the innkeeper's daughter is delivered and commits her crime, aided by her parents; the next day, 22 June, Anne is accused, judged, sentenced, and hanged at one in the afternoon; she remains on the gallows until Saturday 25 June, when, discovered living by her godson, she is taken down, three days and three nights after execution of her sentence, in perfect conformity with the vow she made before being hanged.

A chronology of the sort has two functions. First, by its very precision it helps to persuade of the truth of the event related, as do the specific places and the proper names given. Second, by situating the miracle of the girl saved from hanging at a high point in the traditional calendar of feast days, the midsummer holiday of St John's Day on 24 June, it solicits the imagination, inviting a contrast between Anne, alone on the gallows outside the town, far from its festive hubbub. As in Douai, the brief narrative offers hints that invite readers to go beyond what is written, to invent what is left unsaid in the text, and to complete the story to their liking. There can be no doubt that the mention of St John's Day gave contemporary readers an entire backdrop of ready images as they read the story.

When she is preparing for her execution, the Anne of the Rennes and the Douai *occasionnels* do not perform the same acts or make the same vows. First, although she confesses and prays in both, she does not receive communion in the Rennes version, a detail added in the 1589 text to confound the Protestants by linking the miracle with the Eucharist, and belief in the real presence with the manifestation of divine grace. Next, if Anne invokes the Virgin in the guise of Notre-Dame de Liesse in both versions, a somewhat unexpected saint, St Servais, appears only in the Rennes version. Why indeed this invocation of the bishop of Tongres, who died in 384 after transferring his episcopal seat to Maastricht, having been warned by the Apostles of the imminent destruction of his city by the Huns? In the late sixteenth century, the centre of the cult of St Servais was the church dedicated to him in Maastricht in the Spanish Netherlands, the site of his tomb and repository of his relics. This is where Anne goes when she has been saved and when she undertakes the *voyage de Notre-Dame de Liesse* (in the diocese of Laon and of *Monsieur de sainct Servais*). Although the saint was not unknown in France, his cult, which had spread from the north-east borders of the kingdom, carried by Cistercian liturgy with the advance of the Tartars (who in the thirteenth century seemed new Huns), does not seems to have had a large following in France. Are we to take mention of him in the Rennes *occasionnel* as one more detail to make us doubt that this edition was, in fact, Breton? Might it suggest that the work was printed in the north of France or in the Low Countries? Perhaps, if it is legitimate to link the place of publication – thus of the first distribution – of the pamphlet with the site of the cults mentioned in it. Why, then, should St Servais be eliminated from the *occasionnel* printed in Douai, a city not too far from Maastricht and from the saint's tomb? This might indicate that the Douai pamphlet was destined exclusively for French, even for Parisian, consumption, as it celebrated the shrines of the Virgin in France – Chartres and Paris – visited by Anne.

The manner of telling the tale also changes from 'Rennes' to Douai. The 1588 text gives a large place to the spoken word that is not found in the one printed the following year. To take one example, there is the episode of the broken cord. The Douai version says, 'But the rope broke two or three times.' The Rennes version, as we have seen, says:

Being thrown, her rope broke, and [she] fell to the ground, was picked up, and thrown again with two ropes, the which broke, [and] immediately [she] was raised up and thrown again with three ropes at her neck, which broke again, [and she] was raised up and thrown for the fourth time with six ropes at her neck and left for dead.

This typical motif in the lives of the saints can be seen, for example, in the miracles of St Foy or the legend of St Yves. The 'Rennes' pamphlet, however, states it with the repeated and rhythmic formulas of the oral tradition, whereas the Douai author imposes a laconic written style that has broken with the spoken word. A second difference between the two texts lies in dramatizing dialogue, prominent in the 1588 *occasionnel* and later abandoned. The first exchange, between the young child who in his innocence knows the truth and his incredulous mother, is given in indirect discourse ('The said child returned [and] told his Mother that . . .'; 'To which the said mother replied that . . .'), but the later exchanges between the child and the priest (both the last confessor of Anne and the schoolmaster of her godson) and between the priest and the judge are given in direct address, with the words exchanged fitted seamlessly into the flow of the narrative, introduced only by a comma or a period (since quotation marks were not yet in use).

The Douai text eliminated this dramatization as if it were incompatible with the Christian affirmation of the miracle, carefully distinguished from any suspect form of the marvellous. This led to the elimination of Anne's godchild, replaced by an anonymous and banal *quidam* as the person who finds that Anne is still alive. It also led to a change in the gesture by which she shows that she is still alive. In Douai she rather prosaically moves her legs; in Rennes she raises her hands in a prayerful attitude and turns her eyes towards heaven. Finally, it also led to the elimination of the striking episode of the judge, found 'in good company', who mocks the priest's revelation and loses control of his own actions when he belies his own words ('It is just as true that she is still alive as it is that I am galloping across these viands on the table') by doing just that ('And suddenly the said Judge began to gallop across the said table most horribly').

This scene may contain a transposition or a reminiscence of the scene in the miracles of St James in which a judge, about to sit down to table, declares that he will only believe in the survival of the young pilgrim unjustly hanged if the cock and the hen he is about to consume are resuscitated – which then happens. In any event, the 'Rennes' *occasionnel* is above all a story, a story in which the spoken word can be heard, in which the characters have real existence, emotions, sentiments, and relations, and in which the account of the miracle is accompanied by signs of the marvellous common to medieval hagiography. In order not to corrupt his all-important apologetic and dogmatic aim, the author of the Douai pamphlet rewrote the plot, narrowing what had been a good tale to be told or read to fit the limits of an *exemplum* for use in a sermon.

The 1588 text tells in its epilogue what happened to everyone involved: Anne fulfils her vow and makes her pilgrimage; those who plotted against her face death under a sentence pronounced by *Messieurs de la Justice de Renne* – that is, the judges of the Rennes Parlement. The author has been saving a final effect, however, in the revelation that the innkeeper's daughter is guilty of six infanticides. Thus an *occasionnel* that to the end has balanced the account of a dreadfully cruel crime and that of a miraculous case, to repeat the phraseology of many tract titles of the time, brings the reader back to the 'story' rather than to the 'miracle'.

This unstable and ambiguous status invites two readings, one that sees the story as part of the tradition of the marvellous tale, the other that asks how a tale of the sort came to be put to the service of Church authority at a time of bitter tensions and struggles. Thus we need to investigate two lines of thought. We need first to note, within the corpus of oral literature, the possible matrices of the story that found printed form in 1588; then we need to examine the contemporary scene to ascertain the reasons for the changes made in the 1589 text. This implies that we must consider, at least as a hypothesis, that the plots of the *occasionnels* may have been written or read in relation to very long-term schemata and lasting motifs (those of hagiography or of the oral tale, for instance), but that they also had a purely historical meaning linked to the short-term conjuncture of events surrounding their publication.

Pursuing the Tale

Is there a possible source for the Marian miracle in oral literature? One of the tales repeated by the Brothers Grimm, which they entitle 'Marienkind' (Our Lady's Child) suggests a comparison.[33] This is how the story goes: the daughter of a poor woodcutter is taken to heaven by the Virgin, who goes on a voyage and leaves her the keys to the thirteen doors of heaven but forbids her to open the thirteenth. One day when her guardian angels are absent, the girl disobeys, and she sees the Holy Trinity in the forbidden room. At the Virgin's return, the girl denies her disobedience and is struck dumb and sent away from heaven. The Brothers Grimm end the tale thus: a king out hunting discovers her and marries her. After the birth of her first child, the Holy Virgin appears to her and asks her again about the thirteenth door. The mother persists in denying, and her child is taken away, as are two subsequent children. At this point the people demand the death of the queen as a child-eater and the king acquiesces. On the pyre the queen repents and attempts to avow her error. Rain comes to extinguish the flames and the Holy Virgin gives her back her speech and her children.

What connections could there possibly be between this tale in which no one is hanged and the miracle of Montfort? There are none in the specific details, of course, but there is a similarity in the structures of the two stories, which string together a like series of events. We see an accusation of infanticide, the condemnation to death of the presumed murderess, and a miraculous intervention of the Virgin, who in the end saves the heroine at the moment of her execution.

Any relationship between the late sixteenth-century *occasionnel* and tale 710 in Antti Aarne's and Stith Thompson's canonical classification is evidently distant and reduced to morphological similarity alone, even if in some versions the young queen is sentenced to be hanged rather than burned alive. None the less, the tale combines two elements fundamental to the plot published in Rennes or in Douai: the motifs of infanticide (the child killed by its own mother) and of Marian pardon accorded to a repentant or innocent girl. This of course does not mean that the author of the 1588 text dipped directly into the oral repertory. This would be all the more difficult to prove since the only French version of the tale in which the Virgin, not a fairy, appears in the role of the godmother was collected in the boundary area between the Barrois and Champagne in the late nineteenth century. Still, when it links infanticide and a miracle of the Virgin, the printed text juxtaposes two representations that were among the ones most deeply rooted in the common imagination of people of the sixteenth century, who were at once moved to horror at the most detestable of crimes and comforted by Mary's loving mercy. Under the appearance of novelty – indispensable for success at the bookstalls – the *occasionnels* often drew their strength from particular and local tales based on longstanding motifs that other traditions, both oral and written, had borne along and continued to express in parallel ways, motifs that fashioned and formulated both beliefs and oppressive dreads, fears and certitudes.

Anne and Pierre: The occasionnel in the Service of the League

But the *occasionnel* also operated on another level, that of the current political and religious situation. In both the 1588 and 1589 versions, Anne recommends herself to the Virgin of Notre-Dame de Liesse, who hears her prayer and her vow. The least that one could say about this particular Virgin is that she was allied to the League. It was under her protection that the first League was created in 1576 in the château of Marchais, which was near the sanctuary and belonged to the Guise family.[34] Furthermore, in 1583 Notre-Dame de Liesse was one of the epicentres of penitential and eschatological 'white' processions encouraged by the religious zeal of the League. Pierre de L'Estoile even makes the sanctuary one of the two primary poles of attraction for pilgrims before they set off for Paris. He says:

They say they were moved to make these penitences and pilgrimages because of a number of lights appearing in the air and other signs like prodigies seen in the Heavens and on the earth even towards the zones of the Ardennes from which the first [people] came as pilgrims or penitents, as many as ten or twelve thousand, to Our Lady of Rheims and of Liesse for the same occasion.

In August, September, and October 1583, thousands of penitents from Meaux, Soissons, Laon, Rheims, and Noyon flooded into Notre-Dame de Liesse, a pilgrimage site since the twelfth century.[35] There is nothing surprising, then, in the victorious League celebrating its Marian protectress in 1588–1589.

It is doubtless hazardous to establish too strict a connection between, on the one hand, events unfolding on the barricades of May 1588 that gave Paris to the League, and the end of the year 1589, and, on the other hand, the printing of the two *occasionnels*, given that the exact date of their publication is unknown. Nonetheless, it is sure that both publications reflect the pro-League spirituality that attempted to combat heresy by affirming the cult of the Virgin of pity and intercession, protectress of the upright and the innocent. This is what explains the presence of the *Oraison à nostre Dame de Lyesse* that follows the narrative in the 'Rennes' pamphlet. This *Oraison* is a penitential prayer to be recited by pilgrims making their way towards the sanctuary:

> As [thy] subject to thy holy chapel
> Of Lyesse I take myself, and call thee
> To my aid in my necessity.

The addition of a prayer quite evidently set in type at a different time from the *Discours d'une histoire et miracle* links the pamphlet and the sanctuary more closely and places the *occasionnel* with print materials connected with pilgrimages, such as pilgrim songs, the lives of the saints whose shrines were to be visited, and certificates attesting completion of the devotional voyage. Although it may have been written and printed with other intentions, the 1588 text is thus enrolled in the cause of the League piety encouraged by worship of the *perdurable Lyesse*, as the *Oraison* states. In 1589, a year marked by many processions, in Paris in January, but also in Meaux, Senlis, and Laon (to where the statue of Notre-Dame de Liesse had been transferred to remove it from possible Protestant violence), whoever rewrote the story of the hanged girl miraculously saved celebrated the Marian shrines of Chartres and Paris in particular, but also embraced reverence for 'all the Our Ladies to which one goes on voyages of devotion'. She is still invoked as first protectress, but Notre-Dame de Liesse here becomes only one of the possible figures for the charitable and mediating Virgin, guardian of the destinies of the kingdom and of the people. The pamphlet thus becomes detached from any particular Marian sanctuary to become an invitation to honour and respect the Mother of God the Saviour, offered on the last page as the centre of universal worship.

The connection between the two *occasionnels* of 1588 and 1589 and the Virgin of Liesse does not simply come of the necessity to encourage Marian piety conceived as a rescue and proclaimed against the Protestants. Anne is not the only victim executed on the gallows whom Notre-Dame de Liesse aids. All the sixteenth- and seventeenth-century books and pamphlets that speak of her miracles tell of her miraculous rescue of Pierre. The first occurrence in print of this story that I can find is in a work published by the widow of Jean Bonfons, who worked in Paris between 1569 and 1572. The work is entitled *Les Miracles de Nostre Dame de Lyesse et comme elle fut trouvée et nommée comme pourrez voir cy-après*.[36] The protection of the hanged man opens the series of miracles performed by the Virgin of Liesse after the founding of her sanctuary. The date is 1139 and the text runs thus:

In the year of grace one thousand, thirty nine, [there] was a poor man named Pierre de Fourcy, who did not have great means to provide subsistence for himself, his wife and the three little children that he had. This poor man went every day to the public place where day workers gathered, but he found no one willing to put him to work, at which, seeing that he was earning nothing, he began to moan and to say: 'Alas! Lady of Liesse, come to the aid of your poor servant. Help him to live, he and his family.' This good man went every day from house to house begging the inhabitants to let him earn his living and [saying] that he, his wife and his children were dying of hunger, but he never found that they would give him even one penny to earn. When he saw this, he was most unhappy, and he then began to say, 'Alas! Lady of Liesse, I will die of hunger. I cannot possibly have the courage to go about asking for alms. Alas! Lady, I will have to be a robber, yes I will, if I want to live.' At which, as out of despair, he began to rob his nearest neighbours and those whom he knew had wheat, wine and salt pork. He continued until the neighbours noticed that someone was robbing them and suspected the poor man, at which they agreed to keep such a good watch that they would catch him in the act, for they knew well that he did nothing and had nothing to live on, and that he and all his family seemed in fair health. They then kept such good watch that he was caught in the granary of one of their number filling his sack with wheat. They grabbed him and beat him well, saying they would have him hanged. The poor man, seeing himself caught, was so bewildered that he knew not what to say, except to call upon the fair lady of Liesse most devoutly in his heart. When the two neighbours had tied him up they had him put in prison. And when the Provost interrogated him, he confessed all, for which he was sentenced to be hanged and strangled. In order to accomplish this thing, he was led to the gallows, where he called upon the fair lady of Liesse, most devoutly saying beautiful prayers and praying her to save his life. When these prayers were said, he mounted the ladder and was hanged by the hangman, who took great care to have strangled him. When he was hanged everyone went away, and he remained on the gallows the space of three days without dying, complaining of the pain that he suffered and endured. There passed by a shepherd from the fields who heard this poor hanged man lament. When he raised his eyes to him, he saw that he was not yet dead. When the hanged man saw him, he called to him and said to him, 'Alas! my friend, go get the Provost and tell him to send me the hangman to finish me off and let me die, for I languish in great martyrdom.' Then the shepherd out of pity ran towards the town, where he met the two neighbours who had had him hanged. He said to them, 'Alas! my lords, are you not people of justice?' 'Yes,' they answered, 'What is amiss?' 'My lords,' said he, 'There is a poor man hanged at the gallows for three days who has begged me to go tell the Provost to have someone come finish putting him to death.' Then were the two neighbours much bewildered to hear these news. They said to the shepherd, 'Do not worry, we are coming.' They went immmediately to the gallows, and when they had climbed the ladder, they drew their knives and gave five or six blows to the body of the poor long-suffering man, but never could they manage to kill him. When the shepherd, who was at some distance from them, saw that they were so mistreating this poor sinner, he began to cry out, 'Alas! Those men are finishing off this poor man, and I thought that they were taking him down from the gallows.' He then ran to the town, where he found the Provost, to whom he recounted the case, which much astonished him. Immediately he mounted on horseback and went with the shepherd to the gallows, where he found the two neighbours still there, unable to leave, and who said, one to the other, 'By Our Lady, it seems to me that I am bound by my feet with iron chains.' 'Just what has happened to me', said the other. 'I can neither flee nor leave.' When the Provost saw them, he said to them, 'Ha! my lords, what have you done to this poor man?' 'Aha!, my lord,' said the hanged man, 'they have hurt me a hundred times more than the hangman: they have wounded my body in more than six places.' When the Provost asked him who had kept him from dying, he answered that it was the fair lady of Liesse. Then the Provost commanded the neighbours to cut him down or lose all their possessions. He charged them to nourish him, his wife and his children as long as he should live, and if they would not consent, he would confiscate their houses and their goods. At which they agreed to nourish him all his life. Then they lowered the poor man, carried him to their houses and got him cured. And when he was well he went to visit and thank the fair Lady of Liesse who had saved his life.

Pierre de Fourcy, the thief caught in the act and sentenced to the gallows, obviously calls to mind the even older robber protected by the Virgin cited in anthologies of *exempla* and in the *Golden Legend*. As with the older robber, Pierre is saved twice, once from the rope, which fails to strangle him, and next from his neighbours' knife blows, which fail to pierce his body. Like Ebbo, Pierre owes this grace to his devotion to Mary, whom he reveres and implores. Again like Ebbo, when he is freed he gives her thanks. But between the tale connected with the sanctuary and the medieval *exemplum* there are clear and significant differences. First, the miracle of Liesse particularizes the story by dating and situating it. It also moralizes it by insisting on Pierre's absolute destitution. He is without possessions and without work, and for that reason he is forced to rob. The miracle also condemns the cruelty of Pierre's accusers, who are punished at the end. Finally, it omits the description of the actual operation of Mary's mercy. Here the Virgin of Liesse neither supports her protégé in mid-air nor parries the blows directed at him.

This account, printed *before* the 1588 and 1589 *occasionnels*, can be found in all collections of the miracles of the Virgin of Liesse. It figures in the various editions of the book published by the widow Bonfons, by Pierre Mesnier in Paris at the very end of the sixteenth century, and by Blaise Boutart in Troyes at the beginning of the following century. Next, in various and often abridged forms, it figures in works probably sold at the sanctuary itself,[37] for example, the *Histoire et miracles de Notre Dame de Liesse* of G. de Machault, published in Paris in 1617, a *Sommaire* of which was printed in Troyes in the same year by Claude Briden. The latter work dates the story in 1539, thus bringing it closer to the present. A woodcut portrays the procession leading Pierre de Fourcy to his execution, rather than showing the miracle being carried out, as in the medieval texts. This picture shows Pierre kneeling in prayer before an image of Notre-Dame de Liesse as the wagon carrying him makes its way to the gallows. While the title pages of the Douai *occasionnels* show the discovery that Anne is still alive, a scene that takes place *after* the miracle, the Troyes pamphlet of 1617 shows what was happening *before*. Both pictures, like the texts they illustrate, avoid showing the supernatural act itself, the representation of which might encourage superstitious thoughts.

These closely related images suggest another connection. It seems that the *legenda* attached to Notre-Dame de Liesse in fact constitute the meeting point between the medieval miracle of the Virgin's mediation in favour of Ebbo, her faithful devotee, and the story of Anne, saved by her faith in the lady of the Laon sanctuary. The tale of the hanged man miraculously saved, in the widow Bonfons's version and perhaps in others before hers, attenuates the discrepancy (discussed above) between the *exempla* and the *occasionnels*. An important difference remains: the hanged man of the miracle of Liesse becomes a hanged woman in the Rennes and the Douai versions. This feminization of the victim, which constitutes the fundamental novelty of the *occasionnels*, has its own precedents, however. Like Anne des Grez but before her, Pierre de Fourcy is the victim of men's wickedness, and his sins are so constrained as to approach innocence. Like Anne, he is aided by a simple person who reports the miracle (in one instance a 'shepherd of the fields'; in the other a small boy). Like her, he makes the pilgrimage to Liesse to thank his protectress.

Does this mean that the narrative included in the collection of the miracles of Notre-Dame de Liesse is the direct 'source' of the 1588 *occasionnel*? Perhaps not, but it seems plausible to postulate that the author of the text given as printed in Rennes was acquainted with the accounts of the miracles attached to the royal sanctuary in the diocese of Laon. In any event, this is surely the case with the publisher of the pamphlet, since the *Oraison à nostre Dame de Lyesse* that follows the *Discours d'une histoire et miracle* is directly taken from the work published by the widow Bonfons. The 'Rennes' *occasionnel* is thus grafted onto an ancient and local tradition, from which it borrows its central motif (the miraculous hanging) and its formulas (for example, the dramatization of the dialogues).

What the text invents, perhaps by drawing on other texts, perhaps by taking details from an actual event, is the story of an innocent woman unjustly accused of infanticide and rescued on the gallows. This added a new miracle to the repertory of Notre-Dame de Liesse, a miracle that moved readers to horror before an abominable crime, and that feminized innocence in a time in which women participated *en masse* in penitential processions and pilgrimages. The story might in its turn have been taken over as the basis for other miracle tales. Thus in 1632, when he published his *Image de Nostre Dame de Liesse, ou son histoire authentique*, René de Ceriziers first gives the story of Pierre de Fourcy, which for him happened 'less than a century ago'. He then adds that of a serving girl made pregnant by her master. She is accused of infanticide after the accidental death of her infant, but 'those who knew of the misfortune of our miserable [girl] thought that there was design and malice.' Sentenced to be hanged, she implores the mercy of Notre-Dame de Liesse and promises to go barefoot to her sanctuary. She can then leave her prison without help and without punishment, and like Anne she fulfils her vow by visiting Liesse.

Establishing the meanings and the uses of texts such as the ones analysed here is an almost impossible challenge. No corroborating record exists of the two *occasionnels* printed in Douai and Rennes – if indeed the second imprint can be trusted. To solve the enigma they pose with only the physical object – the pamphlet – and its text in hand requires fragile and risky hypotheses that may soon be contradicted by more rigorous or more fortunate research. We know for certain, though, that printed texts of the sort attracted the greatest number of readers and brought stories, images, and beliefs out of the narrow world of the more literate elites.

The account of the miracle of Montfort, in its two successive versions, bears signs of a tension found in nearly all the *occasionnels* that combined formulas inherited from tradition with current happenings. The story of Anne's miracle is thus partly a new variation on an old hagiographical and pastoral motif, and partly a celebration of a devotional practice closely tied to current events involving the League. This made it possible, within the flexible form of the widely distributed chapbook, to rewrite the story, keeping the general plot outline, and to redirect the message. This is just what the Douai publishers did in 1589. It also made possible multiple uses of the same printed piece, offered and received simultaneously as a *canard* telling of an extaordinary event, as an apologetic text reaffirming the Catholic credo, or as a pamphlet linked to a specific pilgrimage and an object to be used in pious practices. The *occasionnels* gave printed form to stories that in former times had been told, preached, or recited, and this plurality of possible readings, which was both organized by the text and produced spontaneously by its readers, is doubtless one of the major reasons for their lasting success.

Notes

I am indebted to Daniel Vigne, the director of the film, *The Return of Martin Guerre*, for my introduction to the *occasionnels* studied in this essay. It was he who got me to read these sixteenth-century texts that tell the story of the woman hanged and miraculously saved, a story that he was using for a film script. My sincere thanks to him and to all who have helped me with their special knowledge, in particular Alain Boureau, Jacqueline Cerquiglini, Jean-Claude Schmitt, Michel Simonin and Catherine Velay-Vallantin.

1   Douai, Bibliothèque municipale, 1589/4, octavo, 15pp., A-B⁴.

2   Jean Pierre Seguin, *L'Information en France avant le périodique. 517 canards imprimés entre 1529 et 1631* (Editions G. P. Maisonneuve et Larose, Paris, 1964); Roger Chartier, 'Lectures populaires et stratégies éditoriales', in Henri-Jean Martin and Roger Chartier (gen. eds) *Histoire de l'édition française*, (3 vols, Promodis, Paris, 1982), vol. 1, *Le Livre conquérant. Du Moyen Age au milieu du XVIe siècle*, pp. 585–603, esp. pp. 596–8 ['Publishing strategies and what the people read, 1530–1660', in Chartier, *The Cultural Uses of Print in Early Modern France*, tr. Lydia G. Cochrane (Princeton University Press, Princeton, 1987), pp. 145–82].

3   Annie Parent, *Les Métiers du livre à Paris au XVIe siècle (1535–1560)* (Droz, Geneva, 1974), pp. 137–40.

4   A copy of this edition is described in the sale catalogue of P. and A. Sourget, *Cent Livres précieux de 1469 à 1914* (Chartres, 1985), no. 40.

5   Denis Pallier, *Recherches sur l'imprimerie à Paris pendant la Ligue, 1585–1594* (Droz, Geneva, 1976), p. 430, no. 865.

6   Denis Pallier, the best specialist in publishing in Paris at the end of the sixteenth century, has verified this point for me, for which I thank him wholeheartedly. The only mention of Geoffroy du Pont is in Philippe Renouard, *Répertoire des imprimeurs parisiens, libraires, fondeurs de caractères et correcteurs d'imprimerie* (Lettres modernes, Paris, 1965), which gives him as a bookseller in Paris in 1589, an indication that perhaps comes from Renouard's acquaintance with Jean Bogart's edition and his imprint on the title page.

7   The catalogue of their publications is given in Albert Labarre, *Répertoire des livres imprimés en France au XVIe siècle* (Bibliotheca Bibliographica Aureliana, Baden-Baden, Librairie Valentin Koerner, 11e livraison, 53 Douai, 1972).

8   I might mention, however, the listing in Bogart's catalogue of the publication in 1586 of a *Discours admirable et véritable des choses advenues en la ville de Mons en Hainaut, à l'endroit d'une religieuse possessée, et depuis délivrée* (though the work,

octavo, is 136 pages in length, which much exceeds the usual length of *occasionnels*), and the publication in 1588 of a *Récit sur l'advénement de la Royne d'Algérie en la ville, de Rome et comment elle a esté baptizée avec six de ses fils et plusieurs matrones qui estoient en sa suite* (octavo, 7 pp.).

9   Albert Labarre, 'L'imprimerie et l'édition à Douai au XVIe et au XVIIe siècle', *De Frandse Nederlanden / Les Pays-Bas français*, 1984, pp. 98–112; Anne Rouzet, *Dictionnaire des imprimeurs, libraires et éditeurs des XVe et XVIe siècles dans les limites géographiques de la Belgique actuelle* (B. de Graaf, Nieuwkoop, 1975).

10   Claude Brémond, Jacques Le Goff and Jean-Claude Schmitt, *L'Exemplum* (Brepols, Turnhout, 1982); Jean-Claude Schmitt (ed.), *Prêcher d'exemples. Récits de prédicateurs du Moyen Age* (Stock, Paris, 1985).

11   See Michel de Certeau, *La Possession de Loudun* (Gallimard-Julliard, Paris, 1970), pp. 314–9.

12   Jean Delumeau, 'Les réformateurs et la superstition', *Actes du colloque L'Amiral de Coligny en son temps (Paris, 24–28 octobre 1972)* (Société de l'Histoire du Protestantisme Français, Paris, 1974), pp. 451–87.

13   Marie-Claude Phan, 'Les déclarations de grossesse en France (XVIe–XVIIIe siècles). Essai institutionnel', *Revue d'Histoire moderne et contemporaine*, 1975, pp. 61–88.

14   Jean-Louis Flandrin, 'L'attitude à l'égard du petit enfant et les conduites sexuelles. Structures anciennes et évolution' (1973), in his *Le Sexe et l'Occident. Evolution des attitudes et des comportements* (Le Seuil, Paris, 1981), pp. 61–88.

15   Seguin, *L'Information en France*, no. 11, p. 70.

16   ibid., no. 40, p. 74; nos 68 and 70, p. 77.

17   On the *Golden Legend*, see Alain Boureau, *La Légende dorée. Le système narratif de Jacques de Voragine (1298)* (Editions du Cerf, Paris, 1984).

18   Albert Labarre, *Le Livre dans la vie amiénoise du seizième siècle. L'enseignement des inventaires après décès, 1503–1576* (Editions Nauwelaerts, Paris and Louvain, 1971).

19   Parent, *Les Métiers du livre*, p. 232.

20   Jacobus de Voragine, *The Golden Legend, or Lives of the Saints as Englished by William Caxton* (J. M. Dent, London, 1900), vol. 5, p. 107.

21   Frederic C. Tubach, *Index Exemplorum. A Handbook of Medieval Religious Tales* (Akademia Scientiarum Fennica, Helsinki, 1969), no. 2235, p. 179. The *exemplum* is summarized thus: 'A thief, devoted to the Virgin, is held up by Her on the gallows, and is thus saved from death.'

22   Albert Lecoy de la Marche (ed.). *Anecdotes historiques, légendes et apologues tirés du recueil inédit d'Etienne de Bourbon, dominicain du XIIIe siècle* (Librairie Renouard, H. Loones, successeur, Paris, 1877), pp. 102–3.

23   Andrew George Little (ed.)*Liber exemplorum ad usum praedicantium ...*, (Typis Academicis, Aberdeen, 1908), pp. 24–5.

24   See Albert Poncelet, 'Miraculorum B.V. Mariae quae saec. VI-XV latine conscripta sunt. Index', *Analecta Bollandiana*, 21 (1902), pp. 241–360, who mentions nineteen occurrences of the story.

25   Gautier de Coinci, *Les Miracles de Nostre Dame*, ed. V. Frederic Koenig ((Droz, Geneva, and Minard, Paris, 1970), I Mir. 30, pp. 285–90.

26   Alexandre, comte de Laborde, *Les Miracles de Nostre Dame compilés par Jehan Miélot ... Etude concernant trois manuscrits du XVe siècle ornés de grisailles* (Société française de reproductions de manuscrits et peintures, Paris, 1929), pp. 85–86.

27 Paris, B.N., n.a.f. 24541, reproduced in Abbé Poquet (ed.), *Les Miracles de la Sainte Vierge traduits et mis en vers par Gautier de Coincy* (Paris, 1867), p. 501, and in Henri Focillon, *Le Peintre des Miracles Notre Dame* (P. Hartmann, Paris, 1950), pl. xvii.

28 Paris, B.N., fr. 9198, reproduction in Laborde, *Les Miracles de Nostre Dame*, pl. 4.

29 Baudouin de Gaiffier, 'Un thème hagiographique; le pendu miraculeusement sauvé' and 'Liberatus a suspensio', in his *Etudes critiques d'hagiographie et d'iconologie* (Société des Bollandistes, Brussels, 1967), pp. 194–226 and 227–32.

30 Paris, Bibliothèque de l'Arsenal, 8° J 5521]5, octavo, 12 pp., A⁴ B². The pamphlet is reviewed in Seguin, *L'Information en France*, no. 331, p. 106. Another copy is conserved in the Bibliothèque Municipale, Lille, 3546.

31 Labarre, *Répertoire des livres*, 19e livraison, 124 Rennes, by J. Betz, 1975.

32 Edouard Vigoland, *Montfort-sur-Meu; son histoire et ses souvenirs* (H. Caillière, Rennes, 1895), p. 196.

33 On this tale and its Canadian, French, Irish, and Flemish versions, see Nancy Schmitz, *La Mensongère (conte, type 710)* (Presses de l'Université de Laval, Québec, 1972). [The tale is cited here from *Grimm's Household Tales*, tr. and ed. Margaret Hunt (George Bell & Sons, London, 1884, reissued by Singing Tree Press, Detroit, 1968), vol. 1, pp. 7–11].

34 Emile and Aldoin Duployé, *Notre-Dame de Liesse. Légende et pèlerinage*, (2 vols, Brissart-Binet, Rheims, 1862).

35 Denis Crouzet, 'Recherches sur les processions blanches, 1583–1584', *Histoire, economies, sociétés*, 1982, no. 4, pp. 511–63, in particular pp. 525–6; Denis Richet, 'Politique et religion: les processions à Paris en 1589', in *La France d'Ancien Régime. Etudes réunies en l'honneur de Pierre Goubert* (Privat, Toulouse, 1984), pp. 623–32.

36 This edition is described in the *Catalogue des livres composant la Bibliothèque de feu M. le Baron James de Rotschild* (Paris, 1893), vol. 3, no. 2709. I cite from the editions of Pierre Mesnier (Paris, n. d.) and Blaise Boutart (Troyes, n. d.) conserved in the Bibliothèque Nationale.

37 Among others, *Histoire de Nostre Dame de Liesse extraict des oeuvres de Jacques Bossius, de l'Ordre de Saint-Jean de Hierusalem* (Blaise Boutard, Troyes, n. d.; privilege of 1601); Walrand Caoult, *Miracula quae ad invocationem Beatis Virginis Mariae... ac dominam gaudiorum in Picardia vulgo n'e Dame de Liesse, dictam effulsere ab anno 1081, ad annum usque 1605* (Charles Boscard, Douai, 1606); G. de Machault, *Histoire et miracles de Nostre Dame de Liesse* (Paris, 1617); de Machault, *Sommaire de l'Histoire et miracles de Nostre-Dame de Liesse* (Claude Briden, Troyes, 1617, and Pierre Sourdet, Troyes, 1617); René de Ceriziers, *Image de Nostre-Dame de Liesse, ou son Histoire authentique* (Rheims, N. Constant, 1632); *Histoire miraculeuse de Nostre Dame de Liesse. Avec un receuil des grâces que Dieu opère par l'intercession de sa Sainte Mère. Ensemble les figures de ladite Histoire* (Claude Briden, Troyes, 1645).

3

Tales as a Mirror:
Perrault in the *Bibliothèque bleue*

CATHERINE VELAY-VALLANTIN

Who wrote Perrault's *Contes*? Was it a children's book? If so, what was its pedagogical stance? Did it simply reflect a fashion, or was it a popularizing endeavour, like Fontenelle's during the same period? How was it connected with the Quarrel of the Ancients and the Moderns?

Literary historians periodically analyse data and compare dates. Although the identification of sources for the *Contes* has until recently prompted anachronistic representations that have played a role in the great variety in the uses of this collection, it is not my intention to discuss them here. To summarize, I might note the tenacious persistence of the legend, piously repeated in many a biographical introduction, that Pierre Darmancourt, Perrault's son, was the author of the *Contes*. Confronted with the constant and irritating problem of attribution that surrounds all of Perrault's literary and architectural works, historians have long opted, where the *Contes* were concerned, for the 'thin, frail voice' (Pierre Darmancourt was nineteen years old!) that 'was to carry over the centuries to repeat, unchanged, to later generations of rapt children, the old tales of his nurse'. This is an excellent illustration of a soothing romanticism common among folklorists of the nineteenth century, among them Champfleury, who established, in his *Histoire de l'imagerie populaire* (Paris, 1869), the special connection between the notions of the people, childhood, and the primitive.

The nineteenth century thus saw the start of a long series of characterizations of the *Contes* based on the shaky definitions of 'the people' and the 'popular' commonly accepted until then. At the same time, the child as interpreter of the people (in Michelet's formula) became both the object of special interest and a frequent theme in publishing strategy. As a child's book, the *Contes* of Perrault merited a place in the domain of books for children. Was this an a posteriori justification of a search for a new readership and a new sales market? Or did it reflect the resurgence of a real pedagogical interest in the service of the multitude? For the moment, I might simply note the publishers' mixed motives, their renewed but narrow interest in the moral and thematic polyphony of the *Contes*, and their passing but deliberate choice of the most conventional representation possible. With the twentieth century, an imprudent use of new methods explains the reserve, even the mistrust, with which most literary historians regarded the constructions of 'Indianist' or 'ritualist' folklorists and Jungian psychoanalists. Research on Perrault was for the time being abandoned and publishers turned to an abundant supply of children's literature.

In 1953 there came dramatic news: an unpublished manuscript of the *Contes* had been discovered. Made by a professional copier, dated 1695 (or two years earlier than the first published edition) and bound in Morocco with the arms of Mademoiselle, the King's niece, the copy offered a fully worked out state of the text. Jacques Barchilon edited the manuscript, studied the variants, and pronounced Perrault as the work's author. The folklorist Paul Delarue suggested a more nuanced opinion, imagining Pierre Darmancourt collecting oral sources re-elaborated by his father. That *Cendrillon*, *Riquet à la Houppe*, and *Le Petit Poucet* were absent from the manuscript was seen as evidence of collaboration and fashionable competition between Perrault, his niece, Mademoiselle L'Héritier, and her friends, Madame de Murat and Madame d'Aulnoy. According to Mademoiselle L'Héritier:

> There was talk of the good upbringing that he [Perrault] was giving to his children. It was said that they all show much wit, and finally talk fell on the naïve tales that one of his young pupils has recently put to paper with such grace.
>
> Several of these were told, and that led imperceptibly to the telling of others. I had to tell one in my turn. I told the one about Marmoisan, with some embroideries that came to my mind as I went along. It was new to the company, which found it much to its taste and judged it so little known that they told me I must communicate it to that young tale-teller who so wittily occupies the leisure of childhood. I took pleasure in following this advice, and as I know, Mademoiselle [Perrault's daughter], the taste you have and the attention you pay to things in which some sense of morality enters, I will tell you this tale more or less as I told it. I hope that you will share it with your amiable Brother, and you may together judge whether this fable is worthy of being placed in his delightful collection of Tales.

Mademoiselle L'Héritier, who here keeps up the fiction of the 'young tale-teller', later herself published *Marmoisan*, doubtless rejected by Perrault.

Perrault, in Manuscript and in Print

The 1659 manuscript offers an important indication of Perrault's original project. That the collection is incomplete and was not conceived of as a whole lends support to the thesis of a pedagogical work designed to be added to or transformed as the author gained increasing command of his materials. The reconstruction of each state enables us to define the author's personality better – his 'personal equation', as Marc Soriano calls it in his now standard work, *Les Contes de Perrault. Culture savante et traditions populaires*. After Mary Elizabeth Storer's *La Mode des contes de fées*, a pioneering work unique in its time, recent studies on Perrault, freed from personal and ideological viewpoints, show a rigour and eclecticism lacking in literary historians of the nineteenth century.[1]

Before continuing we need to review the chronology of the various publications of the *Contes*:

1691: *Recueil de plusieurs pièces d'éloquence et de poésies présentées à l'Académie française pour les prix de l'année 1691. Avec plusieurs discours* . . . A Paris, J.-B. Coignard. It contains *La Marquise de Salusses ou la Patience de Griselidis. Nouvelle* (pp. 143–94).

La Marquise de Salusses. . . was republished alone in the same year by J.-B. Coignard.

1693: *Les Souhaits ridicules. Conte*. In *Le Mercure galant*, November 1693 (pp. 39–50).

1694: *Grisélidis, nouvelle. Avec le Conte de Peau d'Asne, et celui des Souhaits ridicules*. Seconde edition. A Paris, chez la Veuve de Jean-Baptiste Coignard, Imprimeur du Roy, et Jean-Baptiste Coignard Fils, Imprimeur du Roy, rue Saint Jacques, à la Bible d'or. 1694. Avec privilège.

Recueil de pièces curieuses et nouvelles, tant en prose qu'en vers, Tome I. A La Haye. Chez Adrian Moetjens, Marchand-libraire, près la Cour, à la Librairie française. The first volume contains, among other works of Perrault, *Peau d'Ane*, *Les Souhaits ridicules*, and *Grisélidis*.

1695: Jean-Baptiste Coignard published the fourth edition (the third has not been found) of *Grisélidis*, *Peau d'Ane*, and *Les Souhaits ridicules*. The *Préface* figures for the first time here.

Contes de ma Mère Loye, manuscript, 118 unnumbered pages octavo, made available in 1956 by Jacques Barchilon. The binding bears the arms of Mademoiselle, Louis XIV's niece, to whom the *Contes* in prose are dedicated. The collection includes the *Epître à Mademoiselle* signed Pierre Perrault Darmancourt, *La Belle au bois dormant*, *Le Petit Chaperon rouge*, *La Barbe bleue*, *Le Maître chat ou le Chat botté*, *Les Fées*.

1696: *La Belle au bois dormant. Conte*. In *Le Mercure Galant*, February 1696 (pp. 75–117).

Recueil de pièces curieuses et nouvelles. . . , Tome V. A La Haye. Chez Adrian Moetjens. It contains *La Belle au bois dormant. Conte*.

1697: *Histoires ou Contes du temps passé. Avec des Moralités*. A Paris, chez Claude Barbin, sur le second Peron de la Sante Chapelle, au Palais. Avec privilège de Sa Majesté.

Recueil de pièces curieuses et nouvelles . . . , Tome V. A La Haye. Chez Adrian Moetjens. The text is that of the first printing of the Barbin edition.

Histoires ou Contes du temps passé. Avec des Moralités. Par le Fils de Monsieur Perreault de l'Académie François. Suivant la copie, à Paris. This was a Dutch pirated edition that had several reprintings.

Histoires ou Contes du temps passé. Avec des Moralités. A Trévoux. De l'Imprimerie de S.A. Seren. Mons. Prince Souverain de Dombe.[2]

The earliest editions of the *Contes* in prose show the signs of their dual genesis. On the one hand, they emphasize the transcription from an oral delivery to a written text; on the other, Perrault as author hid behind the anonymous narratress as he did behind the fictional 'Pierre Darmancourt'. In this manner, the 1695 title, *Contes de ma Mère Loye*, is reflected in the frontispiece of the manuscript, which shows an old nurse telling stories to children of good family, as suggested by a lady in a fashionable Fontange bonnet. The repetition of the collection's title in pictorial form in the frontispiece states the obligatory protocols: this was how the work was to be read. Other details to guide reception accompany the strong image of oral presentation: the door latch figures the closed space of the home, and the candle indicates night-time and evening amusements. The nurse's hand is raised in a recounting gesture – the medieval gesture of reckoning – that confirms her open mouth. A pictorial version of the popular oral tale, the frontispiece – Antoine Compagnon called it the *périgraphie du livre* – prepares the elaboration of the text, legitimizes the narrative forms that the reader is to encounter, and suggests that he or she recreate the same circumstances for reading or retelling the tales.

The author was not forgotten, however. His name was obliterated from the title (reflected in the frontispiece) and the work bears the signature of Pierre Darmancourt, but in reality this absence referred to Perrault's personal situation, which was common knowledge – to his disfavour with Colbert, his widowhood, and his dedication to the upbringing of his children, and to the polemic function of the *Contes* in the context of the Quarrel of the Ancients and the Moderns. The lithograph announcing the 'Contes de ma Mère Loye' thus summarizes the representations suggested to the reader, and it can be seen as a strategic framework for imposed readings. It seems to say: this is the status of the concealed author; this is how he transcribed what he heard; this is how this work should be read.[3]

In 1697, Clouzier, the illustrator of Madame d'Aulnoy's *Contes de Fées* (also published by Barbin in April 1697), used the same frontispiece in a copperplate. Perrault, on the other hand, changed his title. *Les Contes de ma Mère Loye* still appeared in the engraving but disappeared from the title page, ceding to *Histoires ou Contes du temps passé*. It is true that the historical dimension had previously been neglected, and perhaps we should see in this new title the influence of the ideas (at times confused) of Mademoiselle L'Héritier on the transmission of oral literature. In 1695, in her *Oeuvres mêlées*, she advanced the hypothesis of tales 'invented by the troubadours', 'our romancers of olden times', which had been transmitted by the people and 'pitilessly disfigured in blue paper books' (1705).

Did Perrault himself postulate the survival of oral literature from one age to the other? Although he proclaimed his attachment to Christianity, which he perceived as a better solution for continuity than paganism, as an academician he had to respond to the arguments of the Ancients. He had already been induced to suggest that Homeric poetry was a learned elaboration of 'old wives' tales' (*Le Parallèle*, vol. 3, 1692, and *Le Mercure galant*, January 1697). With the new title of his *Contes*, Perrault took the risk of glossing over both the destabilizing and the civilizing influence of Christianity. Arguing the necessary secularization of these tales permitted him to affirm the continuity of autochthonous popular traditions. Nonetheless, three years after his public reconciliation with Boileau, Perrault was using the *Histoires ou Contes du temps passé* to continue the battle.

An examination of the earliest editions of the *Contes* (see plates V to VIII) enables us to follow the evolution of Perrault's thought. The successive books of tales, from the hand-copied manuscript to the Barbin edition, show to what extent, in the early stages of their elaboration, the author still considered his tales as transcriptions and how their printing reflected his strategy as a 'modern' academician. The text itself also speaks to this transformation. Jacques Barchilon, who edited the 1695 manuscript, notes the following corrections (as does Marc Soriano):

1695 text	*1697 text*
[These tales] all contain a most reasonable moral, which will be discovered to greater or lesser degree, according to the penetration of *those who listen to them*. (*Preface*)	. . . *of those who read them.*

1695 text	*1697 text*
The better to eat you with. These words should be pronounced with a loud voice to frighten the child as if the wolf were going to eat him. (*Le Petit Chaperon rouge*)	The better to eat you with. [The remainder is omitted.]

According to Marc Soriano, these omissions in the 1697 edition prove 'that in 1695 the author of the dedication thought of these tales as oral works and speaks of them as stories noted down'. The omissions also enlighten us on the relationship between procedures for casting the tales in written form and for putting them into print. We can reconstruct the meaning of the texts chosen and transcribed in a particular fashion by examining the transfer from what is said to the way in which it is expressed: details unique to the Barbin edition (the new title, in combination with the old frontispiece) and changes in the contents attest to Perrault's determination to raise the tales – be they popular or children's literature – to the status of a litarary work. Perrault presents his readers with the meanings they are to retain: whether we examine the choice of his tales, which always refer to previous versions, the vignettes placed above the stories' titles that predetermine their meaning, or the use of such typographic procedures as refrains given in italics, Perrault the academician clearly indicates scenarios for reading his *Histoires ou Contes du temps passé*. These are indications for reading them, but also for viewing them, since the tales were illustrated; for reading aloud, eventually for recitation rather than retelling, since the nurse of the frontispiece stands not only for the oral sources, but also for the exercise now suggested to the reading public. All these were tactics for reaching a probable and a desired reader.

The Literary Editions

How were such indications perceived? What remained of Perrault's original ambition to compose a work all the more complete for being offered for reading, for viewing, and for listening? Even before Perrault's death, a large number of pirated editions were printed in Holland (in 1698, 1700, 1708, 1716, 1721, 1729). Both the Adrian Moetjens edition and the reprint of 1708 by Jacques Desbordes used the text of the Barbin edition, reversing the frontispiece and the copperplate vignettes. In the Seren edition the text was identical but the frontispiece was missing, and the vignettes heading each tale, showing seasonal tasks and the signs of the zodiac, are probably woodcuts taken from almanacs. Desbordes, a bookseller, first attributed Mademoiselle l'Héritier's *L'Adroite Princesse* to Perrault (in 1716) and added it to the collected *Contes*. Was this done because the publisher found it convenient or did it reflect the constant problem of attribution that accompanied the works of the Perrault family? Be that as it may, the error persisted until the late nineteenth century.

The work's many editions may be misleading. That was current practice in bookselling, as the following dialogue demonstrates:

The bookseller: It is a truly fine book, however, and has had several editions in very little time.

M. de Frédeville: Do you think you can make me believe that? Without reprinting a book a second time, you can make six consecutive editions only by changing the first sheet.

The bookseller: Ha ha! Sir, you know all our secrets. (*Carpentariana ou Remarques de M. Charpentier de l'Académie française* [Paris, 1724], *Le libraire du Palais, dialogue*).

The Seren edition does not seem to have been republished, but the Dutch pirated editions copied the Barbin edition, even to the woodcuts, and they kept up the fiction of Pierre Darmancourt's authorship ('par le fils de Monsieur Perrault, de l'Académie française'), even after Perrault's death in 1703. The Parisian editions, on the other hand, were the first to change the title. In 1707, the widow Barbin offered for sale the *Contes de Monsieur Perrault. Avec des Moralités*, dating them, in the interests of publicity, from that very year, or four years after the author had died, but she reproduced the Claude Barbin edition of 1697 page for page, with the same frontispiece and the same vignettes. In 1724, Nicolas Gosselin republished the original edition as well, still with the title, *Contes de Monsieur Perrault. Avec des Moralités*, the title that was repeated thereafter. How could this fail to crown Perrault's ambitions? There was only one collection of tales, and it was Perrault's. All polemic allusion to the 'troubadours' had been forgotten, even in those days of the 'Quarrel over Homer' that reanimated hostilities from 1715 to 1717.

The last edition of the Parisian bookseller Nicolas Gosselin was dated 1724, and the Dutch pirated edition of the bookseller Jacques Desbordes 1729. We have to wait until 1742 for Coustelier to reprint in Paris the *Histoires ou Contes du temps passé. Avec des Moralités, par M. Perrault*. It would be mistaken to see any break in the genre in this thirteen-year gap, since the fairy-tale as a genre remained homogeneous in its contents during the eighteenth century. As early as 1705, Hamilton spoke of the time of fairy-stories as past, and fifty years later the comte de Caylus referred to Perrault's and Madame d'Aulnoy's days as a decidedly bygone epoch, but these false regrets were certainly for publicity purposes. Jacques Barchilon has clearly shown (as have a number of folklorists) that it proved convenient to ignore the oriental tales, which he considers simply a variant on tales of the marvellous. According to Barchilon, from 1690 to the end of the eighteenth century all tales produced were 'fairy-tales' written in French or oriental guise as the moment dictated.

Although this position requires correction concerning the origin of some tales – Gueulette, for example, masked his stories as 'Tartar tales', but Pétis de La Croix produced true Turkish tales – it is nevertheless true that the thematic resources of the tale were conceivable only within stability. In comparison to a permanence of this sort in the genre and its contents, the actual production of tales underwent fluctuations, and such things as innovation, change, imitation, and increased production (or their opposites) were closely associated with fashion. Periods of increased production, both continuous and sporadic, corresponded to moments 'of rupture, of progress and of innovation', in Jean Baudrillard's phrase, that worked to enhance modernity. The fashion for tales of the marvellous in the eighteenth century can be summarized as falling into two periods:

1690–1700: the first enthusiasm for the *conte de fées;*
1730–1758: a new enthusiasm taking as a reference point the preceding vogue thirty years earlier. Simultaneously, an ironic reaction to a taste for a genre judged to be ridiculous and old-fashioned, and a fashion for parodic and licentious tales.

The 'trough of the wave' (from 1700 to 1730) shows a saturation that began with Hamilton and, above all, with Dufresny's bawdy satirical comedy, *Les Fées ou Contes de ma Mère l'Oye* (1697), which was aimed both at Perrault's concern for *bienséance* (propriety) and at the mechanics of fairy-tales such as metamorphoses and supernatural aid.

Were the reprintings of Perrault's works typical of the production of tales in general? This was by no means true for the first period, as has been seen. Editions followed one another in close succession, whereas production was falling off in other sectors. Seeming disorder was even clearer in the incoherent selections made from the works of Madame d'Aulnoy. The Coustelier edition of Perrault in 1742, on the other hand, came at a peak moment in book production and was immediately imitated by Jacques Desbordes in Amsterdam. Publishers thus supposed that they would profit from the second vogue. A potential public existed for Perrault's tales. The reprintings, which might seen anarchical in the context of evolutions in printing in other sectors, clearly refute the simplification that the oriental tale dethroned the fairy-tale in readers' preferences. Perrault in no way suffered from this long process of disaffection. On the contrary, repeated editions prove that people remembered a quite specific genre.

Despite authors' efforts to bring new life to a literary genre by spectacular opposition to older models, the booksellers correctly gauged their public's unchanging tastes. I do not mean to minimize the particularity of each stage in a shifting fashion, but rather to note the basic unity in their narrative procedures, and to show that the new publications, in spite of their differences in form, expressed the same needs of the social group and aimed at the same objectives as the old. The nearly regular reprintings of Perrault and the uniformity in editions up to the mid-eighteenth century suppose the public's unmediated appropriation of the text in a communion as total as in 1697 – unless of course the publishing world was simply judging the public's reading skills and tastes to be unvarying. Occasionally Perrault's publishers took advantage of a new fashion to give their editions a new look, but in general their stategy ignored passing enthusiasms, and their public was as unvarying as the reprinted text, in all points identical to the original edition.

The Coustelier edition introduced an innovation, however, when it sacrificed Clouzier's vignettes – long reproduced and imitated – to a renewed emphasis on the text. The frontispiece and the vignettes engraved by Fokke after De Sève dress the characters in antique garb, and it is hard not to see this Graeco-Latin portrayal as an accentuation of the classicism of the work. Perrault's *Contes* had been treated as a masterpiece of French literature for more than forty years (more by the work's publishers than by its imitators). When compared to the oriental tales, they were incorporated into the classical heritage by their ancient origins – a move that had a certain piquancy if we remember Perrault the 'modern' and his polemic with the 'Ancients'! In spite of this change in form, De Sève's vignettes faithfully illustrate most of the same scenes that Clouzier had chosen.

There are three exceptions to this fidelity, however. First, *Le Petit Chaperon rouge* gained an extra illustration. De Sève offered the reader the traditional view of the wolf devouring the grandmother, a half-page vignette that in certain of Barbin's and Gosselin's editions included the title. However, he also used a full-page illustration showing the meeting of the little girl and the wolf on the path to the grandmother's house. Arguments in favour of a slacking of dramatic tension and an attenuation of the scabrous scene of the wolf and the grandmother were later to prevail, and, as we shall see, certain *Bibliothèque bleue* editions chose to print only the new vignette (see plate VI). A second change occurred in the 1697 edition: the illustration of *La Barbe bleue* shifted the focus of the suspense in the final scene. The original engraving illustrated the passage that ran: 'Then taking hold of her hair with one hand, and holding up the cutlass with the other, he was going to cut off her head. . . . "Recommend thy self to God", and raising his arm . . . At this very instant there was such a loud knocking at the gate . . .' [tr. Samber]. The new vignette accompanied the scene immediately following Bluebeard's death, when he was lying on the ground and his wife threw herself into the arms of her rescuers. The change countered Perrault's objectives, since he had taken care to base the dramatic progression of his tale on chance, delayed expectation and increasing emotion. It also betrayed the literal sense of the text, which states that 'the poor lady was almost as dead as her husband, and had not strength enough to rise and embrace her brothers' (see plate VII). For the third change, Puss in Boots was shown parading like a major-domo before the marquis of Carabas, married to the princess at last. De Sève's wholly imaginary figuration of a 'happy end' replaced an illustration of the cat's wily tricks of a cleverness approaching racketeering.

These three recastings had several characteristics in common. They indicated discomfort at the violence, crudity, and amorality of certain episodes, and they transfigured the stories by their representation of a new sensibility, a new affectivity and a new morality. These edifying vignettes show three aspects of living in society – the couple (*Le Chat botté*), the family (*La Barbe bleue*), and sociability (Little Red Riding Hood's meeting with the wolf is a perfect model of civility!). Was Coustelier, the publisher, adapting to the tastes of a more refined and well-read public? Had he taken on a moralizing, even a civilizing mission? The changes tended towards gentler and more regulated behaviour. Perhaps both explanations are true. In any event, successive literary editions (Bassompière, Lamy) used the same vignettes throughout the eighteenth century. This shift in iconography raises real problems when it is imitated by the *Bibliothèque bleue* editions, as we shall soon see. I might also note that the Coustelier edition was the first to change the order of the tales: *La Belle au bois dormant* lost its place at the head of the collection to *Le Petit Chaperon rouge*, which explains the second full-page vignette illustrating that tale. This was a first step towards autonomous illustration, which became current practice a century later.

The discrepancy between the conscientious fidelity of the title (*Histoires ou Contes du temps passé*) and the liberties taken with the stories themselves merits examination. If publishers left the author's status unchanged and if they thus spared him becoming stereotyped ('Perrault's tales' and 'fairy-tales' were synonymous terms), it was to enhance use of his characters for moralizing purposes.

To continue with the editions of Perrault's tales: they showed a spectacular increase starting with 1770s. There was the Liège edition of Bassompierre (1777), and the Lamy edition (1781); there was also the publication (1775) of the first volumes of the Marquis de Paulmy's *Bibliothèque universelle des Romans*, preceeding that of the *Cabinet des Fées* (beginning in 1785). Careful reading of printers' announcements and introductions shows a new and revelatory interest in the tales for historical and bibliophilic reasons. This was why the tales in verse were to be rescued from oblivion. *Les Souhaits ridicules* found a place in *Les Amusements de la campagne et de la ville* (Amsterdam, 1747), as did *Peau d'Ane* in the *Bibliothèque universelle des Romans* (January 1776). The latter verse tale was published in its entirety as a courtesy to the 'men of letters' who, 'having learned that [the Marquis of Paulmy] did not have this little work readily available, have deigned to provide [him with it]'. The 'rarity' of the document and its literary qualities justified its selection by Bastide, one of the collaborators responsable for the 'originals'. When the *Histoires ou Contes du temps passé. Avec des Moralités par Monsieur Perrault* appeared in October 1775 in the form of 'miniatures', however, the editor limited himself to recalling 'the moral aim of the author in writing them'.

The success of the *Bibliothèque universelle des Romans* was unsure, at least at the start. For what public was it destined? 'Like the collaborators, it was a well-to-do public, curious intellectually and open to innovations, but whose material wealth rested on a perfect integration into the monarchic system.'[4] In point of fact, this public had not waited for the *Bibliothèque universelle* to read fiction elsewhere, and the project was a failure. Having finally determined the 'taste of the public' from readers' correspondence, Bastide and Tressan changed their techniques of selection, which ensured the success of the anthology, even in the provinces, where it had good distribution. The real public for the work was relatively wealthy (with an annual income of at least 12,000 *livres*), closely resembling that of *Le Mercure de France*. Roger Poirier characterizes them as,

the newly rich or those who were on their way to becoming so, who had social ambitions, and the financial means to satisfy their ambitions but possessed neither the education nor the literary or critical baggage to be demanding concerning the literary contents of the texts presented to them.

Women had a preponderant place in this public, and remarks such as, 'the ladies will appreciate . . .' and 'our *lectrices* will excuse us . . .' are frequent in the introductory remarks to the 'miniatures'.

Thus, up to and including the *Cabinet des Fées*, late eighteenth-century publishers felt the need to return – or so they say – to texts threatened with definitive disappearance and oblivion. In reality, historical perspective on the texts of the tales was governed by didactic and moral preoccupations and by the discovery of a new sensibility. The success of the *Magasin des Enfants* of Madame Leprince de Beaumont benefited Perrault, to whom certain of her laboriously edifying tales were attributed. Beyond their iconographic tampering and their intellectualizing claims, the publishers also smoothed out the thematic material and even the literal meaning of the *Contes*, to the point of giving the collection the lulling homogeneity of a heavily stereotyped yet pragmatic morality. The illusion was that the text, thought to be untouchable, was indeed untouched. In reality, readers were given a different story to read, guided by illustrations that told a different tale from the text. The literary homage of the editors and anthologists was stage dressing designed to mask sleight of hand and alterations.

How did the public react? Did it accept easily, for example, seeing the dreamlike erotic appeal of *Le Petit Chaperon rouge* channelled into banality? The tale is unique in its ability to call upon its many previous occurrences in order to remain open enough, in spite of the violence done to it, to permit several versions and multiple motifs to coexist in the imagination and the memory of the reader. Claude Brémond speaks of *le meccano du conte*, comparing the tale to a construction set that can be taken apart and put back together again as often as needed. Brémond says:

The tale's means of survival, whether we envisage it on the individual scale . . . or in collective memory, is *re-use*. Thematic elements of proven worth enter into combination with others at the heart of new configurations, which are put to the test, ready to be forgotten if their welcome is not good, destined to be repeated and imitated should they meet with success.[5]

In this perspective, combining two apparently incompatible motifs (as in *La Barbe bleue*) was not impossible, if one supposes that the reader knew another ending for the tale. Exclusive meaning was all the less likely if this ending occurred in another folk genre – song – which operated in different ways and in another context.

A tale cannot be read, heard, or seen unless it is situated within the complex system of folk references, which means that it necessarily elicits a multiplicity of readings. Ethnologists and folklorists have studied the tensions (things excluded or coexisting) and the reminiscences inherent in the moment of oral transmission. The moment of reading a text or viewing an image also exists only in a dialectic relation, even if it is tempting, when faced with copies seemingly in such perfect conformity with the original, to opt for a functionalist interpretation. I might also note that Perrault's publishers continued to aim at individual, private reading: the compact, tightly spaced text of the original typography and page layout continued to preclude oral retelling. It is not certain, under these conditions of reference, that the moralizing impact of the tales was exactly what was intended.

The first edition of collected tales in both verse and prose tales – Lamy's edition of 1781 – was followed by the publication of the *Cabinet des Fées*, but it appeared when publishers, the printers of the *Bibliothèque bleue* in particular, were beginning to publish the tales separately. The Lamy edition is of particular interest for its apocryphal prose version of *Peau d'Ane*. The tale ends, however, with the last stanza of the verse version, presented as a moral in an evident attempt to bring consistency to the collection by following the model of the tales in prose. The success of *Peau d'Ane* in its prose version encouraged the addition of Madame Leprince de Beaumont's *Les Trois Souhaits* (*Magasin des Enfants*, 1757) to the collected *Contes*. The publishers went through the motions of attributing the short, watered-down version of the tale to Perrault, as though it were an adaptation of his *Souhaits ridicules*. At the beginning of the nineteenth century, the mixing of the two authors was to encourage the annexation of Perrault's *Contes* into the pedagogical domain occupied by Madame Leprince de Beaumont, and to make the tales a favoured instrument for the moral education of children – an unexpected and perverse effect of the prose adaptation of *Peau d'Ane*, which, although it represented a drastic change in form, was not a transgression. The publisher's only aim was to facilitate understanding, a move doubtless judged to be all the more urgent since the tale was well loved and well known.

What explanation can be offered for the learned editions of the tales republished throughout the eighteenth century? Only one: recognition. Only those who already knew Perrault's *Contes* were capable of reading them. Comprehension of the stories was subject to previous knowledge and made of recollections, reminiscences, juxtapositions, and interrogations. Even those reading the *Contes* for the first time recognized them immediately. The printer-booksellers acknowledged the privileged status of the collected tales, but they also turned it to advantage. Moralizing and romanticizing emendations proliferated in total editorial licence. Different readings were all the more admissible because the book could not stand on its own. Comprehension was doubtless difficult if not impossible; the uses to which the work was put were impossible to direct, but those were the very conditions for the consumption of the work as an object.

I have read the tale from end to end four times with sufficient attention, nevertheless my mind can fix on no idea of Peau d'Ane in her disguise. At times I imagine her begrimed and black as a Gypsy, with her donkey skin that serves her as a scarf; at other times I imagine that the donkey skin is a sort of mask over her face . . . ; but after all that, the poet himself, who has not taken the trouble to tell me what this disguise consists of, destroys in a few passing words all that I was trying to imagine on the subject. This story is told in just as obscure and confused a fashion as [when] my nurse used to tell it to me long ago to make me go to sleep.

These are the terms in which a critic of Moetjens's collected tales deplored the inconsistencies of the story, but they also explain its many re-editions. Perrault had carefully worked out the connections between his tales and folklore: from direct filiation to adaptation to independent existence, his *Contes* managed to evoke a multiplicity of possible representations, even with an unchanging text. How, then, did the publishers of the *Bibliothèque bleue* exercise their editorial prerogatives in this fertile domain of the re-use of an established text?

Bibliothèque bleue *Editions: An Inventory*

Was the emergence of 'blue' editions of Perrault's *Contes* determined by the fortunes of 'learned' editions? The first permissions, one to Pierre Garnier, the other to Jean Oudot, printer-booksellers in Troyes, both date from 23 July 1723 and appear in two collections of tales dated 1734 and 1737 (see plate IV). The permissions came sixteen years after the first edition of the widow Barbin, two years after the last Dutch pirated edition of Jacques Desbordes, and one year before the new edition of Nicolas Gosselin. It is not impossible that both Oudot and Garnier in Troyes, and Gosselin in Paris, decided to counter the Dutch offensive, judged all the more harmful since the Desbordes editions contained a number of consistent misspellings and errors of attribution (*L'Adroite Princesse* was attributed to Perrault in 1716). The fact remains that both Oudot and Garnier, on the one hand, and Gosselin, on the other, based their editions on the Barbin edition, and that their choice seems to have been deliberate. As the table below shows,[6] the Garnier and Oudot heirs and the widow Béhourt in Rouen republished the *Contes des Fées* until around 1760.

TABLE 3.1 *Les Contes des Fées*: eighteenth-century collections

'Learned' editions	'Blue' editions
1697	First Barbin edition First pirated editions from Holland	
1707	Last Barbin edition	
1716	Desbordes Dutch pirated edition (*Contes* plus *L'Adrote Princesses*)	
1721	Desbordes	
1723		Permission to Pierre Garnier and Jean Oudot
1724	Nicolas Gosselin (new edition based on Barbin)	
1729	Desbordes	
1734		*Les Contes des Fées*, Troyes, Jean Oudot (Morin, no. 183)
1737		*Les Contes des Fées*, Troyes, Pierre Garnier (who died in 1738) (Morin, no. 181)
1740?		*Les Contes des Fées*, Troyes, Garnier le Jeune (permission to Pierre Garnier. The younger Garnier died in 1743) (Morin, no. 182)
1742	Coustelier	
1750?		*Les Contes des Fées*, Troyes, Madame Garnier (Morin, nos 173, 183 bis)
1756		*Les Contes des Fées*, Troyes, widow of Jean Oudot (Morin, no. 184)
1760?		*Les Contes des Fées*, Rouen, widow Béhourt (1759–63)
1776	*Bibliothèque universelle des Romans*	
1777	Bassompierre	
1781	Lamy (the first *Peau d'Ane* in prose)	
1785	*Cabinet des Fées*	
1793		*Les Contes des Fées*, Orléans, Letourmy (Oberlé, nos. 144, 145)

Time was needed to feel the effects of the publications of the *Bibliothèque universelle des Romans*, of Lamy's edition, and of the *Cabinet des Fées*. First Letourmy in Orléans (around 1790), then Caillot in Paris, Gangel in Metz, Madame Garnier in Troyes, Abadie in Toulouse, Deckherr in Porrentruy (and others) all republished the collected *Contes* from 1800 to about 1820, in small print runs, however. The printers of the *Bibliothèque bleue* were beginning to apply profitably a formula that had proven its merits with more loquacious authors like Madame d'Aulnoy, whose works could not easily be published in their entirety. Increasingly, tales were published separately during the second half of the eighteenth century, usually by recently founded firms (Chalopin in Caen, Lecrêne-Labbey in Rouen, and, in the south, printers in Carpentras) rather than by the heirs of the old families of Troyes. In Troyes itself dissension arose between the champions of tradition – Madame Garnier, then Baudot, to whom she sold her print shop – and the more recently established André family. Thus an anthology including *La Barbe bleue*, *Cendrillon, ou la petite pantoufle de verre*, *Les Fées*, *Riquet à la Houppe*, and *La Vertu* was published in Troyes by 'la Veuve André, 'Imprimeur-Libraire et Fabricante de papiers, Grand-Rue, près l'Hôtel de Ville' (Morin no. 43), in competition with the integral edition reprinted by Baudot.

Editorial strategies became more homogeneous after 1830. Widow André, in association with Anner, returned to the complete *Contes des Fées* (Morin no. 180), and Baudot now opted for publishing the tales separately, becoming a major promoter of such publications with *La Belle au bois dormant* (Morin no. 49), *Riquet à la Houppe* (Morin no. 993), and even *L'Adroite Princesse* (Morin no. 15), still attributed to Perrault. Each of the tales received its separate edition, but *La Barbe bleue*, *Cendrillon*, *La Belle au bois dormant* and *Peau d'Ane* were reprinted more frequently and illustrated more often in late eighteenth-century engravings. Prudence is advisable on this point, however, and any quantitative estimate is subject to caution because it is conditioned by the publication of incomplete collections.

Like the 'learned' editions, the *Bibliothèque bleue* neglected the *Contes* in verse, waiting for prose adaptations by various hands before adding the tales to their catalogues. Thus the *Histoire de Peau d'Ane*, 'à Troies, chez Antoine Garnier, Imprimeur-Libraire, rue du Temple' (Morin no. 536) seems to have been the first edition to appear after the work's publication by Lamy in 1781 (it was probably published around 1790, when Jean-Antoine Garnier took over from his mother). *Peau d'Ane* reappeared in Toulouse in 1810, published by the 'imprimerie de Desclassan et Navarre vendue chez L. Abadie' (Oberlé no. 163), and in Rouen, published by Lecrêne-Labbey at the same date (Oberlé no. 164 and Hélot no. 187). In the early nineteenth century, the tale was incorporated gradually into the collections of the *Contes* in prose, as if Perrault had written nine tales instead of eight.

Our panorama would be incomplete if it omitted the definitive and persistent inclusion of all the tales of Madame Leprince de Beaumont among those by Perrault, as in *La Belle et la Bête, Le Prince charmant, et Les Trois Souhaits. Contes tirés des Fées, avec des moralités par M. Perrault*, published in Troyes by J.-B. Lepacifique at the end of the eighteenth century (Oberlé no. 160), the edition of the same title published by 'l'imprimerie de Madame Garnier' (Morin no. 175), or, even later, the editions of the Deckherr brothers in Porrentruy in 1813 (and in Montbéliard in 1822), and of Pellerin in Epinal in 1827.

Can we reconstruct the reasons for including Perrault's *Contes* in the *Bibliothèque bleue* during the eighteenth century? There is no doubt that they found their niche, and it was not by chance that the verse *Contes*, close as they were to the known texts, were not included in the publications. Indeed, an episode from the second book of *Huon de Bordeaux*, the story of Yde and Olive, prefigures the prologue of *Peau d'Ane* in the father's incestuous desire for his daughter. As for *Les Souhaits ridicules*, the text was imitated from a story (no. 78) of the *Cent nouvelles nouvelles* of Philippe de Vigneulles, extracts of which were printed by the Troyes publishers. The most telling case was that of *Grisélidis*, however.

The Example of Grisélidis

Boccaccio was the first European author to recount the adventures of the marquis of Saluzzo (*Decameron*, X, 10), and in 1374 Petrarch translated the tale into Latin, Griselda becoming Griseldis. Not only Chaucer's *Clerk's Tale* but several French adaptations were based on Petrarch's text: Philippe de Mézières's translation (1389), an adaptation for the stage (1395), and another 'translation' in prose in the fifteenth century. The 'example' of Griselda found a place in didactic and moralizing works such as Christine de Pisan's *Le Livre de la Cité des Dames* (1405) Martin Le Franc's *Le Champion des Dames* (1442) and Olivier de La Marche's *Le Triomphe des Dames* (1493). In 1484, and again in 1491, a prose version of the 'pacience de Grisélidis' was published on which a number of editions were based during the sixteenth century; in the seventeenth century, *Grisélidis* became one of the best sellers of the *Bibliothèque bleue*. Charles Sorel testified, in his *Remarques sur le Berger extravagant* (1627), that village folk read the work and oldsters retold it to children. A character in the *Berger extravagant*, Adrian, cites 'the story of Griseldis', along with the *Ordonnances royaux* and Pibrac's *Quatrains*, among the books one might read 'to enjoy oneself on feast days'.

Perrault never concealed his use of the *Bibliothèque bleue* volume. He wrote,

If I had given in to all the different opinions offered me on the work that I am sending you, nothing would be left of it but the simple, bare tale, and in that case I would have done better not to have touched it and to have left it in its blue paper, where it has lain these many years. (*A Monsieur*... [probably Fontenelle] *en lui envoyant Grisélidis*).

Perrault had occasion to read the Troyes publications during his stays at nearby Rosières, where his brother-in-law Guichon used to invite him for the hunting season. Two poems, the 'Ode à l'Académie française' (1690) and 'La Chasse' (1692) are dated from Rosières. The composition of *Grisélidis*, which Perrault read to the Académie Française on 25 August 1691, may have occurred during that period. In any event, it is certain that Perrault knew the publications on Oudot's list. He states, 'He [Boileau] may do what he will to boast of the extraordinary sales of his satires; their distribution will never come close to that of *Jean de Paris*, *Pierre de Provence*, *La Misère des clercs*, *La Malice des femmes*, or the least of the almanacs printed in Troyes at the Chapon d'Or'(*Apologie des femmes*, Veuve Coignard et Coignard fils, Paris, 1694). According to Abbé Dubos, Perrault did careful research for the story of Griselda:

He searched in vain in all the appropriate historians which marquis of Saluzzo had married Griselda; he knows of no book in which mention is made of this event other than the *Decameron* of Boccaccio, of which the blue paper [version] is an abridgement. Mr Perrault has embellished Boccaccio's narrative, and he gives a lover to the princess so that, after being set on the road to marriage, she does not return to the solitude of the convent. (letter of 19 November 1696)

In this quite special case, then, it is possible to determine exactly what changes Perrault made from his sources – from the *Decameron*, which he knew, and, above all, from the *Bibliothèque bleue* version. His alterations were of limited scope, but they can be discerned by a sharp-eyed reader. For the most part, Perrault took the *Bibliothèque bleue* version as his model. Not only are the main events the same, but he copied certain techniques of exposition. In other developments he 'amplified' his sources, filling out dialogues and rounding out portraits. The psychological motivation of his characters is less bookish, more carefully studied, and avoids the obvious meaning given in the pedlars' books. The anonymous author of the *Bibliothèque bleue* version insisted both on the submission that a woman owed her husband, and on the social distance that separated Griselda and the marquis. Perrault insisted instead on the piety of his heroine: 'You needed to make the Patience of your Heroine believable, and what other means had you than to make her regard the poor treatment of her Husband as coming from the hand of God? Otherwise one would take her for the stupidest of women, which assuredly would not make a good effect' (*Lettre à Monsieur*... *en lui envoyant Grisélidis*).

In a second change, Perrault eliminates Griselda's son, at the same time gratifying her daughter with a 'lover'. It is probable that such changes in details did not upset readers familiar with the *Bibliothèque bleue*. How, on the other hand, did they take the omission of the scene of the shift? As the price of her gift of her virginity, Griselda asks the return of 'one of the chemises that I had when I was called your wife'. Here Perrault failed to make use of the 'parting gift', an authentic folk motif and a much-appreciated touch. In folklore, the heroine, rejected by the man she loves but permitted to take away a gift of her choice, chooses the man she loves and from whom she does not want to be separated. Perrault was perhaps responding to canons of propriety when he eliminated the motif of the chemise, symbolic of the sexual and conjugal life of Griselda, but when he did so he told quite another story, making a martyr or a saint out of a woman in love. He also made it impossible for his version to supplant that of the *Bibliothèque bleue*.

It may have been out of simple convenience that the Troyes publishers continued to reprint their chapbook of *La Patience de Grisélidis*. Why, then, if that title had obviously encouraged sales, did they omit the phrase '... de l'Académie Française' from their publicity? In reality, they found Perrault's alterations too learned. Still, Perrault had taken pains to confuse matters, both in his title, which was almost identical to that of the *Bibliothèque bleue* chapbook, and in the name of the heroine. At first he had named her 'Griselde', a more 'poetic' name and one that he preferred to 'Grisélidis', which was 'a bit sullied by the hands of the people'. With the second edition (1694) he substituted 'Grisélidis' for 'Griselde' throughout, offering no explanation for the change. The reason could only have lain in an attempt to reach a larger public whose familiarity with the salient features of the text had to be respected. Imitation of the formal aspects of pedlars' books fooled no one, however. Nisard reproduced in its entirety the 1656 chapbook, which closely resembled the Jacques Oudot edition (1679–1711): *La Patience de Grisélidis, jadis femme du Marquis de Saluces, Par laquelle est démontré la vraye obédience et honnêteté des femmes vertueuses envers leurs maris* (Morin no. 870). These were undoubtedly the publications that Perrault knew. However, when the Oudot firm declined, Pierre Garnier republished these same chapbooks, showing little interest in the Perrault version (see Morin no. 871, permission to Pierre Garnier of 1736). Their success held firm throughout the eighteenth century, the publishers carefully avoiding a confusion that might have been fatal for their sales. There is little doubt that their readers would have found it difficult to appropriate a text in verse; furthermore, the narrative would have become totally inaccessible if readers failed to find in it the themes that lent it authenticity for them. On the contrary, when the Troyes publishers took over the *Contes* in prose it was to fill a gap in their lists. How, then, were the tales launched?

Perrault's Contes *in the* Bibliothèque bleue
Fidelity and Change

Can we reconstruct the publication of the tales? One thing is obvious when we examine the first five editions in Troyes (1734–56): Perrault's *Contes* were reprinted with perfect fidelity, which supposes a high degree of professional competence. At the most I might remark that in the first Pierre Garnier edition of 1737 the affirmations of the heroine of *Les Fées* are spelled *oui-dea*, which suggests that the compositors may have had archaic leanings, since this spelling, common in the sixteenth century, was no longer in use after the late seventeenth. Soon after, in 1740, Garnier le Jeune took the liberty of spacing out the pages of *Le Petit Poucet* by adding paragraph breaks, perhaps because he found the tale long in relation to the others in the collection. The change seems minor, but it none the less indicates a desire to facilitate reading, one of the strategies of the publishers of the *Bibliothèque bleue*. The same could be said of the elimination of italics in *La Barbe bleue* in the response of the heroine's sister: 'I see nothing but the sun that makes a dust, and the grass that grows green.' Italics were retained, however, for 'Anne, sister Anne, dost thou see nothing coming?', giving it the status of a refrain. The earliest changes in both text and layout, minor as they were, thus facilitated reading, monitored the rhythm of the tale, and encouraged reading aloud by their insistence on a refrain that was already familiar to the readers, as we shall see.

Changes in the title are of another order. The permission of 23 July 1723 accorded to Pierre Garnier and Jean Oudot mentions 'Contes des fées' and this was the title regularly used thereafter. Garnier and Oudot were thus the first to have given this title to Perrault, since Nicolas Gosselin in 1724 was content with *Contes de Monsieur Perrault*, in imitation of widow Barbin (1707), a title that already bore signs of dissatisfaction with the former title, *Histoires ou Contes du temps passé*. The absence of the frontispiece engraving presenting 'Contes de ma Mère Loye' reinforced the novelty of the title and lent meaning to the change. The title *Contes des Fées* was not new, however. In 1697 and 1698, Madame d'Aulnoy had published her tales under this title, and they had been republished under the same title in 1710 and 1725. Madame de Murat had also published a *Contes de Fées* in 1698, and a *Nouveaux Contes des Fées* in 1710 and 1724. These reprintings must have been known to the Troyes publishers, since they were published (or issued) by the Compagnie des libraires, but this is simple conjecture, since the first permissions concerning Madame d'Aulnoy date from 1757 (Gumuchian no. 409), and 1758 for Garnier and Jean Oudot's widow (Morin nos 176, 178). It is thus only a possibility that the use of the same title for Perrault and

Madame d'Aulnoy shows that the printer-booksellers were attempting to create a collection.

The title was not chosen without reason, though. The central position of the tale entitled *Les Fées* in Perrault's collection was part of a strategy of composition, since *Les Fées* was a synechdoche, the part standing for the whole. The theme of that tale, as Louis Marin put it, was the passage from oral transmission by the living word to other forms of communication. But *fée* was also a modifier, as in the *clé fée* (enchanted key) of Bluebeard, or as Gilles Ménage recalls in his *Dictionnaire étymologique* (vol. 1, p. 307). The tale *Les Fées* thus enabled the reader to 'approach the Eloquence of "enchant-ment", which, mixing word and magic, sets aside the [art of] speaking well of learned and academic rhetoric in order to attribute the art of pleasing, instructing, and touching emotions to a mysterious "gift".'[7] The title *Contes des Fées* attests to a phenomenon of overdetermination. Starting with the effect on the senses produced by the tale title *Les Fées*, and by its central position in the collection, the repetition of the word *fée* in its multiple meaning suggested both 'tales told by the fairies' and 'enchanting tales'. The connection between orality and the supernatural becomes obvious when one recalls the scene described by Noël du Fail in his *Propos Rustiques*:

Goodman Robin . . . would begin a fine story about the time when the animals talked . . . of how Renard the fox stole a fish from the fishmongers; about Mélusine; about the Werewolf; about Anette's hide; *about Fairies*, and how he often spoke with them familiarly, even at vespers as he passed through the hedgerows and saw them dancing the *branle*. . . . [He] said that they came to see him, assuring [him] that they were good wenches.

It was as if *raconter les fées* was all it took to tell a tale. In 1730, the title *Contes des Fées* was possessed of at least as much wealth of evocation for the public of the publishers in Troyes as was *Histoires ou Contes du temps passé* for Perrault's Academy entourage. Reiterated and imprecise use of the term *conte des fées* reduced it to a platitude by the end of the eighteenth century, until it reached the stereotyped meaning it has today.

The engravings used as vignettes surmounting the title of each tale also underwent a change in the early collections from Troyes. Only one vignette showing Riquet's meeting with the princess was a clumsy copy of Clouzier's illustration. Even there, the Troyes engraver took pains to underplay the tuft of hair on Riquet's head and accentuate his small stature and his deformity. We shall see that this choice was repeated throughout the eighteenth century and was made for a reason. The other illustrations (which are far from clear) used backgrounds and costumes from classical antiquity well before Fokke's engravings of De Sève's illustrations. The Troyes publishers' use of ancient culture shows that reference to classical antiquity as an aesthetic model and an example of virtue was not restricted to the initiate or the elite of the Academy, for whom the Coustelier edition of 1742 was destined.

Although most of the vignettes were probably re-used woodcuts, this was not true of the illustration for *La Barbe bleue*, which clearly figures the scene of the interrupted decapitation of the heroine. That an original illustration was used implies an unhesitating choice of iconographic representation, which was not the case for *Les Fées* or *Cendrillon*. Not only were the two tales illustrated by the same marriage scene – two women crowning a newly wed couple with laurel wreaths; they were also connected by the appearance of the title, *Cendrillon*, placed as a catchword at the foot of the last page of *Les Fées*. The image thus went beyond exact representation of a particular episode to move towards allegory and even towards symbolic synthesis (Morin no. 181; see plate 5).

Garnier the Younger (Morin no. 182) cut the connection between the two tales by gratifying *Les Fées* with a separate and not particularly appropriate illustration: a banquet in honour of someone, possibly a king. The younger Garnier's rectification of his father's edition of the work shows an interest in decorative aspects to the detriment of both thematic characterization and comprehension of the tales. Madame Garnier, the widow of Pierre I and mother of the younger Garnier (with whom she did not work, however), collaborated with her other son, Jean, to republish the works that had brought success to her husband. Madame Garnier's *Les Contes des Fées, Avec des Moralités; Par M. Perrault* (Morin no. 173) presents analogies with the first edition of Pierre I Garnier. The text is identical, as is the selection of the morals. There is a curious difference, however. The first words spoken by Sleeping Beauty as she awakens, 'Is it you, my Prince? You have waited a great while', are in italics, not so much to call attention to the spoken word (since other dialogues are not similarly underscored), as to mark the high point of the tale in a timid attempt at pedagogy. Only *La Barbe bleue*, with its decidedly unambiguous picture (plate VIII), and now *Cendrillon* were illustrated adequately.

The title vignette for *Le Petit Chaperon rouge* merits a remark. A woodcut shows a man killing an ox, perhaps an ass, before the eyes of a seated personage (plate VIII). Apparently unrelated to the story, this scene of the killing of an animal may have referred to an epilogue to the tale found in the folk cycle in which the wolf is killed by a hunter, after which Little Red Riding Hood and her grandmother are released, alive, from the stomach of the wolf. This happy ending persists to modern days in versions of the tale from the Morvan, Touraine, and the Alps. It occurs less frequently than the better known ending, and it attenuates the main function of this 'cautionary tale', of warning children of dangers. *Le Petit Chaperon rouge* belongs to a cycle of tales that end badly and in which sympathetic characters die a violent death. Since some children found this ending hard to bear, oral tradition offered other solutions: that of the hunter, or that of the little girl using the pretext of an urgent need to leave the house, allowing the wolf to attach a cord to her from which she frees herself once she is outside and escapes. Similar ruses are attested in written tales before the seventeenth century in the Orient and the Far East, and Bernier and Galland may have suggested it to Perrault, who rejected it. In any event, the survival of this motif in nineteenth- and twentieth-century versions owing nothing to Perrault suggests that it existed in France in the eighteenth century.[8] It is possible that Madame Garnier, in reaction to the cruelty of the ending and in the interests of uniformly happy endings, may have used the illustration to attenuate the ending Perrault had chosen.

To close this examination of the first edition of Pierre Garnier (Morin no. 181), I might stress the increased coherence that a shift in the position of *Le Petit Chaperon rouge* brought to the collection. The new plan grouped the tales by pairs, *La Belle au bois dormant* and *La Barbe bleue*, *Les Fées* and *Cendrillon*, *Le Petit Chaperon rouge* and *Le Petit Poucet*. Three major thematic axes – love, marriage and childhood – underlay the restructured collection. *Riquet à la Houppe* and *Le Chat botté* were left in their original places, and they seem suddenly marginal in this new arrangement. I should note, finally, the nearly consistent use, for reasons of space, of a single brief moral. On the level of physical appearance, the Troyes editions differed considerably in their covers. Pierre Garnier used gilt-edged pages with a board binding and the title lettered in gilt; Madame Garnier and the younger Garnier used a chequered blue, green, and yellow paper binding (the latter with the imprint 'A Orléans, chez . . .', indicating commercial relations between print shops).

Until the mid-eighteenth century, the Troyes publishers printed the text of the *Contes*, and of course the *Epître à Mademoiselle*, in complete conformity with the original. The widow Béhourt ('imprimeur-libraire, rue du Petit-Puits à Rouen', 1759–63), seems to have been the first to add the following paragraph to the *Epître*:

Several Roman Emperors, among others Marcus Aurelius, who needs only to be mentioned to be praised, various Kings of France, such as St Louis, Louis XII, Henry IV, etc., moved by a like desire, occasionally abandoned their suite to learn how their least fortunate Subjects lived in their thatched cottages, and they regaled them with gifts.

This apocryphal addition followed Perrault's flattering remark in the *Epître* concerning the 'Heroes of your Race' who condescended to 'look more closely at what of the exceptional takes place in huts and hovels'.

At first sight, the widow Béhourt chose her models to reinforce the edifying example of princely behaviour. Her choice of figures could not be more traditional: Marcus Aurelius, St Louis, Louis XII, and Henry IV were among the exemplary figures cited in almanacs or in the *Histoire de France avec les figures des roys, depuis Pharamond jusques au roy Henri IV ... revue et augmenté de la chronologie des Papes et des Empereurs*, first published in Troyes by Jean Oudot in 1608. The widow Béhourt's paradigmatic use is not far from the *exemplum*.

Deciphering the reasons for this addition is a more delicate matter. The collected tales were published in Rouen at a moment when the legend of Good King Louis XV was gaining ground, aided by the sanctimonious horror inspired by Damiens's attempt to assassinate the king in 1757. The *Messager boiteux de Bâle* of 1758 was full of expressions such as '... the marks of Divine Protection of that Monarch, too worshipful not to be remarked'; or, '... that Monarch who received the praise and the testimony of his People that he personally had never done the least harm, even to a child, and that he is the most amiable Lord who has even ascended the Throne' (this on the occasion of a detailed description of Damiens's execution); or again, '... the humanity of Louis XV, which leaves to other sovereigns only the glory of imitation' (1766). Although it was principally the *Messager boiteux* that purveyed remarks of the sort as historical impartiality, Jean Oursel of Rouen did so as well in the almanac that he published at regular intervals. Are we then to see the exemplary trilogy of St Louis – Louis XII – Henry IV as a justification for the proclaimed symbiosis between Louis XV and his people? We must remember that the addition was made to a letter of introduction to a collection of tales. The *Messager boiteux* of 1758, on the other hand, indulged in a veritable polemic profession of faith:

We have neglected nothing, up to this point ... in order to compile the historical relation of our veritably limping Messenger in such a fashion that it agree ... with the title of Veritable. We pay no more heed than does the judicious reader to old wives' tales; we leave them to charlatans and fast-talking vendors. All our attention will be turned, then, to giving the Public true, real, and impartial things. This is what our readers can expect, and it is what they will find in the following pages.[9]

It is possible that widow Béhourt attempted to respond by lending legitimacy to her edition, bringing up to date the author's dedication by referring to all the 'great men' of this world.

Further examination of the collected *Contes* would only confirm what we have seen thus far: these collections were on the whole presented as a homogenous corpus as far as the establishment of the texts was concerned. The publishers were visibly seeking recognition of their professional qualities by stressing their respect for the author and their fidelity to the text they printed. In contrast, they worked with total latitude in picking images, in the order of the tales, and in the title (since no one was fooled by Perrault's own coyness). The written narrative was fixed, then, but the pictorial representations and the manner of reading were left open. The only firmly closed domain was that of the wording.

Widow Béhourt's edition broke with the monolithic consistency of the Troyes editions. When it was exported into Normandy and into other provincial cities, the *Bibliothèque bleue* freed itself from textual constraints. Changes, additions, and alterations began to show in the collections, at first timidly, but later the editions of separate tales varied from the original text more openly. The pendulum swung in the other direction: late eighteenth- and early twentieth-century publishers in Troyes, the Garnier heirs, A. P. F. André, the widow André, her son-in-law Anner, and finally Baudot neglected the virtues of the Oudot and Garnier printings and their fidelity to Perrault in favour of less carefully produced books that were better adapted to the reading skills of the greater public, and thus sold more widely. Each of these 'popularizing' publishers established his or her own estimation of the reading skills and the needs of the public, so that each book bears the mark of its printer. There were constant exchanges between the printing firms, however, each printer appropriating from the others and putting a personal stamp on the result, to be pirated in turn. A few basic and shared principles can nonetheless be drawn from their publishing procedures.

Facilitated Reading and Intelligible Narrative

Mechanisms aimed at facilitating reading first appeared in the collected tales; they came later, much amplified, in the editions of separate tales. The first changes were semantic. Terms judged to be out of date were replaced by 'modernized' and more explicit synonyms; explicative relative clauses were added to clarify meaning. At least those must have been the publishers' views: in reality, Perrault's concise style was broken up by a weightier syntax. The elimination of certain modifiers and the 'correction' of verb tense use (in disregard of tense sequence) produced mixed results, but they do show proof of the printers' interest in making the stories intelligible.

Professional intrusion and distortion of the text, which became increasingly visible with time until the mid-nineteenth century, was akin to an operation of the unconscious on narratives that by then were part of a common patrimony. Perrault and the respect that the publishers in Troyes had shown him in the first half of the eighteenth century now seemed somewhat forgotten: he may have been perceived more as a transcriber or a link in the 'chain of tradition' than as an author. In short, Perrault was returned to his status as a teller of tales. The publishers' appropriation of the tales – each to his or her own measure – was more a recuperation than a theft, a recuperation all the more justifiable for being coupled with an exploitation of earlier competing editions with which publishers juggled, choosing a type of illustration from one and a characteristic from another. When the text and its editions were seen as one and the same, it seemed to legitimize all sorts of

encroachments, which were taken as simple professional operations authorized by exchanges or competition between print shops, and presented as
simple physical modifications of typography or layout.

The first half of the nineteenth century saw the appearance of a new page
layout with increased space and more paragraphs – seven in the edition of
Cendrillon of A. Hardel, rue Froide, Caen, in place of the three in the original
edition; nine in an edition of *La Belle au bois dormant* from Baudot in Troyes,
instead of the five in the original edition. Spaces to breathe were more
attractive to the reader, and new paragraphs inserted into the narrative at
strategic moments facilitated comprehension. The publishers did not limit
themselves to such purely physical changes, however, but were concerned
with clarifying the meaning of certain episodes as well. They were not totally
in the wrong, since the passages on which they concentrated were illogical
or psychologically contradictory.

La Belle au bois dormant provides a good example. Historians of literature –
recently, Ester Zago[10] – have questioned the lack of psychological verisimilitude in the conduct of the prince. Although he fears his mother, for
'she was of the race of the Ogres, and the King would never have married
her, had it not been for her vast riches', he waits until his father dies to make
his marriage and the birth of his two children public. Then, in what seems
the height of stupidity, he gives the ogress queen command of the kingdom
and puts his wife and his children in her care while he goes off to war. When
she is caught redhanded, the queen mother abruptly decides to kill herself,
thus freeing the young prince from the responsibility of punishing her.

The queen mother does not figure in medieval versions of the story. In
Basile's tale, *The Sun, the Moon and Talia*, she is the wife and not the mother of
the prince, thus presenting the three characters in a thoroughly middle-class
adulterous relationship. In Basile's version, as Ester Zago states, 'Perrault
changes the basic triangle by giving the role of the "other" woman to the
Prince's mother, and, precisely in order to avoid the complications that a
jealous mother would have involved, the wickedness of the Queen is justified
by the fact that she descends from a race of Ogres.' The improbability of the
Prince's conduct, the jealous suspicions of the Queen and the embarrassed
silences of her son (whose thoughts were the object of psychological analysis
in the early draft of the tale published in the *Mercure Galant* of 1696) all
indicate survivals in Perrault of earlier versions of the tale – notably Basile's.
Perrault's publishers in Normandy – widow Béhourt and J. J. L. Ancelle –
were unaware of these precedents, and the prince's relationship with his
mother obviously seemed odd to them. This explains the elimination of
several remaining indications of maternal/conjugal jealousy, such as 'she
began to suspect he had some little amour, *for* he lived with the Princess
above two whole years.' Even more telling is the addition of an explanation,
at first sight unnecessary, of the Prince's filial indifference: 'He soon

comforted himself with his beautiful wife, and his pretty children, *when he learned that she had wanted to devour them.'*

Late nineteenth-century editions in Metz and Paris eliminated the second part of the tale, ending the story with the marriage of the prince and Sleeping Beauty. It is possible that the devourment theme weighed heavily in their decision. Still, Pellerin in Epinal, in his 'brilliant series', says that a kitchen-boy 'went to fetch the poor little things. He killed a young kid and a little lamb, and when they had been cooked on the spit, he presented them to the queen, who found them wonderfully good.' This is an embellishment taken by the adaptor from the first part of the tale, where pheasant and partridge are prepared to be served to the princess at her awakening. The threat of cannibalism, latent throughout the second part of the tale, does not seem to have created discomfort or pedagogical distress. Furthermore, the culinary motifs are stressed less out of black humour than to concentrate attention on the children. Thus, Baudot was the first to replace the usual image of the discovery of Beauty asleep with a picture of her son, little Day, 'a little foil in his hand, fencing with a large monkey' (Morin no. 49).[11] It is as if Baudot was aiming at a new public – children – and as if a new activity – skill with arms – was to be learned just like reading and writing. It is perhaps not by chance that this tale is followed in the Baudot edition by an anecdote entitled *La Signature*, telling a lamentable tale of an engagement broken because Eléonore did not know how to write.

Three overall tendencies can be seen in *La Belle au bois dormant*. The first is simplification and reduction: the tale was truncated and the second part dropped. Second, certain episodes were reinterpreted by means of brief interpolations or encroaching illustrations: the text was on the whole respected but its meaning was twisted. Third, imagery provided wide latitude for rewriting: in the *Belle au bois dormant* of Wentzel in Wissembourg, the king sends away 'his wife', the queen mother, after her attempt to murder the princess and her children, which restores the adultery of the Basile version. The publishers took a tale told with truly classical economy, but that contained certain psychological failings evidently unique to Perrault's version, and drew from them a richness of meaning and a plurality of levels of reading that seem paradoxical in terms of the formal simplification that they sought in other domains (as the Wissembourg illustrations show). All shared a concern for verisimilitude, however. Although it is true that each publisher put his or her own stamp on the text that he or she printed, certain areas – notably those involving sexuality – proved more susceptible to rewriting than others.

Moralizing Emendations

Psychoanalysts have noted the allusion to the dreams that Perrault gave his Sleeping Beauty: 'It is very probable ... that the good fairy, during so long a sleep, had given her very agreeable dreams.' Confirmation of the erotic sense of this sleep is found in the first part of the moral (the entire moral in the 1696 version in the *Mercure Galant*):

> But then to wait an hundred years,
> And all that while asleep ...
> Not one of all the sex we see
> To sleep with such profound tranquillity.

The key words of this moral resurface in the dialogue of a comedy of Regnard, *La Baguette de Vulcain*, performed in 1693, three years before the manuscript text of *La Belle*:

(Roger awakens the sleeping Bradamante)
Bradamante: What? Has it been two hundred years since I saw the light of day?
Roger: You were thus a maiden when you fell asleep?
Bradamante: Truly, yes.
Roger: And are you yet?
Bradamente: Assuredly.
Roger: That is problematic, and I think you would not have slept so tranquilly.

The allusion is obvious. Perrault knew this comedy and of course was aware of the sexual possession of Beauty and her childbirth during her sleep in Basile's tale. These texts reflect a prevalent folk attitude concerning the virtue of beautiful sleepers. In point of fact, this motif (475.2 in Stith Thompson's *Motif Index*, henceforth referred to as T) is part of the structure of the tale 'The Sons on a Quest for a Wonderful Remedy for their Father' (tale type 551 in Antti Aarne and Stith Thompson, *The Types of the Folktale*, henceforth referred to as AT), and it exists in latterday France in versions of the tale from the Nivernais, Brittany, and Poitou. Perrault eliminated the motif and accentuated the burlesque aspect of the moral by using the word *femelle*, which, according to the *Dictionnaire de l'Académie*, belonged to the vocabulary and style of raillery. Suppressed though it was, the folk theme emerges in the allusion to Sleeping Beauty's dreams, and where the *Bibliothèque bleue* versions contain moralizing emendations, this sentence is deleted. This is the case in the late eighteenth-century edition from the south of France, *La Belle au bois dormant. Conte* (duodecimo, 15pp., Oberlé no. 158), the text of which otherwise conforms with the original.

There were also some publishers whose religious vocation made them diffident of folklore. The Périsse brothers in Lyons made their reputation as booksellers specializing in the distribution of works of piety, a tradition that continued up to the nineteenth century. In the eighteenth century, between 1762 and 1767, 29 per cent of the books in their catalogue were works of religion. As Roger Chartier states, 'The Périsse bookshop, religious to excess and by vocation, expresses the persistence of a notable demand for books of religion at the very moment that a cultivated elite was developing more secular interests.'[12] In 1811, the Périsse brothers printed separate editions of four tales, with a permission from the prefect of the département of the Rhône: *La Princesse au bois dormant*, *La Barbe bleue, ou l'Avare attrapé et la curieuse punie*, *Cendrillon, ou la Petite pantoufle de verre* and *Le Petit Chaperon rouge*. These tales were the ones most in demand at the beginning of the nineteenth century, and it was only in 1830 that *Le Petit Poucet*, perhaps initially judged too long, was added to the series.

The titles of the tales give an immediate indication of the Périsse brothers' moralizing tendencies. Certainly turning Bluebeard into a miser changes the thrust of the story considerably! It was their version of *La Belle au bois dormant* that received the most insipid emendation, though. The heroine was gratified with a baptismal ceremony; instead of being gifted with 'the wit of an Angel' she had simply 'great wit'; her aura was 'something luminous and surprising' rather than 'luminous and divine'. The prince was no longer 'amorous' but 'filled with fire'; the term *amourette* is eliminated in favour of *ruse*. And of course Sleeping Beauty no longer dreamed beautiful dreams.

In the same order of ideas and a few years later, Gangel, who ran a print shop, lithography press and image manufactury in Metz, deleted from *Le Petit Poucet* an ironic and highly idiosyncratic remark of Perrault's on woodcutters' tendency to have large families: 'People were amazed, that the faggot-maker had so many children in so small a time; but it was because his wife went quick about her business, and brought never less than two at a time.' One could draw up a long list of narrowly censorious deletions, inspired perhaps by what Nisard called 'the lack of delicacy and the naïveté of the *Contes* of Perrault'. Elsewhere Nisard wrote, in a different vein: 'Their morality is so excellent that it could in certain aspects rival the one that has its source and accompaniment in religion, and its impression would be just as long lasting if, like religious teaching, it had the advantage of being the object of our constant preoccupations from the remotest age of life.'[13]

It is possible that printers deliberately forced the moral tone of the tales and strengthened their morals (the irony in which disappeared) in order to adapt the lessons of the *Contes* to their own estimation of their readers' consciences.

Adaptation could take a surprising turn when a similar cycle of tales existed. This was the case with *Le Petit Chaperon rouge*. We have already seen that the first generation of publishers in Troyes tended to attenuate the unhappy ending of Perrault's tale by using an image showing justice being done to the wolf. Paul Delarue has discussed this problem as it pertains to the outcome of the Brothers Grimm's version of the tale. For Delarue,

The happy ending added to Grimm's *Little Red Riding Hood* by the young narratress, in whose memory the dual tradition, French and German, coexisted, or by the carrier or carriers of the tradition who transmitted the tale to her from print, is the one that most frequently ends another tale, *The Wolf, the Goat, and the Kids* (AT tale type 123), in Germany and in Central Europe.[14]

In the latter tale the wolf, who has eaten the kids in the absence of their mother, falls asleep. The goat, informed by a surviving kid, arrives, opens the wolf's stomach, and liberates her young. For Paul Delarue, this is 'a contamination brought on by the shared motif of the small victims swallowed up by the wolf and by the modern psychological tendency of narrators and listeners who want a tale to end well, their sensitivities adapting poorly to a tragic ending.'

It is possible that if indeed there was contamination here, it was older in western Europe than Paul Delarue thought. The image of the ox being killed in the edition of Madame Garnier (Morin no. 173) could be called upon to support this notion. The publishers were led into this contamination by another route, however, a hypothesis that can be backed up by examination of the editions. Clouzier's illustration showed the wolf in an equivocal position, but, as we have seen, the effect was mitigated in the Bassompierre edition of 1777 by an illustration showing the meeting of the wolf and Little Red Riding Hood. On the whole, publishers of the end of the eighteenth century seem to have been uncomfortable with Clouzier's picture, because they 'chose' to make the mistake of illustrating the tale with the engraving for the preceding or the following tale.

Other publishers of *Le Petit Chaperon rouge* used only the scene of the meeting to illustrate the tale: Blocquel (and Castaiux) in Lille, in their collection of *Contes de Fées* in 1813; widow André in Troyes, who published *Le Petit Chaperon rouge* in an eclectic collection also including *Les Trois Souhaits* of Madame Leprince de Beaumont and several anecdotes (turn of the nineteenth century). Widow André chose to illustrate the moment of the wolf entering the grandmother's house while the little girl is still far behind on her way. This eliminated, or at least neutralized, the ferocity of the devourment scene (plate VIII).) The dominant note in this collection was still pessimistic, however, since one of the anecdotes, *Les enfants égarés dans les bois, histoire attendrissante*, tells of children who die of cold and hunger when their uncle purposely loses them in the Norfolk woods (in England) in order to take possession of their inheritance. It is as if widow André had chosen an actual news item to serve as counterpoint to the tale of *Le Petit Poucet*, usually coupled with *Le Petit Chaperon rouge* to provide two stories about children and addressed primarily to children.

Once the widow André had started the process, her successors turned to another crucial turning moment in the story, the moment when the wolf knocks at the door (Gangel in Metz, and Pellerin in Epinal). The same picture was used during the same period for *Le Loup et les chevreaux*, also published in Epinal. The only change was in the scallop shell symbolizing the wolf's disguise as a pilgrim of St James, omitted in the illustration for *Le Petit Chaperon rouge*. Marie-Louise Tenèze has noted the influence of this imagery on the versions of *Le Loup et les chevreaux* of the later nineteenth century: 'The existence of an introductory Epinal image in the working out of the tale, the motif of the wolf disguising himself as a pilgrim on the advice of the fox (influence of the *Roman de Renart*), and the ending of the wolf being invited to leave by the chimney, [influences] eight versions.'[15] One might thus suppose that this iconographic contamination led to a thematic con-tamination, in the happy ending when the kids or the child are saved from the wolf's stomach. This is still only a hypothesis, however, since, as Paul Delarue rightly remarked, several Asian versions of the story (Chinese and Korean in particular) entitled *The Tiger and the Children* are attested in the seventeenth century and end happily. Is this an ancient element, then, or a later addition?

As far as nineteenth-century, France is concerned, there is scarcely any doubt that printing played a role in the mutual attraction between the two tales. It is probably also true that the imagery portrayed the ending that the readers expected, whether it was reminiscent of a previous state of the text or not. I might add that in the early nineteenth century Flemish printers published images in broadsheet form of twelve or fifteen woodcuts. One-third of them show the happy ending rather than the wolf devouring the child.[16] It is thus possible to reconstruct a process of transformation set off by the moralizing resolve to play down the 'primitive scene' that could be read into the image of the devourment – a process that moved irresistibly towards a motif that was perhaps already inherent in the tale or at least close to it. We should not exaggerate the impact of changes in print on oral custom, however, since only seven French oral versions out of thirty-five present the happy ending and the killing of the wolf.

Emerging Folklore

There were also certain tales that were read differently from the printed text. This was the case with *La Barbe bleue*. As we have seen, De Sève's illustration (Coustelier edition, 1742) showed the heroine throwing herself into the arms of her rescuers, quite contrary to what Perrault's text said. This image was not often reproduced in the *Bibliothèque bleue* editions, which preferred Clouzier's vignettes. In contrast, in the early nineteenth century the scene brought a good many broadsheets of twelve or fifteen woodcuts to a close, such as those produced by the Brépols firm in Turnhout (around 1815) and the series published in Brest and engraved by Mercier in Nantes (around 1820).[17]

A second change (as we have also seen) was the use of italics in only one speech – 'Anne, my sister Anne, dost thou see nothing coming?' – thus giving the line the status of a refrain, which it did not have in Perrault's text, where it was simply part of the dialogue. This change persisted in all editions to the end of the nineteenth century.

A third change appeared between the late eighteenth century (widow Béhourt) and the Baudot and Chalopin editions: the heroine 'would have moved a *tiger*, so beautiful and sorrowful was she', where the word 'tiger' replaced the earlier 'rock'. The change was far from fortuitous, as it instantly recalled the *tigre altéré de sang qui me défend les larmes* (tiger athirst for blood who bans my gloom [tr. Solomon]) in Corneille's *Horace* (IV, 5), which had become a cliché by the eighteenth and nineteenth centuries.

A fourth change occurred in the dialogue between the heroine and her sister Anne, 'a veritable cascade of ambiguities', as Marc Soriano has shown. Some publishers, P. Chalopin in Caen, for example, credited Bluebeard with the final speech: 'God be praised, it is my brothers!', which is patently absurd. Soriano has shown that Perrault deliberately played on the ambiguity of this dialogue by attributing speeches to the two sisters indiscriminately, and adding to the confusion with odd quirks. Each of the sisters speaks as if she was alone, as if the other did not exist, or as if she was at once herself and the other. 'Go up, I desire thee, upon the top of the tower, and see if *my brothers* are not coming', the heroine says, rather than the more appropriate 'our brothers'. Soriano also remarks that many contemporary editions distort the dialogue by attributing the final speech (which Chalopin had assigned to Bluebeard) to the heroine. In short, this dialogue seems to have elicited both bewilderment and attempts to rewrite it. Marc Soriano's interpretation is known: Perrault makes use in his *Contes* of a 'veritable technique of ambiguity . . . in order to orient the reader's mind *both* towards the masculine and the feminine', in which Soriano sees the influence of Perrault's twin birth and his parents' preference for his twin. But how did the publishers deal with the incomprehensible aspects of this dialogue and the artificial doubling of the heroine by her sister Anne?

In reality, all these changes point to the emergence of a parallel folk tradition, expressed in a song. The folklorist Paul Delarue has noted (as have musicologists) the similarity between the tale of *La Barbe bleue* and a ballad, *Renaud, le Tueur de femmes*, which was known throughout a good part of

Europe. The theme – Renaud kills all his wives – closely resembles the initial situation of the tale, but there is a basic difference because Renaud's last wife, wilier and braver than the others, tricks him, disarms him, and drowns him. There is another song, however, *La Maumariée vengée par ses frères*, perhaps of southern French origin and certainly of long-standing tradition. The song was published in the sixteenth century under the title *Romance de Clothilde*, and it spawned a dance tune known as *la mal maridade* that figures in the list of dances following the supper in Lanternland in Rabelais's *Pantagruel* (V, 33 bis). More than forty versions of this song have been collected in Quebec, which would weaken the hypothesis of the musicologist George Doncieux concerning its origin in the south of France. In any event, it was an extremely old and widely known song.

In the song, three brothers pay a surprise visit to their sister, who is married to a hangman who beats her and pricks her to bleeding with a knife or a pin. The wife's blood is collected in a flask and is drunk, according to the version, by the husband or the heroine herself, or it is sent to one of the brothers. The brothers ask their sister for her news, and she answers loudly that she is well, but softly that she is mistreated and her husband has killed their children. (In fact, some Quebec versions entitle the song, *Maumariée, parle tout haut, parle tout bas*.) The brothers kill the husband with their daggers. In the oldest Midi versions, the unhappy wife has to protect her remaining children from the brothers' avenging daggers. Negotiations follow: a girl is put in a convent, the oldest boy is left with his mother, and the brothers take away the younger children.[18]

The song throws light on a good many aspects of Perrault's version of the story, and on publishers' interpretations of it. The ambiguous dialogue between the heroine and her sister Anne can be read as the responses of one person alternately speaking in a whisper (the heroine 'cries out softly'), then out loud. It is possible that the publishers, relying more on the song that filtered through the story than on Perrault's problems with twinship, presented the dialogue as a monologue. There are other versions of the song (from the Nivernais) that give a dialogue between the heroine and her husband in which *he* is the one who sees the brothers coming and does not know where to hide. This perhaps explains the 'he cried' in the Chalopin editions.

As for the ending that was illustrated, it seems obvious that it might show the heroine throwing herself at the feet of her brothers to protect her children from them. The repeated and consistent use of the word 'tiger' in place of the word 'rock' might easily evoke the ever-present blood motif, apparent in the husband's gory death, to be sure, but also in the heroine of the song constrained to drink blood. Finally, the refrain, 'Anne, my sister Anne . . .', recalls the repeated questions that the brothers in the song put to their sister: 'Jeanne, ma soeur Jeanne . . .'. All these shifts of motif and meaning are merely indicative, since no documentary proof exists connecting the tale and the song. The comparison needed to be made, however. It allows us to underscore, once again, strong links between a known theme in the oral tradition, rooted deep in collective memory, and a written text that was both close to it and different from it.

Every tale, in its own way and varying with its publisher, illustrates this emergence of folklore. Alterations – even incoherences – in some editions are not all survivals nor appeals to tacit references, but many are nothing but that. That was the case in the recurrent illustration for *Riquet à la Houppe* that eliminated the kitchen boys and the table shown in the original illustration (depicting the somewhat exaggerated demonism of the protagonist). In the Troyes edition of 1734, Riquet is figured as short and deformed, which fits the known versions of the tale, 'The Name of the Helper' (tale type 500), which Perrault's story closely resembles. Moreover, according to Littré, the word *riquet* in Norman meant 'deformed, hunchbacked'. It is also true that some copies of the second printing of the Barbin edition show the vignette of *Riquet à la Houppe* in the place of the vignette for the dedicatory *Epître*. Not only does that mean that the illustration for *Riquet* appeared twice in the volume; the subject, Riquet saluting the princess, also took on a somewhat piquant cast when it accompanied the *Epître*. That may have contributed to the accentuation of the grotesqueness of Riquet in later editions, but it in no way explains the disappearance of his tufted top-knot, a detail particular to Perrault and not found in any 'popular' version of the story.

Perrault's version of the story of the children lost in the forest also contains references to folklore. He calls his protagonist Poucet, a name that closely resembles *Pouçot*, a minuscule character in another cycle of tales in which the hero's adventures are shaped by his small size: he is born in a cabbage, he lodges in a horse's ear and visits a cow's stomach, and so forth. It was Pouçot's tale that the Epinal firms chose to illustrate, at least in some woodcut broadsheets (Pellerin, no. 701; Pinot and Sagaire, no. 471). The firm of Jacques Henri Le Tellier, printer-bookseller in Lierre (established 1779, died 1809), had already published the tale of *Pouçot* and distributed it in broadsheets in Belgium and Holland, where it was reprinted by other Flemish firms. Arnold Van Gennep remarked that firms outside Holland, such as the Metz publisher Gangel, worked for the Dutch trade around 1840–50, simply adding Dutch text to their usual broadsheets. It was perhaps Le Tellier's success that persuaded French publishers to adapt their broadsheets to a Flemish public, more familiar with the tale of *Pouçot* than with *Le Petit Poucet*.

Finally, there is a reference to a practice that may be fictional. *Cendrillon, ou la petite pantoufle de verre*, subtitled *conte moral* under the influence of Marmontel's *Contes moraux*, appeared in a hard-cover collection of the latter eighteenth century, without mention of place of publication or date, but presenting typographical similarities to the chapbooks published by Gaudibert Penne of Carpentras. The collection also contained *Le Sermon de Bacchus*, *Comédie ou Le Devoir des savetiers*, *Le Passe-temps des gens d'esprit*, *La Malice des Femmes*, *L'Imperfection des Femmes*, *La Malice des Filles*, *Le Catéchisme des grandes Filles* and *Le Jardin d'amour*. *Cendrillon* appears between *Le Devoir des savetiers* and *Le Passe-temps des gens d'esprit*. The text of the tale (with no illustration) conforms completely to Perrault's original. It is probable, however, that *Cendrillon* owed its inclusion in the collection by chance to its position following *Le Devoir des savetiers*. In that text, a dessert offered for the wedding banquet of an apprentice shoemaker and his master's daughter includes 'an entire service of Lemon pudding accompanied by Pumpkin juice from Cinderella's Pumpkin'. *Le Devoir des savetiers* is a burlesque text composed for the reception of an apprentice into journeyman status. It dates from the latter seventeenth century, and was reprinted in Troyes under different titles and with different contents. The edition in question tells of the meeting of the apprentice and the master's daughter and their resulting marriage in a burlesque and bawdy key. Several allusions are made to shoes, the symbol of conjugal bliss, and occasionally to Cinderella's slipper.

Folklorists and ethnologists have taken pains to detail possible links between the story of Cinderella and rites of marriage. In several regions of France the fiancée's attempted flight as she goes to the church is an attested rite, as is the theft of the bride's shoe during the nuptial meal. Recalling the juridical role of the shoe as a sign of ownership and domination, notably in Christian marriage ritual in the Middle Ages, Dorothée Kleinmann even concludes that 'in all phases of the tale, the two major aspects of the symbolism of the foot and the shoe, the socio-juridical and the sexual, are mixed.'[19]

I should note, finally, that Provence and the Comtat Venaissin had a flourishing artisan tradition in the ceramics of Marseilles and Moustiers from the end of the seventeenth to the early nineteenth centuries. Among the decorated objects produced there were faience slippers, marriage slippers and later *sabots de Noël*, which a young man would offer his fiancée on the eve of their wedding, and which the bride, to prove her fidelity, was to guard carefully against repeated attempts at ritual theft. That essentially Mediterranean practice is alluded to in Le *Devoir des savetiers*.[20] We have no evidence of a specifically ritual use of *Cendrillon*, but I might at least note the strong presence of the theme of love and marriage in the tale. This was the way the first generation of Troyes publishers interpreted the tale, ignoring the moral maxims Perrault appended to it, which speak to the need for proper upbringing and for 'godfathers and godmothers for advancement' in this world.

Printed Tales and Oral Traditions

Can the effect of the *Bibliothèque bleue* editions of the *Contes* on oral literature be measured? Folklorists Paul-Yves Sébillot, Paul Delarue, and Marie-Louise Tenèze have studied individual tales in an effort to account for and evaluate Perrault's influence on versions of the tales collected after the latter nineteenth century. Paul Delarue, for example, counts only four oral versions (out of twenty-nine) wholly based on Perrault's version of *Le Petit Chaperon rouge*, and he notes that they were collected after 1934. This is representative of the general attitude of narrators towards Perrault: the versions based on his writings are few, and certain of them only borrow details such as the title or characters' names.

On the whole, the oral versions show a mixture of independent motifs and episodes taken from Perrault. *La Barbe bleue*, *Peau d'Ane* and *Cendrillon*, for example, take over themes dear to Perrault; *Le Petit Poucet* appears in versions combining Perrault and Madame d'Aulnoy with episodes of folk origin; original motifs in *Le Petit Chaperon rouge* – Pins and Needles road, for example – seem more resistant to change. Other European collections present similar results: in Germany, where Grimm's Tales have been known to several generations, its influence on oral tradition has remained fairly weak, as the folklorist Kurt Ranke has shown.

Still, the inclusion after 1888 of abridged versions of Perrault's *Contes* in primary school readers is responsible for a growing emergence of those versions – though in altered form – in the oral corpus collected in the twentieth century.

This is how Baudrillard and Kuhn define their pedagogic mission in the preface to the anthology, *Lisons!* (1908): 'To make known some of the words and thoughts that are the honour of humanity: its traditional fables, its time-honoured tales, the famous events and legends that have inspired its poets and its artists, the extracts consecrated as classics by their universal popularity.' The collections of *morceaux choisis* were thus the form in which the culture of university professors enamoured of pedagogy pierced through.[21]

Morceaux choisis were not chosen for recreation but for instruction, which explains the late entry of Perrault's *Contes* (and the Grimms' tales) into such manuals. All the pedagogical institutions' mistrust was insufficient to avoid defining the *Contes* as literature once they bore the seal of scholastic approval: it was by reciting Perrault's versions that the tale-tellers interviewed by the folklorists proved their schooling.

Diffusion of the *Contes* in the schoolroom – where pedagogical authorities disapproved of them but the common people welcomed them and forgot that they belonged to a denigrated folklore – must not be confused with the distribution of pedlars' books, Epinal images, or even children's picture books. It was not the same public who read the same tales in different editions at different times. There were exchanges between publics, though, and probably some sharing. In a quantitative survey of various tales collected in Upper Brittany in 1894, Paul-Yves Sébillot has shown the predominance of tales distributed in *Bibliothèque bleue* versions:[22]

TABLE 3.2 Oral versions similar to the *Contes* of Perrault

	Upper Brittany	*Other provinces*	*Total*
Cendrillon	11	18	29
Peau d'Ane	7	15	22
Le Petit Poucet	6	11	17
La Barbe bleue	4	9	13
Le Petit Chaperon rouge	3	9	12
Le Chat botté	3	2	5
Les Fées	3	2	5
Riquet à la Houppe	1	0	1
La Belle au bois dormant	1	1	2

Although Sebillot's figures correspond roughly to the overall circulation of each tale – it is true, for example, that *Riquet à la Houppe* had few reprintings in the *Bibliothèque bleue* – still, *La Belle au bois dormant*, one of the best known tales, the best-loved, and the most frequent in books, broadsheets, and images, is rarely encountered in the corpus of tales collected by the folklorists. This obviously poses a problem. Rather than offer a solution, I might note the autonomy of oral versions of the tales. It is clear in the case of *La Belle au bois dormant*, and it filters through in all the other tales. An examination of the thrust of the only *Bibliothèque bleue* publication that shows thematic differences from Perrault – *Le Petit Poucet* – proves this sovereignty of the oral tradition.

The Flemish picture plates combine two tales, 'the children lost in the forest' (AT 327), and 'Pouçot' (AT 700). If we add the oral versions collected, only five out of 147 versions reflect this amalgam in its entirety: one from Forèz (1836), a Basque version (1878), a version from La Beauce and Le Perche (1915), one from Brive (1923) and a recent version from Besançon (1957). Prudence is of course required concerning the conclusions to be drawn from those quantitative data. Most of the versions that present anomalies are linked to tale type AT 700, however, and are contaminated by tale type AT 327. If we group these versions together, we get 4.8 per cent of the total. If, on the contrary, we consider only the versions collected between 1860 and 1900, the period in which they may have been contaminated by the amalgam reflected on the picture plates, we get only 10 per cent.

It is possible that the blending of the two stories occurred independently of the broadsheets, as spontaneous creations of isolated tale-tellers. If they had an influence it was weak. Nonetheless, it indicates appropriation in a real situation of a contamination in print. Tenuous though it may be, an indication of the sort is not negligable, since it attests to a practice of oral re-use of a written amalgam of folk materials that tale-tellers – or at least some of them – found to their liking.

Readings of the Tale

It goes without saying that telling and reading a tale are of two separate domains. The publics, the immediate stimulations, and the functions of the two activities declare their difference. The exercise of one, however – reading – is subordinate to oral references and to recognition of those references. The autonomy of the other – oral relation – can on occasion be infringed upon by previous knowledge of a text transformed by the alchemy of contamination. One constant remains: in this perpetual circuit of exchanges in which multiple ricocheting influences coexist, each tale risks its own evolution – which in spite of all is independent of both editorial interests and practices particular to one place or time. The case of *La Belle au bois dormant*, which had no editorial adjuncts and was exempt from the vicissitudes of public favour, is exemplary in this sense.

The problem of the share of popular tradition and learned culture in a composite text, and of the relation between the two, could be raised concerning many texts difficult to define as clearly the one or the other. Marc Soriano was thinking in similar terms when he wrote of Perrault's *Contes* that they are

a work that rapidly and lastingly reached the largest public. But at no moment was this a miracle. This success was due to a fertile and knowledgeable collaboration between the written art and the oral art. By a conjunction of historical and personal circumstances, Perrault, with no apparent preparation for the role, put to the service of popular art, its freshness and its profundity, the resources of learned art.

Marc Soriano was discussing the manifest success of Perrault's *Contes*. The disparity included in the *Contes* cannot be resolved into a clear dichotomy that would have allowed the popular public and the more literate public to evaluate the work differently by taking from it what they sought. For Soriano, the contrary was true: collaboration between the work's publics assured the unity of the work before its audience and made its success a sign of its universal worth. Moreover, by noting that the resources of learned art are put to the service of popular art, Soriano inverts the hierarchy traditionally accepted. He sees the rehabilitation of popular tradition in Perrault's *Contes*, rather than its annexation by learned culture. What does such a 'collaboration' mean, regarding Perrault? What was the relative importance of characteristics from written culture or oral culture? And what interaction took place between them in both learned editions and *Bibliothèque bleue* editions?

Soriano adopts a paradoxical point of view in order to rehabilitate the *Contes*. His is a radical reversal of the scale of values imposed by cultural tradition. Essentially, legitimate culture reposes on the value of the written word, which it erects as an absolute and necessary point of reference. If only what is culturally important is written, by the same token all that is important achieves written form. Passing into written form records and consecrates a given contents in order to set it up as a fundamental corpus of references. Orality, under these conditions, becomes an allogeneous practice. By opposition to writing, in which established authority finds a stable reference to be invoked – a sort of vulgate – the oral loses caste, stricken with the banality of life itself. This is how a clear break is instituted between written and oral cultures, a break that preaches the dignity of writing and clinches the eviction of the oral from the sphere of legitimate culture.

The difference between written culture and oral culture can also involve the status of the contents that is noted or uttered. In the first instance, a discourse becomes fixed; in the second, it must be practised 'in situation'. Whereas the written text itself becomes the reference put into play in communication, oral culture implies a strict and determining contingency in the conditions of communication. It is from these conditions that the meaning of the message transmitted derives.

This means that the two-way passage between the oral and the written modes radically reverses the characteristics of communication. On the one hand, although oral commentary, elaboration, and criticism of a written text bring that text back to actuality by reintegrating it into a living practice, that is not enough to legitimize completely the transfer of a written message into the oral mode, since the conditions of communication are recognized only peripherally as parts of a public reading aloud or a pedagogical activity. On the other hand, if the text of an oral communication is transcribed, it acquires the authority of a reference, but it loses all it had contained that was specific to actual oral practice – as seen in the passage from the 1695 manuscript of the *Contes* to the 1697 edition. Plato, speaking of the myth of Theuth, judged of the written and the spoken differently. For him, what was written was discredited as a dead letter, in contrast to the living word (Plato, *Phaedrus*, part 4).

The primacy of writing over oral expression accompanied and paralleled the supremacy of learned culture over popular culture. When he turned to writing to echo popular tradition, Perrault could only betray it, unless his *Contes* were inserted – if not by him, then by the publishers who took them in charge after the early eighteenth century – into a new oral practice liberated from the domination of the text. Perrault's didactic aim was recognized from the first appearance of the *Contes*. His stories truly did work to constitute a patrimony of oral culture, but their efficacy was accompanied by a refusal of the hegemony of the text. This is why the *Contes*, totally absorbed into a new practice and yet still attached to orality by the indirect link of images and parallel references, show their real vitality through new variants suggested by the oral tradition or imposed by censorship, voluntary and involuntary. Thus we have a rich assortment of versions of the 'tales of Perrault' and a variety of characteristic transpositions and confusions, occasionally recognized, as in the happy ending to *Le Petit Chaperon rouge* that coexisted with the ending of the author's choosing. When the *Contes* turned once more towards a new and living practice, it was proof that the authentic authorial text had been abolished and (in the late eighteenth century) had lost its status as a classic – at least in the *Bibliothèque bleue*.

This means that Marc Soriano's judgement requires reappraisal. If Perrault's *Contes* enjoyed undeniable success, it was not in the form that their author had originally given them. By re-establishing a connection with an oral practice adapted to the book, both the 'popular' and the lettered public rediscovered – beyond Perrault's deft writing – the interplay within folk matrices of the primordial popular tales and their variants, and even the rites accompanying them. The success of the *Contes* as an instance of oral practice thus easily eliminated the author, and by the end of the eighteenth century the title *contes de Perrault* no longer referred to the author, Charles Perrault, but had become a generic label.

This new orality presented some ambiguous characteristics, however, and its ambivalence finds an echo in an ambivalence in Perrault's text. The 'oral art' of Perrault was in reality carefully constructed. In trying to reconstitute popular tales that were still alive in the oral tradition, Perrault necessarily did violence to this primordial orality by fixing one particular version of a given tale in a precise form based on the artificial conditions of its collection. The academician formed his own 'oral popular literature'. Consequently, his *Contes* became part of a whole that was by definition contradictory and artificial – a whole made of multiple grafts of elements arising from very different cultural domains. If Perrault did indeed resort to using the 'resources' of written and learned culture, it was to fabricate an orality satisfactory enough to be taken for the real thing. It was in fact taken for real, and for more real than real.

How, then, could anyone have seen through this process of creating a factitious orality and concealing the mechanisms of its making? And did anyone perceive the ambiguity of the game that Perrault – a man profoundly rooted in learned culture – was playing with what was marginal in that culture? Doubtless some did, since in order to survive in the face of the subtle practice that introduced into legitimate culture a universe normally excluded from it, and that inspired both fascination and rejection, Perrault's *Contes* turned to re-use in the *Bibliothèque bleue*. The publishers of the *Bibliothèque bleue* invoked the limited cultural worth of their product to take the liberty of creating a different reading of the *Contes*, using all the technical means at their disposal – page layout, semantic emendations, illustrations. Far from denouncing the factitious orality of the tales that Perrault offered it, the public reached by those publishers fell under its spell. Before the blossoming of 'popular art' could be recognized it needed to be transplanted; it needed a graft out of which a new way of narrating emerged that was uniquely different from its predecessors.

How can one explain the success of the *Contes* in more lettered milieux? First, it is obvious that Perrault's reservations concerning his *Contes*, when he reduced them to a utilitarian and moralizing aim in the *Epître* signed Pierre Darmancourt, worked to his favour in cultivated circles. By announcing from the start that he was offering a marginal children's literature, Perrault invited his lettered readers not to take the moral message of his *Contes* seriously – they were not destined for them – but to enjoy the game of literary and cultural exoticism. Doubtless those in literary milieux took pleasure in vagabondage outside the traditional norms. This pleasure could be all the more freely indulged when the tale itself appeared gratuitous. To note that a tale could be reduced to a formula valid for an entire family of stories was, conversely, to recognize in each tale something like a variation on a more or less clearly stated theme, and to appreciate the originality of a detail as a particular choice within a group of variants. When this occurred, the tales opened up to multiple meanings, depending on various symbolic systems of reference.

Reading the tales is to juggle with pieces of a construction set – the *meccano du conte*. This happens all the more freely when the author disappears behind tradition. Then the hegemony of the text fades, permitting the written texts of the tales to serve as a base for a new form of orality. The author's task – and the reader's – is to decide on one fixed form to give to the text. This is how the originality of a particular author's version of a tale is established. Reworking the text in this manner – by writing it or by reading it – always prompts an appeal to an extremely rich intertextuality. The text is conditioned by the existence of other texts, whether it follows them or rejects them, or whether it works to elicit unexpected comparisons. This means that the pleasure arising from the text comes from word play, structural effects, and the setting up of a make-believe universe brought to life by use of a number of formulas. This more aristocratic reading and this sort of pleasure in the written text would be born – as they were initially in the context of orality – from the unique relationship established between the author and the readers, over and above the narrative itself. Perrault constantly played with his more literate readers; he worked to keep them uncertain about the playfulness or the seriousness of what he had written, and he maintained an ambiguous relationship of complicity, even of duplicity, with them. The 'learned' publishers showed that they were aware of the game they were putting into the hands of their cultivated public.

Its relation with its public was all-important to the vitality of the tale. If the oral tale was to be recognized, it had to win over public censorship, which can perpetuate, transform, expurgate, or destroy the narrative produced by an anonymous teller. In the long run, it is the interaction between the narrator and censorship that assures the survival of a collective work. What about the written tale? Like any written work, it survives by its very form. The *Contes* of Perrault were of a hybrid nature that called for two specific methods of approach and appreciation. If the *Contes* have enjoyed recognition as authentic 'classics' of folklore and the oral tradition, it was by turning the hegemony of the authorial text upside-down to produce infinite variations of the matrices that this type of narrative offers.

In the case of the *Contes*, success, both popular and learned, affected the general contents of the work as it was read, recognized, then transmitted with a plethora of variants and in a great variety of forms – different sorts of books, broadsheet images, etc. Its success was of greater breadth than depth, but the study of specific modes of oral transmission, popular literature, learned literature, and their interrelations reveals that the hierarchic reversal of learned culture and popular traditions that Marc Soriano speaks of is illusory. Oral culture no more needed rehabilitation than denigration in readers' eyes. By substituting a factitious orality for true oral culture, Perrault left his publishers and then his readers the latitude to reconstruct another form of orality in which, paradoxically, the book had a motivating and determinant place.

The success of the *Contes* as the work of a specific author was of another nature. Here the problem was no longer to revive the folk heritage in general for use as a language that would lend itself to multiple variants. Quite on the contrary, this time it was as a unique work – a work conceived as one actualization of those games among others – that it was appreciated. This was why the Marquis de Paulmy judged it necessary to republish Perrault's *Contes*. The hegemony of the text restored here implied a reversal of the primacy of the contents. Tales were no longer told to instruct, but for the sake of the telling. The pleasure of the text was then based on a first distancing, on a refusal to take the tale seriously, within a culturally more aristocratic mode of reading.

Notes

1 Mary Elizabeth Storer, *Un épisode littéraire de la fin du XVIIe siècle. La Mode des contes de fées (1685–1700)* (Champion, Paris, 1928); Paul Delarue, 'Les contes merveilleux de Perrault et la tradition populaire', *Bulletin folklorique de l'Ile-de-France*, 1951–3; Delarue, 'Les contes merveilleux de Perrault. Faits et rapprochements nouveaux', *Arts et Traditions populaires*, nos 1 and 3(1954); Jacques Barchilon, *Tales of Mother Goose: The Dedication Manuscript of 1695 Reproduced in Collotype Facsimile with Introduction and Critical Text* (2 vols, Pierpont Morgan Library, New York, 1956); Marc Soriano, *Les Contes de Perrault. Culture savante et traditions populaires* (Gallimard, Paris, 1968); Bruno Bettelheim, *Psychanalyse des contes de fées* (R. Laffont, Paris, 1976) [*The Uses of Enchantment: The Meaning and Importance of Fairy Tales*, Knopf, dist. Random House, New York, 1976]; Raymonde Robert, *Le Conte de fées littéraire en France de la fin du XVIIe à la fin du XVIIIe siècle* (Presses Universitaires de Nancy, Nancy, 1982). These titles are only a few of the works on the genesis of the *Contes* and on Perrault's methodology.

2 For further information on the first editions of the *Contes*, see: Perrault, *Contes*, ed. Gilbert Rouger (Garnier, Paris, 1967); Avenir Tchemerzine, *Bibliographie d'éditions originales ou rares d'auteurs français des XVe, XVIe, XVIIe et XVIIIe siècles* (10 vols, Editions des Bibliothèques Nationales de France, Paris, 1933), vol. 9, pp. 161–87; *Children's Literature: Books and Manuscripts; An Exhibition, November 19, 1954 through February 28, 1955* (Pierpont Morgan Library, New York, 1954), nos 87–116: The 'Perraults'; Gumuchian & Cie, *Les Livres de l'enfance du XVe au XIXe siècle* (2 vols, Paris, 1931), pp. 316–22. [When possible, quotations from the tales are given from Robert Samber's translation in Jacques Barchilon and Henry Pettit, *The Authentic Mother Goose Fairy Tales and Nursery Rhymes* (fascimile reprint of London, 1729, Alan Swallow, Denver, 1960)].

3 This argument partially echoes the conclusions presented by Louis Marin in his paper, 'La trajectoire d'une illustration: un conte de Perrault', presented 1 March 1984 in Roger Chartier's seminar, 'Socio-Histoire des Pratiques Culturelles, XVIe–XVIIIe siècles', Ecole des Hautes Etudes en Sciences Sociales, Paris. See also Louis Marin, *La Parole mangée et autres essais théologico-politiques* (Meridiens Klincksieck, Paris, 1986).

4 Roger Poirier, *La Bibliothèque universelle des Romans: rédacteurs, textes, public* (Droz, Geneva, 1977), p. 100.

5 Claude Brémond, 'Le meccano du conte', *Magazine littéraire*, 'Contes et mémoire du peuple', 150(July–August 1979), p. 13.

6 The numbers refer to the pamphlets that eventually were noted and catalogued in *'La Bibliothèque bleue'. Belle collection de livres de colportage du XVIIe au XIXe siècle*, catalogue by Gérard Oberlé (Cercy-la-Tour, March 1983); Alfred Morin, *Catalogue descriptif de la Bibliothèque bleue de Troyes (almanachs exclus)* (Droz, Geneva, 1974); René Hélot, *La Bibliothèque bleue en Normandie* (A. Lainé, Rouen, 1928).

7 Marc Fumaroli, 'Les enchantements de l'éloquence: *Les Fées* de Charles Perrault ou de la littérature', in Fumaroli (ed.), *Le Statut de la littérature. Mélanges offerts à Paul Bénichou*, (Droz, Geneva, 1982), p. 165.

8 I follow here the conclusions outlined in Paul Delarue, 'Les contes merveilleux de Perrault et la tradition populaire', *Bulletin folklorique d'Ile-de-France*, 1951–3.

9 All citations from the *Messager boiteux* are taken from Geneviève Bollème, *Les Almanachs populaires aux XVIIe et XVIIIe siècles. Essai d'histoire sociale* (Mouton, Paris and The Hague, 1969). The changes in the *Epître* were kept by J. J. L. Ancelle, printer-bookseller in Evreux in the early nineteenth century.

10 Giovanna Franci and Ester Zago, *La Bella Addormentata: genesi e metamorfosi di una fiaba* (Dedalo, Bari, 1984); Ester Zago, 'La Belle au bois dormant: sens et structure', *Bulletin du CERMEIL*, vol. 2, no. 5(February, 1986), pp. 92–6.

11 The appeal of animal allegories showing monkeys dicing, playing cards, or practising with arms may have influenced Baudot's illustrator. Such woodcuts were very popular from the sixteenth century, and they were often included among depictions of the 'world upside down' (see Maurits de Meyer, *De Volks-en Kinderprent in de Nederlanden van de 15e tot de 20e Eeuw* [Standaard-Boekhandel, Amsterdam, 1962], pp. 193, 417).

12 Roger Chartier, 'L'Académie de Lyon au XVIIIe siècle. Etude de sociologie culturelle', *Nouvelles Etudes lyonnaises* (Droz, Geneva, 1969), pp. 133–250, in particular pp. 215–16.

13 Charles Nisard, *Histoire des livres populaires ou de la littérature du colportage depuis l'origine de l'imprimerie jusqu'à l'établissement de la Commission d'examen des livres du colportage. 30 novembre 1852* (2 vols, Paris, 1952), vol. 2, p. 512 (2nd revised edition, G. P. Maisonneuve et Larose, Paris, 1968).

14 Delarue, 'Les contes merveilleux de Perrault' (1951), p. 294.

15 Paul Delarue and Marie-Louise Tenèze, *Le Conte populaire français* (3 vols, Maisonneuve et Larose, Paris, 1976), vol. 3, pp. 384–5.

16 Maurits de Meyer, 'Le Conte populaire dans l'imagerie populaire hollandaise et belge', *Fabula*, 1(1958), pp. 183–92.

17 Pierre-Louis Duchartre and René Saulnier, *L'Imagerie populaire: les images de toutes les provinces françaises du XVe siècle au second empire . . .* (Librairie de France, Paris, 1925), pp. 68ff.

18 George Doncieux, *Le Romancéro populaire de la France: choix de chanson populaires françaises* (E. Bouillon, Paris, 1904), pp. 185ff. and 351ff.; Damase Arbaud, *Chants populaires de la Provence* (2 vols, Makaire, Aix, 1862), vol. 1, p. 83; Fernand Benoit, *La Provence et le Comtat Venaissin. Arts et Traditions populaires* (Aubanel, Avignon, 1975), pp. 291–3). For the Quebecois songs, see Conrad Laforte (ed.), *Catalogue de la Chanson folklorique française* (4 vols, Presses universitaires de Laval, Quebec, 1977), vol. 2, pp. 22–6, 31–3.

19 Dorothée Kleinmann, 'Cendrillon et son pied', *Cahiers de Littérature orale*, 4(1978), pp. 56–88.

20 See Edward Westermarck, *Histoire du mariage*, tr. Arnold Van Gennep (6 vols, Payot-Mercure de France, Paris, 1934–45), vol. 4 (1936), pp. 190, 297 [*The History of Human Marriage* (2 vols, Macmillan, London and NY, 1891; 5th edition, 3 vols, Macmillan, London, 1925; Allerton, NY., 1922)]); Martine Segalen, 'Le mariage, la quenouille et le soulier. Essai sur les rites du mariage en France', in *Naître, Vivre et Mourir* (Musée d'Ethnographie, Neuchâtel, 1981), pp. 135–47; *Faïences de Marseille* (Paris, 1985), pp. 39–66.

21 Emmanuel Fraisse, Jean Hébrard, Hélène Mathieu, *L'Invention d'une 'littérature scolaire': les manuels de morceaux choisis de l'Ecole Primaire (1872–1923)*, forthcoming. Fraisse has summarized this in an article with the same title, in *Etudes de linguistique appliqué*, 59(July–Sepptember 1985), pp. 102–9.

22 Paul-Yves Sébillot, 'Contes de la Haute-Bretagne qui présentent des ressemblances avec des contes imprimés', *Revue des Traditions populaires*, vol. 9(1984), pp. 36–105.

PART II

Religious Uses

INTRODUCTION

Three uses of religious printed matter – liturgical, ritual, and heretical – interest us here. The problem is to show, through an analysis of particular materials and situations, how the work of Christianization used the book and the printed image to cement community, propose correct devotions, and inculcate the teachings of the Church, but also how written objects could help perpetuate religious beliefs disapproved by orthodoxy. Printed matter (and before it and with it the manuscript book) was the instrument of a religious acculturation controlled by authority, but under certain circumstances it also supported resistance to a faith rejected, and proved an ultimate and secret recourse against forced conversion.

Our choice has favoured certain objects, the book of hours first, because it was incontestably the commonest of all books of religious practice from the age of the manuscript and at the scale of all Christendom. Next, the marriage charter, which in some dioceses at least (such as Lyons) was both handled in ritual and owned in the home. Finally, the 'evangelical books' inherited, bought, and copied (and hunted out by Catholic inquisitors) in Bohemia during the Counter-Reformation. Each of these materials was, in its own way, imbued with a basic tension between public, ceremonial, and ecclesiastical use of the book or other print object, and personal, private, and internalized reading. When the print piece was licit and approved, this duality was expressed in alternate uses (as was the case with books of hours, in which certain prayers were to be said aloud when the faithful gathered and others read individually in silence), or it extended the moment of the rite into a perpetual lesson, as was true of the *chartes de mariage*, whose imagery recalled Christian truths and models for good marriage throughout their owners' lifetimes. When the book was forbidden, its use in private in the home became the target of all suspicions and inquisitions, because it provided an intolerable refuge beyond the reach of visible institutional disciplinary procedures.

At the heart of this investigation into the different uses of printed matter to encourage faith, we thus find the notion of appropriation. It permits us to understand how different ways of reading grasped the same materials differently (for example, hours read aloud and silently); how a secular use and a strong affective investment could be deposited in a ritual object like the nuptial charter; how rebellious readers found the strength to preserve their identity and their history by reading, singing, and memorizing the written word. Although the task is difficult and beset with uncertainties, it is complex and contradictory practices such as these, either inscribed in the objects themselves or described by the witnesses or the subjects, that we are interested in rediscovering.

4

Books of Hours
and the Reading Habits
of the Later Middle Ages

PAUL SAENGER

Books of hours are the most widely known of the many genres of medieval manuscript. Their frequently attractive and often radiantly beautiful illuminations have made them the object of lavish facsimile editions, and reproductions of individual illuminations are to be found in almost every general survey of medieval culture. Despite their alluring visual properties, books of hours, even in Catholic countries, are among the least understood of the written artefacts of the Middle Ages. In particular, little consideration has been given to how these books were actually read and used in daily life. The problem of their use is reflected in the very name we have given them. The label 'book of hours' itself conceals the variety of texts, in addition to the various offices of the canonical hours properly defined, which these books contained.[1] Books of hours have often been grouped in manuscript catalogues with liturgical books, yet many of the texts found in books of hours exist elsewhere exclusively in literary compilations.[2]

If attention is shifted from textual content to format, books of hours are generally considered to represent a specific variety of relatively small and portable books pertaining to private piety. Indeed, the portable format of books of hours is one of their most remarkable traits. When their peculiar combination of format and content is placed in a larger historical context, books of hours may be fairly regarded as a unique product of the late medieval Latin West. In the tenth century, a Greek equivalent for every genre of Latin book pertaining to liturgy, prayer, and devotion existed.[3] In the fifteenth century, when books of hours were being produced by tens of thousands in manuscripts and subsequently in printed edition, in western Europe, the Byzantine world remained content with the same genres of liturgical and devotional texts that had sufficed half a millennium before. In the medieval Hebrew and Islamic traditions as well, there evolved no equivalent to the books of hours, the largest quantity of which were produced in France, the Low Countries, and Italy.

The emergence of a new and distinctive type of portable book of prayer was closely tied to the evolution of reading habits in the late medieval period. The proliferation of books of hours was a result of the advent of silent reading following the systematic introduction of word separation throughout western Europe in the first half of the eleventh century. This graphic innovation ultimately permitted the entirely private fusion of the previously public acts of reading and prayer.[4] By the fourteenth and fifteenth centuries, the habit of reading silently had devolved from scribes and university scholars to an ever larger portion of the lay population.[5] It is therefore important to emphasize that books of hours proliferated in a new milieu composed of two types of reading ability that have often been obscured confusingly under the modern term 'literacy'. One reading ability I shall term phonetic literacy. Phonetic literacy was the ability to decode texts syllable by syllable and to pronounce them orally. Such reading was closely related to oral rote memorization, and has its equivalent today most notably in Islamic countries where Arabic is the language of prayer but not of daily discourse.[6] Although the readers often had from extraneous sources a general appreciation of the sense of the text, they were not competent to comprehend its precise grammatical meaning. In fifteenth-century France, to read a Latin prayer aloud or to recite a written text from memory was a pious act that could be performed by many monks and laypeople insufficiently literate in Latin to be able to translate devotional prose or verse phrase by phrase into the vernacular.

Alongside the ability to read phonetically, a second type of literacy existed. This facility, which I shall term comprehension literacy, was the ability to decode a written text silently, word by word, and to understand it fully in the very act of gazing upon it. Certainly, many clerics could read Latin with this degree of comprehension, and even a greater number the of laity and clerics who possessed only phonetic literacy in Latin had comprehension literacy in the vernacular. Modern studies of the history of the book that seek to understand the intricate and often bilingual structure of books of hours must begin with the awareness that this genre of books for the laity developed in a milieu in which these two levels of reading ability existed side by side. Major clues to detecting which mode of reading specific portions of a book of hours were designed to serve lie in the rubrics that accompanied the texts. Unfortunately, these rubrics have often been ignored by manuscript cataloguers; they have only rarely been reprinted and they have never been edited critically.[7]

To understand the relationship of the rubric to the mode of reading prayers, we must begin by analysing the late medieval vernacular nomenclature for reading and praying. In late antiquity, prayer was typically oral from a written text.[8] In contrast, in the modern world praying is often entirely unrelated to reading. Although oral prayer in the Catholic religion is usually from a printed text, among modern Protestants both oral and especially silent prayer are often impromptu and following from the individual inspiration of the person praying.[9] In the fifteenth century, this dichotomy between programmed and spontaneous prayer was only incipient and the relationship between text and prayer was universally much closer than the one that exists today. It is clear in innumerable instances that – at least among orthodox Catholics – the injunction of theologians to say a prayer meant, in fact, to read it aloud or silently. Indeed, in rubrics the term *dicere* was customarily used in conjunction with written prayers in books of hours where *legere* would have been used in rubrics of secular texts. *Legere* was used relatively seldom in Latin and vernacular derivative forms to describe the reading of prayers in books of hours.[10] Like *legere* when used in conjunction with secular texts, *dicere* did not indicate that the reading was oral rather than silent.[11]

The fifteenth century had its own vocabulary, which only partially corresponds to our own, to distinguish oral prayer from silent prayer. This vocabulary had only limited precedent in earlier centuries, and its full development at the end of the Middle Ages was of great significance for the history of private piety. In the liturgy of the western Church before 1300, truly silent prayer was unknown. All prayer pertaining to the celebration of the Mass and the canonical offices was oral, and even in private masses all prayers had to be pronounced in order to be valid. The prayer *Oramus te, domine per merita sanctorum* and the Canon and the Secretum of the Mass, which were referred to in ordinals and pontificals as prayers to be said *in silentio* or *sub silentio*, were in fact pronounced in a lowered voice so as to be audible to the priest but not to the congregation.[12] Such prayer in a hushed voice was likened to the clearly oral prayer of Jesus in the Garden of Gethsemane during the night preceding the Crucifixion.[13] When, in the twelfth century, Hugh de Saint Victor described the act of private prayer in his widely circulated treatise *De modo orandi*, he unambiguously endorsed vocal prayer not to benefit God, who knew the intimate thoughts of all, but to stimulate the person praying to a higher state of devotion. Thomas Aquinas agreed with Hugh, adding to his argument that oral private prayer served to edify others.[14]

It was only in the mid-fourteenth century that theologians offered the pious a clear alternative between oral and silent prayer.[15] The Franciscan, Nicolas de Lyra, for example, in his *Postilla* on the New Testament, equated private prayer with silent prayer, and public prayer with vocal prayer. For those for whom vocal prayer enhanced devotion, it was to be commended, but those who found oral prayer distracting were urged to indulge in private mental prayer.[16] In about 1400 Jean Gerson, in describing the hierarchy of prayer, distinguished between a higher form – mental prayer – which took place entirely in the soul, and a lower form – vocal prayer – which pertained to the body.[17] A brief treatise by an anonymous author on the state of the mind during prayer, included in a book of hours compiled for an unknown woman of central France (which today forms Poitiers, Bibliothèque Municipale MS 92), distinguished between two forms of prayer: verbal or oral prayer and prayer by fervent mediation and contemplation within oneself.[18] Albertus de Ferrariis of Piacenza, in his treatise *De horis canonicis*, sanctioned oral prayer for the public observance and silent prayer for the private observance of the canonical hours.[19] In about 1450, Pierre de Vaux composed a *Vie de Sainte Colette*, the celebrated reformer of the Franciscan orders, in which he stated precisely that Colette prayed in two manners, *vocalement* and *mentalement*, and that it was from the latter mode that she entered into the highest stage of religious ecstasy.[20] De Vaux also made it clear that books played an essential role in St Colette's prayers, noting that she prayed at night with the aid of candles and that when a candle fell, burning her book, she cried not because her book was ruined but because her prayer had been rendered imperfect.[21] The anonymous Burgundian treatise entitled *Sermon sur le Pater noster*, translated in 1476 or 1477 in the same milieu by Jean Miélot for Philippe le Bon, similarly divided all prayer into *oeuvre meritale* [*mentale*] and *oeuvre vocale*.[22]

The Paris-trained Franciscan theologian, Pierre des Gros, in his *Jardin des Nobles*, a vernacular summa of theology, law, and history written for the laity in 1464, provided perhaps the fullest fifteenth-century description of the various types of prayer. He listed three distinct modes of prayer: first, silent prayer, as he termed it, prayer by the heart only without expression of the external voice; secondly, prayer by the mouth only without internal attention; and thirdly, mixed prayer, simultaneously by the heart and the voice.[23] Des Gros correctly recognized that mixed prayer was the form recommended by Augustine, and following de Lyra, he unequivocally believed that orally pronounced prayer was required for the public liturgy of the Mass and for prayers of obligation imposed by the sovereign, such as prayers of penance and the chanting of canonical hours required by the secular clergy and by members of religious orders according to their rules, which held force through papal approbation.

Silent prayer, however, was recommended as a valuable aid for private devotion. To illustrate silent prayer, des Gros used the example of Hannah, the mother of Samuel, whose silent prayer, described in I Samuel 1. 12–15, elicited the suspicion of the high priest Eli that she was intoxicated, an indication of how foreign the notion of silent prayer was to the mentality of ancient Israel. From patristic times to the thirteenth century, this passage of the Vulgate, which referred to prayer in the heart ('*loquebatur in corde suo*'), served as a locus for commentaries warning against vain ornamental loquacity in prayer.[24] In the mid-fourteenth century, however, Nicolas de Lyra abandoned this argument and stated tersely that Hannah's mode of prayer with the heart alone, which he identified with the *via contemplativa*, was to be commended highly.[25] In the mid-fifteenth century, Denis the Carthusian's moralization of this passage in his *Enarrationes* on the Old and New Testaments argued explicitly for silent prayer's advantage of direct and unbridled communication with God. Only prayer that was entirely within oneself was truly private and not susceptible to interception by evil spirits.[26] A mid-fifteenth-century compilation, entitled *Livre de devotion* and atttributed to frère Bonaventure of the Observance, similarly suggested that through silent prayer one gave expression to more intense sentiments than in oral prayer and that only in isolation could one communicate freely with God.[27] John Calvin's exegesis of I Samuel 1. 12-25 (in his *Homiliae in primum librum Samuelis*) also identified silent prayer with private devotion and communication between the individual and God. Calvin used the example of Eli to argue that private prayer was properly outside the bounds of official priestly supervision.[28]

The fifteenth and sixteenth century sources I have enumerated (and others) reveal no consistent terminology for true silent prayer. Since prayer *in silentio* was pre-empted for the secretly pronounced prayers of liturgy, private silent prayer was referred to variously as prayer said in thought, mental prayer, meditative prayer, contemplative prayer, and prayer of the heart. The formulation 'prayer of the heart' (*prière de coeur*) was perhaps the expression most frequently used, and it is certainly the most problematical in the context of the modern vocabulary of devotional practices. To the modern speaker, prayer of the heart means sincere prayer;[29] for writers of the fifteenth century, prayer of the heart meant that the act of praying transpired within the mind of the person praying. Prayer of the heart could be accompanied by the voice, as in des Gros's *prière mixte*, but participation of the voice was not necessary, and some authors, notably Denis the Carthusian, believed that the vocal expression of prayer was to be entirely suppressed except in public prayer.[30]

The late medieval use of the phrase 'prayer of the heart' can be explained by a conception of cognition that was physiologically different from our own. In the twentieth century, cognition is known to be the exclusive function of the brain. In the fifteenth century, cognitive function was thought to be divided between the brain, which according to Galen was the locus of sense and memory, and the heart, which according to the Bible, Aristotle, and numerous Latin patristic authorities was the intangible seat of the rational soul.[31] The anonymous *Sermon sur le Pater noster* stated specifically that the head was the seat of the sense of the physical body in the same manner that the heart was the seat of the thoughts of the mind.[32] Robert Ciboule believed that the brain received sensation from the eyes and ears via nerves; he regarded the blood flowing from the heart as the source of understanding and the beginning of all sensation.[33] Guillaume Fillastre, second chancellor of the Burgundian Order of the Golden Fleece, postulated that the emotion of fear was generated by the interaction of the brain and the heart.[34] Although fifteenth-century theories of cognition and psychology were exceedingly varied and complex, most respected the authority of Aristotle and Scripture in recognizing the heart as the ultimate receptacle of sensation and the seat of abstract understanding as well as of the emotions that such understanding naturally generated. For Jean Gerson, the *bouche* was the organ of speech; the *coeur* the organ of thought.[35] In the treatise of the *Douze perils d'enfer*, translated between 1446 and 1461 for Queen Marie of France, wife of Charles VII, the *coeur* was the organ of imagination and cognition.[36] For precisely this reason, medieval man originated the belief that the cessation of the movement of the heart marked the moment of death, the disengagement of the spiritual soul from its abode within the physical body.[37] The special funerary monuments erected during the Renaissance for the hearts of great noblemen reflected the reverence bestowed upon the organ that had served as the corporal sanctuary of the intangible soul and mind of the deceased.[38]

This vocabulary for the function of cognition provided formulations for the description of both silent internal reading and prayer. For example, in the early sixteenth century Octavien de Saint Gelais translated Ovid's explicit reference to the silent reading of an erotic love letter (*Heroides* XVI, 3) as 'And so I read it *en cueur* without pronouncing it aloud.'[39] Analogous references to the heart as the organ of prayer were frequent in the rubrics of fourteenth-century books of hours. For example, the prayer of the *Trois verités* had the following rubric, which circulated with the text: 'Comment par dire de bouche ou de cuer trois veritez que nous nous mettons hors de peche mortel et en estat de grace' ('How by saying by mouth or by heart three truths we put ourselves out of mortal sin and in a state of grace').[40] The instructions for reading the prayers contained in a book of hours copied for an unidentified queen of France *c* 1400, stated that the prayers therein were to be said '*plus de ceur que de bouche*' (more by heart than by mouth).[41]

The shift from the mouth to the heart as the primary organ of prayer was of great importance. In the seventh century, Isidore of Seville had stated that the etymology of *Oratio* was *oris ratio*, and this definition of prayer had been repeated by Dhouda's *Manuale*, the principal Carolingian treatise on lay piety.[42] In contrast, a late fourteenth-century treatise on prayer stated boldly that prayer is conducted by the heart and not by the lips, for it is preferable to pray silently with the heart than only with words and without thought.[43] For Denis the Carthusian, Isidore's definition of prayer pertained to sermons and preachers.[44] The instruction to pray with pure, good, or contrite heart, common in Latin and vernacular rubrics of devotional prayers in books of hours, was virtually unknown in the formal liturgical books for public oral prayer. The texts of the devotional prayers themselves, excluding those of the offices contained in books of hours, referred far more frequently to the *coeur* of the person praying than to the *bouche*. For Robert Ciboule, prayer in the heart ('*contemplation, oration, et sainte meditation*') was the principal remedy to '*ordes et viceuses cogitations*'; the second remedy was, for those who knew how to read, private study and reading.[45] The term 'cordial attention' used in conjunction with silent prayer clearly implied that cognition took place in the heart and not in the brain.[46] Therefore, when prayer book rubrics refer to the mind of the supplicant, it is correct to conclude that by the mind they meant the heart rather than the cerebrum. It is in this sense that one must interpret Chicago, Newberry Library, MS 104.5, an English portable prayer book of the second half of the fourteenth century, which promised protection from sudden death to whoever annunciated or held within his mind a specific prayer.[47]

References to the eyes and vision become more frequent in the rubrics of fifteenth-century prayers. As the primary organ of reading, the eyes were regarded as channels by which external impressions passed directly to the heart.[48] A prayer in Paris, BN, MS fr. 13168, equated vision with understanding: 'S'ensuit autres oreson en francais qui son de grande devotion comme l'en peut veoir en les lisant' ('There follow five other prayers in French which are of great devotion, as one can see by reading them').[49] The rubric in MS 104.5 offered indulgence to the penitent who pronounced orally (*dicerit*) or scanned silently (*viderit*) the prayer that followed.[50] A prayer in Chicago, Newberry MS 56, a Flemish-Dutch book of hours of the mid-fifteenth century, began by announcing that the prayer was being 'heard' through the eyes of the person praying.[51] Such formulations in books of hours are consistent with the numerous references to the 'eyes of the heart' in contemporary devotional treatises. Indicative of the new mentality is the incident, related by Denis the Carthusian, that Augustine on his death bed had the Penitential Psalms painted on the walls so that he could rest his eyes upon them and thereby contemplate them.[52]

Fifteenth-century devotional literature offers many clues as to which portions of the texts contained in books of hours were thought more apt to be read aloud or silently. Perrine de la Roche, in her *Vie de Sainte Colette*, reported that the saint's favourite vocal prayers were the Psalter, the seven penitential psalms and the litany.[53] For the Psalter, Colette may have used a portable Psalter, a Bible, or a breviary, but for the seven psalms and the litany she probably used a book of hours where these texts were conveniently and discreetly transcribed. De la Roche's text also strongly suggests that Colette prayed vocally during the canonical hours, in accordance with Franciscan practice.[54] Pierre des Gros, in the *Jardin des Nobles*, provides further evidence that the hours were also often said orally by the laity. His examples of ineffectual prayer '*de bouche seulement*', in which the lips moved and the voice was heard but the thoughts of the supplicant were not on God, was that of a person who said his hours perfunctorily and inattentively.[55] The brief exposition of prayer accompanying the book of hours forming Poitiers, BN, MS 95, unambiguously grouped the recitation of the canonical hours with psalms and 'other verbal prayers'.[56]

The rubrics of the canonical office of the Virgin and other prayers for oral recitation in many French, English, and Spanish books of hours confirm this hypothesis in that they were often partially in the vernacular so as to be fully comprehended by the person praying, while the texts themselves, which obtained force through pronunciation, remained in Latin.[57] Moreover, other evidence associated with vernacular prayers for the hours which were not translations of the canonical offices indicate that the former may have been subsumed under the practice of private silent reading. For example, the mid-fifteenth-century hours of Jean de Montaubon contained traditional Latin hours, with the liturgical responses historically associated with oral use, and hours of the Passion in French, without liturgical responses, preceded by the rubric 'Cy commence une maniere de penser en la passion de nostre seigneur Iheucrist.'[58] The word *penser*, referring to the activity of the person praying, also occurred repeatedly within the text. Similar prayers for the hours as well as suffrages to the saints without the traditional oral format of liturgical responses are to be found in other late fifteenth- and sixteenth-century Latin and vernacular manuscripts, one of which, dating from about 1500, included in addition to 'Les Heures de notre seigneur' a 'maniere de dire sept fois Ave Maria a la beinoiste vierge Marie par maniere de contemplation'.[59] Frère Robert of the Carthusian Order, the late fourteenth-century author of the *Chastel perilleux*, seems to be referring to this distinction between the older oral mode and the newer mode of silently reading prayers for the canonical hours when he warned his female cousin, to whom he addressed this handbook of devotion, that while in church she ought never to fail in her obligation to say her hours orally by saying them silently, a practice that he identified with the desire to give expression to her intimate thoughts.[60] For Frère Robert, devotional prayer of the heart was an internal activity entirely separate from the chanting of the Holy Office.[61]

In the late fourteenth and fifteenth centuries, whether one prayed one's hours or other texts orally or silently determined the character of concentration and devotion that was brought to the prayer. Oral public prayer of the canonical hours and the Mass, with the possible exception of certain problematical liturgical compilations perhaps intended for use by lay members of religious third orders, was in Latin.[62] Although some of the laity could read with comprehension these Latin texts, typically drawn heavily from the Psalter, most could not. The exposition on prayer in Poitiers, BM, MS 95, which described precisely the state of the mind during the act of prayer, recognized this fact by stating that the *attention mentale et cordiale* for reading hours and other verbal prayers need only be the actual or habitual desire to serve God. To this end, it was not necessary to have full comprehension of all or each portion of the text as it was recited. Similarly, for prayers recited for one's own salvation or to obtain indulgences for the benefit of the dead, the person praying was only required to begin praying with correct intention and to maintain a modicum of attention during the prayer.[63] Jean Gerson, in his letter treatise *De valore orationis et de attentione*, compared such prayer to the habitual labour of a skilled craftsman or to the movement of a boat carried forward by its own momentum even when the sailors ceased to row.[64] For this type of prayer, the written text served only as a prompt script for a prayer which might ultimately be retained by rote memory through frequent repetition.

Pierre des Gros, in his *Jardin des Nobles*, listed three types of attention possibly pertaining to vocal prayer: attention to the order and correct pronunciation of the words, attention to the significance and understanding of the words, and attention to the end or object for which one prayed.[65] The first two attentions not only were unnecessary for efficacious oral prayer but were even potential impediments, for two specific reasons. First, because the simple and unlettered were not able to achieve them, and second, because the complexities of Latin syntax and grammar often caused the mind to wander away from holy thoughts. Modern psychological studies confirm that the distraction of pronouncing correctly poses a problem for the reader who reads aloud and simultaneously attempts to understand a text in a foreign language.[66] However, des Gros found that the third attention – that is, attention to the pious end of the prayer – was both sufficient and necessary for an oral prayer to be effective. Like Gerson, des Gros agreed that the attention to the text of the person praying need not be actual or continual throughout the recitation but could be merely habitual. Des Gros compared the saying of a prayer to the launching of a missile to which the requisite velocity was conferred only once at the beginning of its flight. Citing Hugh de Saint Victor's *De modo orationis*, he asserted that the force of this form of prayer might be so great as to cause the person praying to forget all else. Des Gros thus described in positive terms the obliviousness that the rote repetition of oral prayers often engendered, a state of mind that troubled Gerson and would offend subsequent Protestant reformers, particularly John Calvin.[67] However, for Perrine de la Roche, the fact that St Colette, old and sick, prayed vocally and apparently without full mental presence (for she could not keep track of her place in the sequence of repetitions) in no way vitiated the holiness of her prayers.[68] Elsbet Stagel, the fourteenth-century chronicler of the convent of Dominican sisters at Töss, near Zurich, described with approbation the type of prayer in which the number of repetitions and the 'mind set' of the supplicant were more

important than his or her comprehension of the text.[69] Certain texts and
rubrics in books of hours and other prayer books presuppose just this kind of
prayer.[70]

A far different kind of concentration was required for internal silent
prayer. For it, the very effort to understand that des Gros deemed to be a
distraction in oral prayer was a prerequisite and, in fact, the essence of this
devotional experience, inseparable from the act of recitation. The external
voice that was central to vocal prayer was seen as a potential impediment to
the attention of the heart needed for private devotion.[71] Internal silent
prayer reflected a new aesthetic ideal, which equated silence with holiness
and viewed all sound, including sermons and singing, as obstacles to the
highest levels of spiritual experience. The *Douze perils d'enfer* demanded that
the devout monk maintain silence in reading, in prayer, and at Mass.[72] The
author of the brief exposition on prayer in Poitiers, BM, MS 95, defined
silent prayer as the means by which the praying individual was made deaf to
the outside world, to seek by the grace of God inner spiritual reflection
through 'devout and fervent meditation and contemplation'. To arrive at this
state of grace, it was necessary 'to have understanding in the heart' – that is,
full mental comprehension of the text of the prayer. The author insisted that
this mode of praying was appropriate for ecclesiastics and particularly for
members of religious orders, but that it was not necessary for every
individual.[73] In the fifteenth century, different modes of praying were clearly
deemed appropriate for different levels of society, reflecting to an important
degree their respective levels of literacy. Thus, when Jean Gerson endorsed
the traditional monastic repetition of oral Latin prayer with pious intent, he
felt obliged to note that even some learned theologians were sometimes
unable to grasp the meaning of text through this external mode of prayer.[74]

One of the alluring qualities of silent prayer was the higher state of
spiritual awareness it was believed to offer. While not rejecting oral prayer,
Wessel Gansfort, writing in Latin to a learned audience a decade later,
extolled the spiritual superiority of silent mental prayer and meditation over
the distracting oral recitation of the canonical hours, the Psalms and the
rosary.[75] However, to offer spiritual grace to the elite, capable of reading
Latin or vernacular prayers fluently and with full comprehension, was in
contradiction to the conception of Christianity as a religion open to the
lowest levels of society. Thus, des Gros recommended the suppression of
orality for those for whom the voice impeded the attention of the heart in
individual and voluntary prayer, but he recommended oral prayer for the
unlettered.

The qualified preference of fifteenth-century theologians as spiritual
counsellors for silent prayer was consistent with their preference as scholars
for visual private study as opposed to oral public lectures.[76] The diminished
esteem for oral prayer in the fifteenth century was thus paralleled by the

reduced status of the sermon, described by Ciboule as a simplified form of communication tolerated by the educated because it was appropriate for addressing diverse audiences, including the simple.[77] Furthermore, Ciboule attached a certain stigma to sermons since their orality was itself a breach of silence.[78] Similar attitudes would have been foreign to St Bernard, for whom sermon collections such as those on the *Canticum canticorum* formed an important genre for scriptural exegesis. In an attempt to bridge the gap between learned visual and popular oral culture, Jean Gerson advocated the contemplation of pictures in books as a substitute for reading for the unlettered who sought a private devotional experience.[79] Books of hours served the needs of the unlettered as well, for they frequently contained illuminations of the 'Mass of St Gregory', which were specifically intended for stimulative contemplation to accompany the recitation of the Pater noster, the Ave Maria, and the Credo, short texts traditionally recited from memory, even by the totally illiterate.[80]

At the end of the fifteenth and in the sixteenth century, as comprehension literacy increased and new habits of private prayer evolved, silent prayer could be avocated with increasing zeal. García de Cisneros, writing in the vernacular for Benedictine monks, considered silent prayer to be an essential part of the reformed life of Monserrat.[81] In a brief vernacular consideration of prayer that served as a preface to Francis I's personal manuscript book of hours, prayer was defined as a spiritual colloquy of the soul with God during which the faculties and functions of the physical body were superfluous.[82] For Ignatius Loyola, silent prayer played an important role in the *Spiritual Exercises*, a manual for meditation incorporating contemplations of the life of Christ for the canonical hours, which were clearly inspired by the tradition of the silently read contemplative hours found in the personal prayer books of the fifteenth century.[83] At the end of the sixteenth century, the Jesuit St Alphonsus Rodríguez, whose treatise on silent prayer was widely circulated in the original Spanish and English translation, boldly asserted that 'all are capable of mental prayer, and there is none who may not use it.'[84]

Early Protestants, like Catholics, saw silent prayer as an important adjunct to vocal prayer. John Calvin recommended an alternating pattern of silent and vocal prayer for personal devotion, criticizing the mechanical vocal repetitions of prepared texts.[85] A small vernacular portable Protestant manuscript prayer book, dating from the first half of the sixteenth century and bearing the arms of Anne de Montmorency, blamed the corruption of the Church on the negligence of pastors who had taught prayers in Latin that produced only a fluttering of the lips without comprehension of the text. This reformist condemnation of the rote vocal repetition of Latin prayers was a conscious rejection of the late medieval doctrine of virtual or habitual attention in oral prayer. The anonymous author recommended that one 'pray spiritually and internally to the Lord in the sky and not vainfully and carnally to creatures of the earth'.[86] He described his prayer book as small in format so as to be easily carried, for he intended it to serve as a vernacular key to scripture. It included many texts closely related to fifteenth-century books of hours, among them a *maniere de méditer* the Passion of Christ.[87]

The spread of silent reading of prayers also affected the external etiquette of prayer. Denis the Carthusian, in his moralizations of I Samuel 1. 12–15, emphasized that silent prayer would neither offend God nor interrupt the devotions and prayers of confreres by noise from the mouth and undisciplined gesticulations. In the presence of others, all obtrusive conduct was to be avoided.[88] In oral prayer, des Gros, following Augustine, considered it desirable to accompany the external voice with movements of the body in order to increase the fervour of the prayer; swaying during oral prayer still remains an important part of both Islamic and orthodox Jewish religiosity. Movement in medieval oral prayer was evidence of the physiological link, existing to a greater or lesser degree in all cultures, between oral activity and body movement.[89] In contrast, Denis the Carthusian's remark suggests that a more static posture of praying accompanied the shift from oral to silent private prayer, and in another passage he specifically stated that the hands should be kept clasped or joined during prayer in order to stimulate increased devotion.[90] Similarly, in his *Modus orandi et meditandi*, Ludovico Barbo recommended that the gesture of joined hands accompany contemplation before secret verbal prayer.[91]

It was precisely during the fourteenth and fifteenth centuries that the representations of prayer in manuscript illuminations changed dramatically. Instead of showing the supplicant with his arms conspicuously raised, fourteenth- and fifteenth-century depictions of prayer and piety often portrayed the individual's hands joined with palms and fingers touching or, less frequently, clasped with fingers interlocked, the positions recommended by Denis, which left the hands at rest close to the body in a position creating an area of restricted space symbolic of the activity of prayer within the heart.[92] In the second half of the fourteenth century and in the fifteenth century, the scene of the Annunciation was represented by two distinct iconographic traditions. In one, one or both of the Virgin's arms were extended with open palms while she turned from a book and listened to the words of the angel.[93] The gestures of the angel and Virgin were those of oral communication frequently used, for example, by artists to represent the preacher and the listener. In the second iconography of the Annunciation, suggesting a silent infusion of Divine grace, the Virgin was depicted in the act of reading, light beams rushing to her heart, her hands either placed with palms and fingers touching or with arms crossed upon the chest, another static posture of prayer signifying internalization.[94] Both scenes of the Annunciation regularly prefaced the Hours of the Virgin and were a reflection of a society in transition from an oral recitation of prayers to their silent contemplation. Similarly, in the frontispiece of Margaret of York's copy of Pierre de Vaux's *Vie de Sainte Colette*, Colette was shown in silent prayer, as described by her biographer, kneeling at a prie-dieu, her hands joined palm to palm and finger to finger, with radiating beams of light passing to her head.[95] It was this position of prayer that also became standard in donor portraits in books of hours and French aristocratic funeral effigies after 1350, an apparent statement that the dead were in a state of silent prayer.[96] It is not without significance that the Protestant leaders of the sixteenth century were regularly represented in portraits with their hands in these new prayer positions that emphasized the internal quality of their private devotion.

The fact that in the fourteenth and fifteenth centuries a significant portion of vernacular prayers could be read silently from small portable codices dramatically affected the relationship between the celebrants and the laity in the performance of the public ceremonies of the Church, particularly the Mass. By the High Middle Ages, the Mass had become a priestly monopoly from which the laity and many clerics were barred by their inability to comprehend Latin.[97] As a substitute for not being able to understand the oral prayers, the practice of private prayer during the Mass, especially at the elevation of the Host, developed. Such prayers were apparently recited softly, from memory, with hands held in the palms-touching posture.[98] For Holy Offices, illiterate lay brothers of the early thirteenth century were only able to participate by the rote recitation of a prescribed number of Pater nosters at the apposite moment.[99] The spread of comprehension literacy in the fourteenth and fifteenth centuries revolution-ized the role of silent prayer in the Mass and other Holy Offices. Specifically, the advent of silent reading allowed for a structured and sequential synchronization of the silent prayers or contemplations of the laity with the oral prayers of the celebrants of the Mass. In the early fourteenth century, various series of Latin prayers to be read during the Mass began to be copied as a regular segment within books of hours, In the second half of the century, translations of these Latin prayers and a great variety of original vernacular prayers were similarly incorporated into books of hours or circulated separately in small prayer books. They were brought to the Mass not only by the laity but by members of religious orders as well. These fifteenth-century French books of piety constituted what might properly be termed lay ordinals for contemplation and they contained vernacular prayers for the Mass with rubrics linking them by cues to the apposite oral Latin prayers. Using these books, laypeople with only phonetic literacy in Latin could read with comprehension the French prayers at the appropriate time. The rubrics stating this linkage specified that these vernacular prayers were to be said *en pensé*, *de coeur*, or *en coeur* – that is, silently. Tracts for the Mass, some included in books of hours and some in other collections of small and portable format, provided an array of such prayers as well as less formal chains of topics to be contemplated at various points during the ceremony.

Silent reading in the vernacular allowed the devout to vary the pace of their reading and to intersperse reading and intense prayer from written texts with intense meditation upon assigned subjects loosely related to the content of the Mass.[100] By providing a programmed alternation between readings, prayers, and meditations, these vernacular books followed a model of reading developed in Latin in the early fourteenth century by James of Milan's *Stimulus amoris Christi*, the exceedingly popular Franciscan book for private devotion.[101] One mid-fifteenth-century vernacular tract for the Mass, in a small codex (Paris, BN, MS fr. 19247), advised that it was a more noble act to think of the Lord during the Mass than to recite hours or other texts.[102] Another small prayer book dating from *c*.1370–80, produced in Paris by artisans close to the royal court, included a forty-page *Oroison contemplative* for the Mass as well as specific prayers to be thought and said *de coeur* at particular junctures of the Eucharist.[103] Clearly, the reader here was expected to use the techniques of silent reading to skim and read selectively a portion of the texts, which were too long to be fully exhausted during a single Mass. Paris, BN, MS fr. 1879, dating from the fifteenth century, included a treatise on 'How to hear the Mass' with prayers to *penser et dire* during the Credo.[104] Paris, BN, MS fr. 190, a tome written for the library of Louis of Bruges between 1480 and 1483, contained a treatise entitled 'An *ordo* for whoever would like to hear the Mass while contemplating the mysteries which are there represented without saying anything with his mouth'.[105] This short tract, undoubtedly inspired by the liturgical *ordines* for oral prayer and external ceremonial gestures, contained directions for contemplations and pious thoughts. It specifically recommended the silent recitation of the Pater noster beginning with the elevation of the Host and continuing until the end of the Mass.[106] Paris, BN, MS fr. 402, a mid-fifteenth-century vernacular lectionary that permitted the laity to follow silently in French the temporal readings pronounced aloud in Latin, was prefaced with instructions on *La Significance comment on doit penser en la messe* (the significance of how one should think during the Mass).[107] An unidentified 'frère Olivier' (probably Olivier Maillard, the fervent advocate of the strict observance of the rule of St Francis during the reign of Louis XI), in a treatise entitled 'Some devout meditations upon which a person ought to meditate during the Holy Mass', declared that the person who heard the Mass profitably was not the individual who listened to the words, but rather the one who remembered with his or her memory and heart the passion of our Lord.[108] The author noted that the more fervently and devoutly these prayers were said 'in the heart', the more one pleased God.[109]

It is important to emphasize that the precise topics and order of meditations as well as the specific prayers for specific parts of the Mass in these books of hours, prayer books, and tracts were all invariably different. This new genre of text became the source of a revolution in the experience of the Mass in the consciousness of those who attended it, and as such was profoundly subversive to the outward liturgical uniformity according to the use of Rome that had increasingly been established by missals from the middle of the thirteenth century onward.[110] It was apparent by the mid-fifteenth century that, while oral prayers of the Mass might be performed according to the single prescribed usage of the papal court, the prayers recited silently in the vernacular from written texts could vary with the private desires of the individual. While the manufacture of public prayer books was subject to exacting textual control set by the approved ordinals, private prayers were disseminated by scribes and *librairies* according to the highly fluid dictates of the market-place.[111]

In the fourteenth-century universities, private silent reading had been forbidden in the classroom; in the fifteenth century, the private reading of prayers and other devout texts during the Mass became a source of concern at least among some scholastics who were, of course, literate in Latin and able to understand the Mass as it was pronounced. Wessel Gansfort, to an audience literate in Latin, specifically forbade the practice of reading or praying during the Mass because it violated communal participation in its meaning, an ideal still held to be valid by modern Catholic liturgical reformers.[112] Among the laity, however, the custom of bringing books of hours, tracts on the Mass and other texts to church spread without any serious attempt by the authorities to impede it. An anonymous English description of the habits of the pious laity, dating from the early fifteenth century, commended the practice of silent reading during the Mass as an edifying spiritual exercise.[113] The great variety of separately programmed religious experiences which in this manner were cloaked beneath a single uniform Roman Mass was therefore most striking, and did not go unnoticed during the Counter-Reformation. Pius V, in addition to completing the medieval effort to standardize the missal and the breviary, attempted, without success, to impose on Catholic Europe a single uniform book of hours entirely in Latin and containing a single set of prayers for the Mass.[114]

In addition to permeating the established oral liturgy of the Church, silent prayer created a new intimacy between the devotee and the book. At the end of the Middle Ages, the ideal of the small book of prayers, always close at hand and inseparable from the reader, replaced that of larger prayer books used in the past for both public and private prayer.[115]

The choice of texts in books of hours was sufficiently wide to permit scribes or *libraires* to assemble the kinds of prayers preferred by the purchaser of the book, and to adjust these prayers to the appropriate gender and name.[116] These books, although mass-produced, were far more personal than those of previous epochs. Spaces could be left for the person praying to insert his or her own requests or desires.[117] The book itself, according to its rubrics, became a talisman that, if always carried on the person of the owner, would protect him of her from disasters merely by the possession of the written word. Instructions in rubrics also promised that a given written prayer, if placed above a woman in childbirth, would insure a safe delivery.[118] Such instructions with their implicit references for the personal disposition of the mute text had no parallel in earlier liturgical compilations for public prayer.

It is not a coincidence that erotic illuminations began to accompany books of hours. Artists took advantage of the privacy afforded by each person's own book of hours to portray erotic scenes unimaginable in public art or publicly displayed liturgical texts. Inspired by the sexually explicit illustrations in secular texts, notably the vernacular translation of Valerius Maximus' *Facta et dicta memorabilia*, artists decorated books of hours with increasingly suggestive erotic scenes, often ostensibly depicting the vices for which penance was required but consciously intended to excite the *voyeur* of the book.[119] Borrowing elements from representations of holy baptism, scenes of Bathsheba in the bath evolved into titillating vignettes depicting auto-excitement.[120] Similarly, the resurrection of the dead was imbued with erotic qualities not present when the scene was depicted in works of public ecclesiastical art.[121] These erotic scenes achieved their zenith in manuscripts produced at the very end of the fifteenth century and in the first decade of the sixteenth century. They even began to permeate printed books, only to be eliminated in the austere new books of hours produced in the period following the Council of Trent.[122]

The ubiquitous spread in the fourteenth and fifteenth centuries of books of hours with their accompanying special rubrics, prayers, and illustrations reflected a new phenomenon in the history of private piety, the fusion of prayer with silent reading. Although sixteenth-century Protestants would turn away from advocating the reading of silent prayers and emphasize the need for spontaneous prayer formulated completely in the words of the individual, such sentiments were still rare in the fifteenth century, when men and women in general attached no stigma to the practice of gazing upon and contemplating prayers written by others. In fact, fifteenth-century laity and clerics were especially attracted by the manner in which the new medium of silent reading allowed literate devotees to associate themselves vicariously with the intimate devotions that, according to the rubrics of books of hours, had been written by the hand of Jesus himself or by great saints and popes. From the pen of the author to the eye of the devout layperson, the new internal mode of prayer, cultivated in the late Middle Ages, created a new intimacy between the praying individual and the book, an intimacy that both stimulated more individualistic practices of devotion, and kindled the desire for control of text and accompanying illustrations that would culminate in the Counter-Reformation's zeal for censorship of the printed page.

Notes

This text is a revision of an article that appeared in English in *Scrittura e civiltà*, 9(1985), pp. 239–69.

1   See Victor Leroquais, *Les Livres d'heures manuscrits de la Bibliothèque Nationale...* Paris (2 vols, Protat frères, Mâcon, 1927), vol. 1, pp. xiv-xxxii.

2   This is particularly true of the devotional prayers. For numerous vernacular examples in non-liturgical manuscripts, see Jean Sonet, *Répertoire d'incipit de prières en ancien français* (E. Droz, Geneva, 1956).

3   For an appreciation of the variety of Greek medieval texts, see Robert Devreesse, *Le Fonds Coislin* (Imprimerie Nationale, Paris, 1945); Devreesse , *Le Fonds grec de la Bibliothèque Vaticane des origines à Paul V, Studi e Testi*, 244 (Biblioteca apostolica vaticana, Città del Vaticano, 1956).

4   See Jean Vezin, *Les Scriptoria d'Angers au XIe siècle* (H. Champion, Paris, 1974), pp. 153–4. This observation has been substantiated by a careful scrutiny of the various catalogues of *Manuscrits datés* now in progress and the standard corpus of nineteenth- and twentieth-century palaeographic reproductions now complete, as well as an examination of the Porcher collection of photographs at the Bibliothèque Nationale (Paris).

5   Paul Saenger, 'Silent Reading: Its Impact on Late Medieval Script and Society', *Viator*, 13(1982), pp. 367–414. For the wide dissemination of books of hours in the sixteenth century, see Roger Chartier, 'Culture as Appropriation: Popular Cultural Uses in Early Modern France', in Steven L. Kaplan (ed.), *Understanding Popular Culture: Europe from the Middle Ages to the Nineteenth Century* (Mouton, Berlin, NY, Amsterdam, The Hague, 1984), pp. 229–53.

6   For a discussion of Qur'anic literacy in contemporary Liberia, see Sylvia Scribner and Michael Cole, *The Psychology of Literacy* (Harvard University Press, Cambridge, Mass, 1981), pp. 68–9.

7   Examples of rubrics that would be worthy of careful collation and edition include the rubric that customarily precedes the *Abbreviated Psalter of Saint Jerome*, the related rubrics that accompany the *Adoro te*, and the rubric accompanying the prayer of the *Trois verités* (see note 39). Leroquais, *Les Livres d'heures*, remains the most complete collection of transcribed rubrics. Another useful source is Christopher Wordsworth (ed.), *Horae Eboracenses: The Prymer or Hours of the Blessed Virgin Mary* (Publications of the Surtees Society 122, London, 1920), which contains rubrics from a number of printings of the *Horae* of York and Salisbury. Malcolm Parkes, *The Medieval Manuscripts of Keble College, Oxford* (Scolar Press, London, 1979); Gerard Achten, Leo Eizenhöfer, and Hermann Knaus, *Die Lateinischen Gebetsbuch-Handschriften der Hessischen Landes- und Hochschulbibliothek Darmstadt* (O. Harrassowitz, Wiesbaden, 1972), give useful additional transcriptions from French, German, and English manuscripts.

8   See, for example, Augustine, *De cura pro mortuis*, v, and *De magistro*, i, where

Adeonadus states that prayer is verbal and oral. Similar views are expressed by Origen, *De oratione*, II, 5, and XIII, 7; Tertullian, *De oratione*, 17.

9 This is particularly true among Quakers; see Friedrich Heiler, *Prayer: A Study in the History and Psychology of Religion*, tr. and ed. Samuel McComb with J. Edgar Park (Oxford University Press, NY, 1958), p. 317; J. Hillis Miller, *The Practice of Public Prayer* (Columbia University Press, NY, 1934), pp. viii, 11-19 and 89-90.

10 Examples of *legere* do exist. In Latin, *legere* is used several times in the rubrics of Newberry MS 56, *The Hours of Margaret de Croy*. Anne of Brittany's prayer book (Newberry MS 83, fol. 49v) uses *legere* in a sense that surely suggests oral liturgical recitation: 'Oratio habita per beatum Leonardum qua liberata fuit regina Francie, que in partu deficiebat nec parere iam desperata; et est maxime uirtutis quoties deuote legitur et attente auditur cum mulier est in partu.' Other rubrics are more ambiguous and may in fact signify intensive silent reading; see Achten, Eizenhöfer, and Knaus, *Die lateinischen Gebetbuchhandschriften*, p. 43 (no. 4, fos 124-134v).

11 *Dicere*, used in the sense 'to state', was a commonplace of the scholastic vocabulary and was only gradually replaced by *scribere* at the end of the Middle Ages; see Saenger, 'Silent Reading'.

12 Denis the Carthusian describes these prayers '*in silentio dicit*' as being prayed '*uoce submissa*'; see his *Opera omnia* 35: 347.

13 Ibid. *loc. cit.* Since the words of Jesus's prayers were set down by Matthew, Mark, and Luke, it is clear that he prayed orally.

14 PL 176: 982D. Hugh's views are a repetition of Augustine, *De cura pro mortuis*, v. Thomas Aquinas, *Scriptum super sententiis magistri Petri Lombardi*, Dist. 15, quaest. &, art. 2, ed. Marie Fabien Moos (4 vols, P. Lethielleux, Paris, 1947), vol. 4, pp. 734 and 737-8. Humbert of Romans in his *Super constitutiones fratrum praedictatorum Opera*, ed. J. J. Berthier (Rome, 1888-9), vol. 2, p. 92, also suggests a vocal component to private prayer. For an instance of oral private prayer serving as an example for others, see Jean-Claude Schmitt, 'Between Text and Image: The Prayer Gestures of Saint Dominic', *History and Anthropology*, 1(1984), pp. 127-62.

15 See the remarks of Thomas Merton, *Contemplative Prayer* (Herder & Herder, NY, 1969), p. 75.

16 Nicolas de Lyra, *Postilla literalis et moralis super totam Bibliam*, GW 4284 (Basel, 1498), V, on Matthew 6.6 proposed (at note e) the following rule for selecting a mode of prayer: 'Putant enim et cetera. Posunt tamen multiplicari uerba in oratione duplici de causa: una est per significationem uerborum deuotorum animus melius eleuetur in Deum. Alia causa est ut homo in Deum eleuet non solum mentem sed etiam corporali uoce, secondum illud Psalmiste, Cor meum et caro meo exultauerunt in Deum uiuum [Ps 83, 3]. Non tamen posunt dari ibi regula certa sed debet homo talibus uerbis uti in oratione quantum percipit homo, quod faciant ad deuotionis excitationem. Si autem econtrio percipiat quod multiplicatio uerborum distrahat mentem eius, debet cesare a uerbis et orare Deum affectibus mentis, hoc tamen intelligendum est in oratione priuata, ut dictum est, quia oratio publica debet esse in uerbis ut posset percipi ab aliis.' A century later, Denis the Carthusian stated that vocal prayer ought to be

restricted to those who did so because of an official obligation (*De oratione*, art. 26, in *Opera omnia*, 41: 53).

17 Jeannine Quillet, 'Quelques textes sur la prière de Chancelier Gerson', in *La Prière au Moyen Age: Littérature et civilisation* (Université de Provence, Aix-en-Provence, distr. H. Champion, Paris, 1981), p. 425.

18 Poitiers, BM, MS 95, fo. 139v. The precise terms used were *oroisons verbales* and *fervente meditacion et contemplacion [en soy]*.

19 Albertus de Ferrariis, *De horis canonicis* (Rome, 1475), ch. 35. For the identification of the author, see L. A. Sheppard, 'Albertus Trottus and Albertus de Ferrariis', *The Library*, ser. 5, vol. 2(1947), p. 159. See also Hieronymous Savonarola, *Dell' oratione mentale* (Florence, 1492) for an Italian commendation of silent private prayer dating from the second half of the century.

20 Pierre de Reims, called de Vaux, *Vie de Sainte Colette*, ed. P. Ubald d'Allençon, *Archives franciscaines*, 4(1911), p. 73.

21 ibid., p. 69.

22 'Mais quant tu feras tes oraisons, entre en ta chambrette, c'est entrer au secret de son coeur. . . . Et est assavoir per le nom de oroision sentend cy toute oeuvre meritale [sic for mentale] ou vocale qui est faitte par devotion et en l'onneur de nostre seigneur, comme chanter en l'eglise, dire messe, vespres ou les autres heures du jour, ouyr sermons preschier la parole de Dieu, estudier les saintes escriptures et penser a nostre seigneur.' ('But when you make your prayers, entering into your cell is to enter into the hidden places of His heart. . . . And be it known that by the name of prayer is meant here all mental or vocal operation that is done out of devotion and in honour of our Lord, such as singing in the church, saying Mass, vespers, or the other hours of the day, hearing sermons, preaching the word of God, studying the holy Scriptures and thinking of our Lord') Brussels, BR, MS 9092, fo. 27v.

23 'Et se peuvent faire ces vi especes ici en trois manieres: La premiere de cuer seulement sans expression de voix au dehors, comme nous lisons au premier chapitre du premier livre des roys que la sainct femme Anne, niece de Samuel, prioit Dieu au temple car ses lieffres se mouvoient mais sa voix n'estoit point oye. La seconde de bouche seulement quant une personne prie Dieu sans affection ou atencion quelconque ne actuelle ne habituelle, ou premise comme quant [u]ne persone dit ses heures et tout son atencion est arrier de Dieu comme dit Dieu par le prophete Ysaie, Ce peule icy me prie de bouche, mais le cueur est loing de moy [Isiah 29.13]. Le tierce est mixte qui est quant on prie Dieu de cuer et de bouche par voix exprimee au dehors.

Et cette oroison de voix avec l'affection de cuer fait quatre biens. Le premier est qu'elle excite la devocion et de celuy qui prie et des auditeurs pource dit saint Augustin ou livre de la cure que on doit faire pour les morts [*De cura pro mortuis*, 5]. Combien que la volonte invisible de celuy qui prie soit a Dieu cogneue et que Dieu n'ait point indigence de voix pour cognoistre le cuer humain, car il cognois la parole du cuer, touteffois l'omme use de voix et des membres de son corps affin que l'affection croisse en priant plus humble plus devote et plus fervent. Le second est congregacion de pensee et de attencion, car la pensee plus se unist quant la voix est avec la devocion car ainsi que les

maulvaises paroles distraient la bonne pensee, les bonnes le unissent et
conioingnent a Dieu. Le tiers est entier service a Dieu, car raison est que l'ome
serve a Dieu de tout ce qu'il a pris de Dieu. Puis qu'il a pris ame et corps, raison
est que de ame par oroison de cuer et de corps par oraison de bouche il serve
Dieu. Le quart est redundance car par telle oroison le devocion de l'ame
redunde au corps car comme dit Ihesu Christ en l'evangile de l'abondance du
cuer la bouche parle [Matthew 12.34] par ce ist le Psalmist, Mon cuer est esjouy
et ma langue a pris exultacion [Psalm 15.9]. . . . Ces quartiers causes, icy dit
Ozee au darienier chapitre de son livre, Rendons a ces beaus [Osee 14.3], c'est a
dire les sacrafices de nos lieffres.

Touteffoiz se la voix empeche l'attencion du cuer, c'est le mieulx de la lesser
en oraisons particulieres et qui son seulement de devocion non pas de
obligacion, non pas en oraisons commune et qui sont de l'institucion de l'eglise
ou de veu ou de charge commes de penitence ou de commandement du
souverain. Car telles de doyvent dire de bouche et es oroisons communes qui
sont de l'institution de l'eglise le peuple doit convenir es festes et dimanches
especialment a la messe depuis le commencement jusques a la fin . . . comme
desent les decrets, Missas et com [ad] celebrandas et omens fideles et qui die
solemni, en la premiere distinction de consecration [chs 64, 65, 62, 66 D.I de
cons.].' ('And the six species can be done in three manners: the first [is] with the
heart alone without external expression of voice, as we read in the first chapter
of the First Book of Kings that the holy woman Hannah, Samuel's niece, prayed
to God in the temple, for her lips were moving but her voice was not heard. The
second [is] by mouth only, when a person prays God without emotion or
attention, actual or habitual, or perhaps as when a person says the hours and
fixes all attention on God, as God says through the prophet Isaiah, ". . . this
people draw near with their mouth and honour me with their lips, while their
hearts are far from me". The third is mixed, which is when one prays God with
heart and mouth, by voice, expressed externally. And this prayer by voice with
involvement of the heart does four good things. The first is that it excites the
devotion both of the one who is praying and of the listeners, as St Augustine
says in the book of the care one must have for the dead. Although the invisible
will of the one who is praying is known to God and God has no need of voice to
know the human heart, for He knows the heart's language, still one uses uses
his voice and the members of his body so that loving emotion grows by praying
more humbly, more devoutly and more fervently. The second is the conjunction
of thought and attention, for thought is united more when the voice is with
devotion, for just as evil words distract good thought, good [words] unify it and
conjoin it to God. The third is total service to God, for it is reasonable that man
serve God with all he has taken from God. Since he has taken soul and body, it
is reasonable that he serve God by soul in prayer from the heart and by body by
prayer from the mouth. The fourth is redundance, for by such a prayer the
devotion of the soul redounds to the body, for as Jesus Christ says in the Gospel,
"out of the abundance of the heart, the mouth speaks" , and the Psalmist,
"Therefore my heart hath been glad, and my tongue hath rejoiced" . . . . Osee
speaks of these four causes in the last chapter of his book: "Take with you

words, and return to the Lord", by which he means "the calves of our lips". None the less if the voice prevents the heart's attention, it is better to leave it for private prayers that are purely of devotion and not of obligation, not for common prayers that are of the Church's instruction, or by vow, or assigned as penance or by the order of the sovereign. For such [prayers] should be said by mouth and the people should join in the common prayers that are of the Church's institution on feast days and Sundays, particularly at Mass, from the beginning to the end, . . . as the decrees say, "Missas et com [ad] celebrandas et omens fideles et qui die solemni", in the first distinction of consecration.' Paris, BN, MS fr. 193, fo. 346v.

24 These views are summarized in the marginal *Glossa ordinaria* on this passage which is based on John Chrysostom's homily on Matthew 6.6 (PG 57: 276–7).

25 Nicolas de Lyra, *Postilla litteralis et moralis* II, at notes g and h and the moralization at the end of the chapter.

26 *Enarratio in cap. I libri primi regum*, art 2; see note 88. Denis expounded on this theme again in his *De oratione*, art. 11, *Opera omnia* 41:33. 'Ostio uero clauso oramus, dum strictis labiis cum silentio supplicamus scrutatori, non uocum, sed cordium. In abscondito autem oramus, quando corde tantum et mente intenta petitiones nostras soli Deo effundimus, ita ut nec aduersae potestates ualeant genus nostrae petitionis agnoscere. Propter quod cum summo silentio est orandum, non solum ne fratres adstantes nostris susurris aut clamoribus impediamus, et orantium sensibus obstrepamus, sed ut ipsos quoque inuisibiles hostes, qui nostris orationibus maxime aduersantur, lateat nostrae petitionis intentio' (*Opera omnia*, 41:33).

27 'Car se aucun uoeult a aucun homme parler en secret, il y parleroit plus volontiers se il le trouvoit seul que s'il estoit en grant compagnie de gens. Ainsy quy voeult a Dieu parler secretement en oroison bien y doit prendre garde que il soit seul aveques Dieu. . . . Car comme monseigneur saint Augustin dist, Oroison est conversion de pensee a nostre seigneur [Pseudo Augustine, *Liber de spiritu et anima*, 50].' ('For if anyone wishes to speak with any person privately, he would speak more willingly if he found himself alone than if he were in a great company of people. Thus whoever wants to speak to God privately in prayer must take care that he is alone with God. . . . For as our lord St Augustine said, "Prayer is the conveyance of thought to Our Lord."') Paris, BN, MS fr. 190, fo. 104.

28 *Homilia III in I lib. Samuel*, in his *Opera quae supersunt omnia*, in *Corpus reformatorum*, 29(1885), pp. 270–1.

29 In modern terminology, the heart is associated with feelings and passion, and therefore heartfelt prayer signifies prayer accompanied by emotion. The use of the word 'heart' in such phrases as 'knowing by heart' and 'memorizing by heart' harks back to the medieval sense of the heart as the organ of cognition as well as of passion.

30 *De oratione*, art. 26, in *Opera omnia*.

31 On ancient and early medieval antecedents to the dichotomy between the physiological functions of the head and the heart, see Margarites Evangelides, *Zwei Kapitel aus einer Monographie über Nemesius und seine Quellen*, Inaugural

Dissertation (Berlin, 1882), and the commentary of William Telfer on Nemesius of Emesa, 'On the Nature of Man', in Telfer, *Curil of Jerusalem and Nemesius of Emesa*, The Library of Christian Classics, 4, (SCM Press, London, 1955), p. 347.

32 'L'unction du chief signefie espirituelle liesse au coeur du jeunent. Par le chief est entendu le coeur. Car si comme par les sens qui font au chief est governe l'homme de dehors. C'est a dire le corps, sembleblement est gouverne l'homme de par dedans, c'est a dire l'ame, et ses affections et ses pensees.' ('The anointing of the head signifies spiritual joy in the heart of he who fasts. By the head is understood the heart. For as by the senses that operate in the head man is governed from without - that is to say, the body – man is in the same fashion governed from within – that is, the soul – and his affections and his thoughts.') Brussels, BR, MS 9092, fo. 30. For a similar identification of the heart with the mind, see, note 23, paragraph 2.

33 'Et est le sang du cueur comme dit Aristote net, cler et chault et de greigneur puissance et moult convenable a l'entendement. Et pource que le cueur est le commencement de la vie de tous les sens et de tout mouvement ... Ains est la fontaine et le principe du sang. Toute delectation sensible et toute doulceur commence au cueur et retourne au cueur pource que sa vertu est extendue a tous les membres.' ('And the blood of the heart, as Aristotle says, is pure, clear and hot and of great strength and most convenient for understanding. And [this is] because the heart is the beginning of the life of all the senses and of all movement ... Indeed, it is the fountain and the origin of the blood. All delectation of the senses and all pleasurable sensation begin in the heart and return to the heart because its virtue is extended to all the members.') Robert Ciboule, *Le Livre de saincte meditation* (Paris, 1510), fo. 44v; see also fo. 40.

34 Paris, BN, MS fr. 139, fo. 21.

35 '... qui pourroit dire de bouche ou cuer le penser' ('which could be said by the mouth or thought in the heart'), Jean Gerson, *La Mendacité spirituelle*, in his *Oeuvres complètes*, ed. Palémon Glorieux, (10 vols, Desclée, Paris and NY, 1960–73), vol. 7, p. 317.

36 Describing the pains of hell, the author declared, '... tu yras sans respit au feu de fer et la seras tourmente plus cruellement que cueur ne pourrait penser ne langue exprimer' ('You will go without delay to the fire of hell and there you will be tormented more cruelly than the heart could think or tongue express') Paris, BN, MS fr. 449, fo. 48v. See also ibid., fo. 68.

37 Lynn Thorndike, *A History of Magic and Experimental Science* (8 vols, Macmillan, NY, 1923–58), vol. 4, p. 289.

38 See Alain Erlande-Brandenburg, *Le Roi est mort: Etude sur les funérailles, les sépultures et les tombeaux des rois de France jusqu'à la fin du XIIIe siècle* (Droz, Geneva, 1975), pp. 95-6 and 118.

39 'J'ay eu certes forte peur et grant crainte / Quant veu ta lettre de divers mots emprainte / Et si l'ay leuhe en *cuer* sans prononcer / Doubtant des dieux n'irer et offenser.' ('I of course took much fright and great fear / When I saw your letter imprinted with certain words / And I read it in heart without pronouncing / Fearing to anger and offend the gods.') Paris, BN, MS 875, fo. 124v. The literal meaning here is consistent with the definition of *coeur* suggested in L. Foulet, 'Pour le commentaire de Villon, *Romania*, 47 (1921), pp. 582-4.

40   Poitiers, BM, MS 95, fo. 41, which also contains the exposition on the modes of
prayer cited above. The same form of this rubric is in Paris, BN, MS fr. 13168,
fols 31–2. In other manuscripts the rubric reads 'de cuer et de bouche'
(Leroquais, *Les Livres d'heures*, vol. 1, nos 181 and 236; vol. 2, nos 28 and 155). In
no. 181 the rubrics of the entire book indicate a distinct orientation towards
oral prayer. A related rubric in yet another manuscript reads: 'Les trois verites
qui fault dire sans mentir de cueur pour soy mettre en estat de grace' ('The
three truths that one must say without lying in the heart to put oneself in a state
of grace'), Leroquais, *Les Livres d'heures*, no. 47 [fifteenth century], vol. 1, p. 126).

41   Paris, BN, MS nouv. acq. lat. 592 fo. 45v.

42   PL 82:81. Dhouda, *Manuel pour mon fils*, eds Pierre Riché and Claude Mondesert
(Editions du Cerf, Paris, 1975), p. 125.

43   'Car si comme il est escript Mt. V^a, Beati qui lungent, quoniam ipsi con-
solabunter, Benez sont ceulx qui plorent, car il [sic] seront confortez. Et mesmes
doibt venir devant Dieu en compunction de cuer cellui qui doibt ourer. Car si
comme dit saint Augustin, Compunctio excitat orationis affectum, c'est a dire
que compunction de cuer esmeut le desir de oreson. Et dist celui mesmes
doctour, Oratio cordis est, non labiorum necque uerba deprecantis dicitur
intendere Dominus, sed orantis cor aspicit, quoniam melius est cum silentio cordis orare quam solis uerbis sine intentu mentis. Internus [?] iudex mentem
potius quam uerba considerat. Cest a dire que oreson est du cueur et non pas
des levres. Et Dieu si n'entent pas en paroles du priant, mais il regarde le cuer de
l'omme, car miex vault ourer o silence de cuer que en soulez paroles sans le
regart de la pensee. Le juge de dedans, c'est Dieu, considère miex la pensee que
les paronc en plusieus maniers que lermez, plors et componction de cuer sont
bonne preparation a oreson.' ('For as it is written, Matthew 5, *Beati qui lungent*
. . ., Blessed are they that mourn, for they shall be comforted. And one who must
pray should even come before God in contrition of heart. For as St Augustine
says, *Compunctio escitat* . . ., that is, that contrition of heart stimulates the desire
for prayer. And the same doctor says, *Oratio cordis est* . . ., that is, that prayer is of
the heart and not of the lips. And God does not understand from the words of
the prayerful, but he looks to the heart of man, for it is better to pray in heart's
silence than in words alone without the regard of thought. The judge from
within, God, considers better thought than words. So it appears thus in several
manners that tears, lamentation and contrition of heart are a good preparation
for prayer.') Paris, BN, MS fr. 24, 748, fo. 33. The reference to Augustine seems
to be to an unidentified Pseudo Augustine text (see note 27) and the
unidentified sources of Bernardo Boils's Gastelian treatises on mental prayer;
see Cipriano Baraut (ed.), *García Jiménez de Cisneros, Obras completas* (2 vols,
Abadía de Monserrat, Monserrat, 1965), vol. 1, p. 16, note 38. The same text was
known but not explicitly attributed to Augustine by Smaragdus, *Diadema
monarchorum*, PL 102: 594–5, and Isidor, *Sententiae*, III, 7, 4 (PL 83: 672), both of
whom used the text to express views similar to patristic authorities (see note
24). Augustine, in his genuine works, assumed prayer to be both internal and
external; see *De cura pro mortuis*, v, and *De magistro*, i.

44   Denis the Carthusian, *De oratione*, art. 1, in *Opera omnia* 40: 14.

45   Paris, BN, MS fr. 1841, fols 141v–142.

46 In the brief exposition on prayer in Poitiers BM, MS 95, fo. 139v, we find the phrase *attencion mental et cordiale*.

47 '. . . et scias quod si istud breue siue ista nomina quinques dixerit uel in mente habuerit, illo die non potest mori mala morte . . .', fo. 28v.

48 The intimate relation of the eye to the heart is indicated in the following passage: 'Garde tes yeaux, car l'oeul est messenger de Dieu. Le premier messangier de luxure vision est. . . . Le cueur et la pensee sont pres de l'oeul, donc regard ton oeul.' ('Watch your eyes, for the eye is the messenger of God. The first messenger of lasciviousness is vision. . . . The heart and thought are close to the eye, so watch your eye.') Paris, BN, MS fr. 402, fo. 5v.

49 Fo. 32.

50 'Santus Leo Rome apostolus fecit hanc et misit ad regem Carolum et dixit quod quicumque hanc orationem dixerit uel uiderit, illo die non dubitabit de morte subtinea . . .', fo. 28v.

51 'Maria, o dilectissima est tibi tale oculum hunc audire uersiculum', fo. 171v. The scribe seems to have been reworking an older form of the prayer: see *Liber meditationum ac orationum devotarum* (Petrus Le Dru, Paris, 1502), fo. 69.

52 'Il n'est chose nulle quelconques qui plus ardamment puist esmouvoir le coeur de l'homme a devotion . . . comme par humble pensee souvent mettre devant les yeulx de ton coeur et de ta pensee la tres deloureuse mort et tres precieuse croix de nostre seigneur Ihesus. Pour plus longement tenir le coeur en suspension devote, piteuse et enflambee de contemplation la personne poeut mediter a toute heure par la maniere qu'il s'ensieult.' ('There is nothing that can more ardently move man's heart to devotion . . . than by humble thought often to put before the eyes of your heart and your thought the most dolorous death and most precious cross of Our Lord Jesus. To keep the heart in devout suspension longer, the person, filled with pity and enflamed by contemplation, can meditate at all times in the manner that follows.' (rubric from the *Livre de devotions* attributed to 'frere Bonaventure' of the Observance: Paris, BN, MS fr. 190, fo. 152). Denis the Carthusian relates Augustine's contemplation of the Psalms: 'Unde S. Augustinus agens in extremis, septem psalmos poenitentiales fecit in pariete scribi et oculis suis opponi; quos legens ac intuens, fleuit uberrime.' (*De particulari iudicio in obitu singulorum dialogus*, art. 38, *Opera omnia* 41: 481).

53 Perrine de la Roche, *Vie de Sainte Colette*, ed. P. Ubald d'Allençon, in *Archives franciscaines*, 4(1911), p. 224; see also the comments of Pierre de Vaux, ibid., p. 73.

54 ibid., p. 224.

55 See note 23.

56 '. . . la personne en disont ses heures, psaulmes ou autres oroisons verbales . . .' ('. . . the person while saying his hours, psalms, or other verbal prayers . . .'), fo. 139v.

57 Good examples of a book combining French rubrics with Latin prayers are Newberry MSS 41 and 47. For a Catalan example, see Newberry MS 39. For an example of English and Dutch rubrics, see Parkes, *Medieval Manuscripts of Keble College*, pp. 35-6. For further English examples, see the rubrics for Salisbury use

recorded from early printed editions in the notes of Wordsworth to the *Horae Eboracenses*. Among the prayers with vernacular rubrics and Latin texts are the prayers endowed with papal indulgences.

58   Paris, BN, MS lat. 8026, fos 243v-49. The verb *penser* referring to the act of prayer appears in the text on fo. 243v.

59   Paris, BN, MS fr. 984, fo. 35v.

60   'Et devez savoir, doulce cousine, que se pour aulcunes oroisons privees, qui sont de graces, soient salus de nostre dame ou aultres oroisons, vous laissez ou tardez aultre heure deue a dire les heures ou aultres choses qui sont de debte, ou vous laissez a aler a l'eglise aux droites heures, ou dictes vos heures privement pour plus dire choses privees, ou quant vous estes a l'eglise et vous devez chanter et porter vostre fais avecques les autres, se vous laissez a faire vostre devoir a l'eglise et ce a quoy vous estes tenu pour dire oroisons privees, en tous ces cas vous pechez moult griefment.' ('And you must know, sweet cousin, that if for any private prayers, which are for graces, either salutations to Our Lady or other prayers, you leave or put off to another due hour saying the hours or other things needful, or you fail to go to church at the proper hours, or you say your prayers privately to say more private things, or when you are in church and you must sing and carry your burden with the others, if you fail to do your duty in church and what you are held to for saying private prayers, in all these cases you sin most grievously.') Marie Brisson, *A Critical Edition and Study of Frère Robert (Chartreux) 'Le Chastel Perilleux'* Analecta cartusiana 19-20, University of Salzburg, Salzburg, 1974), p. 251.

61   Brisson, *A Critical Edition*, pp. 418-19.

62   Paris, BN, nouv. acq, fr. 4412, is an example of a vernacular liturgical book of the fourteenth century clearly related to a breviary, the rubrics of which indicate oral use.

63   'Je vous dy que pour avoir attention mentale a son oraison l'en pervient mieulx a obtenir envers Dieu nostre createur ce que prie et requiert a nostre seigneur qui regarde le cuer et non la parole. Et pour ce qui n'a ceste attencion mentale et cordiale, et celuy ne peut si bien pervenir, ne recevoir, le fruict ou l'effect de son oroison lequel fruict ou effect d'oroison s'entend en trois manieres: la premiere si est que la personne en disont ses heures, psaulmes ou autres oroisons verbales quiert et desire par telz chose servir Dieu son createur et luy plaire et avoir merite et en ce cas n'est pas necessaire l'adrecer mentale attencion en tout et partout ce que l'en dit. Mais souffist ce que dit est maintenant, c'est assavoir actuelle entencion ou habituelle seulement de servir Dieu et cetera, comme dit est. La seconde maniere de fruict ou l'effect d'oroison est que la personne desire et quiert empetrer pour soy ou pour aucun sien parent ou amy aucune grace envers Dieu nostre createur. Et par ce obtenir, souffist encore la premiere entencion du priant comme dit laquelle Dieu regarde principalement en rapportant ou reflectant son entencion a l'utilite d'icelluy pour qui on veult prier, et tousjours soubz la volonte divine.' ('I say to you that to have mental attention to one's prayer, one succeeds better in obtaining from God our creator what Our Lord who regards the heart and not speech begs and requires. And anyone who has not this mental and heart's attention cannot as easily arrive at,

nor receive, the fruit or the effect of his prayer, the which fruit or effect of prayer is to be understood in three ways: the first is that the person, while saying his hours, psalms, or other verbal prayers, asks for and desires by such things to serve God his creator and please Him and to have merit, and in this case it is not necessary to address mental attention to everything one says everywhere. But what is now said will suffice, to wit, actual or habitual intention only to serve God, etc., as has been said. The second manner of fruit or the effect of prayer is that the person desire, and begs for some grace from God our creator for himself or for some relative or friend of his. And to obtain this, the first intention of the praying person suffices, the which God looks at principally in relating or reflecting one's intention to the use of the one for whom one wishes to pray, and always under divine Will.') Poitiers, BM, MS 95, fos 139v-141.

64 Jean Gerson, *Oeuvres complètes*, vol. 2, pp. 185-6.

65 'La premiere [condicion de orayson est] que, se elle est vocale et proferee de bouche, elle doit estre attentive quant a troys actencions, se il est possible. La premiere quant a l'ordre et prolacon des paroles que riens on ne lesse ou on ne erre en la prolacon des paroles. La seconde quant a la significacion et au sens de paroles. Et ces deux actencions icy ne sont point de neccessite de oroyson pour deux causes. Premierement car les simples et sons lettres ne les peuvent avoir. Secondement, car telles actensions causent souvent vagacon de pensee. La tierce actencion est actencion de la fin de oroison qui est Dieu et la vie eternelle pour la quelle avoir on prie et est c'est actencion neccesser a oroyson non pas actuellement. Car ainsi que celui qui va en peregrinacion n'a plus actuelle actencion continuellement a sa fin, ou celui qui donne l'aumone aussi n'a celuy qui prie. Mais au moins fault que l'actencion soit habituelle et que elle soit meritoire en la vertus de l'actencion qui aura este au commencement de l'oroyson, ou devant comme la pierre qui est getee et est hors de la main du getant est menee en la vertu du movement que luy baille le getant quant il la tient. Et dit Hugues de Saint Victor que ceste tierce actencion est auccunefoiz si grande en oroyson que elle fait toutes aultres choses oblier [*De modo orandi*, ii; PL 176:980].' ('The first [condition of prayer is] that, when it is vocal and proferred by the mouth, it must be attentive to three attentions, if this is possible. The first concerns the order and pronunciation of the words, so that nothing is left out and no error is made in the pronunciation of the words. The second concerns the significance and the sense of words. And these two attentions are not needed for prayer for two reasons. First, as the simple and the unlettered cannot have them. Secondly, as such attentions often cause vacancy in thought. The third attention is attention to the end of prayer, which is God and life eternal, for having which one prays, and this attention is not actually necessary for prayer. For just as one who goes on a pilgrimage no longer has actual attention continually on its goal, or one who gives alms, so it is with one who prays. But at least attention must be habitual and meritorious in virtue of the attention that there was at the beginning of the prayer or before, just as the stone that is thrown and is out of the hand of the thrower is led by virtue of the motion that the thrower gives it when he holds it. And Hugh de Saint Victor says that this third attention is at times so great in prayer that it makes us forget all other things.') Paris, BN, MS fr. 193, fo. 348-348v.

66 Harry Levin with Ann Buckler Addis, *The Eye-Voice Span* (MIT Press, Cambridge, Mass., 1979), p. 36.

67 'Sic fideles conspicimus non praemeditata uerba quodammodo proiicere et balbutire inter precandum; atque quo simplicior eorum est oratio, eo certius testimonium non fictae aut simulatae qua Deum inuocant esse fidei.' John Calvin, *In lib. Sam.*, cap. 1, col. 270.

68 De la Roche, *Vie de Sainte Colette*, p. 224.

69 Jeanne Ancelet-Hustache, *La Vie mystique d'un monastère de dominicaines au moyen-âge d'après la chronique de Töss* (Perrin, Paris, 1928), pp. 54, 58 and 61; Elsbet Stagel, *Das Leben der Schwestern zu Töss . . .* , ed. Ferdinand Vetter (Weidmann, Berlin, 1906), pp. 18, 47 and 70.

70 In some books of hours, the serial recitation of the Pater noster or Ave Maria is incorporated within the text by means of cues typically inserted in red ink. In other instances, instructions for the recitation of these and other *formulae communes* were indicated either in the rubrics or in instructions following the principal text. The use of cues in books of hours defines the commonly expected limits of oral recitation and rote memory and gives a clue to the relatively small body of simple texts that, in the late Middle Ages, even the totally illiterate may have retained through rote memory.

71 See note 23.

72 'Et [tu] as garde silence en lieu et en temps [tu] as vacque en lecon, oroison et au service divin sans toy distraire ne habandonner a quelconques autre plaisir .' ('And [you] have kept silence in the place and the time that [you] have spent in reading, prayer and at divine service, without distracting yourself or abandoning yourself to any other pleasure.') Paris, BN, MS fr. 449, fo. 104v. For an example of an aversion to singing, see the English reformer Thomas Bilney, DNB, II, p. 503.

73 'La tierce maniere du fruict et effect d'oraison est quant la personne bien humble quiert par la grace Dieu avoir en soy refection spiritualle, c'est a dire sentement spirituel par devote et fervente meditacion et contemplacion. Et pour parvenir a ceste grace est de necessite avoir son entendement cordial en son oroison en tout et partout. Et a ce fruit se doivent plus exciter gens d'eglise et encore plus gens de religion, car y parvenir et y devouer ou y'estre bien habitue au milieu du monde soudoit fort. Aussi, il nest pas de necessite a chacun.' ('The third manner of the fruit and effect of prayer is when the person, most humble, asks by the grace of God to have in himself spiritual renewal – that is, spiritual sentiment by devout and fervent meditation and contemplation. And to accede to this grace it is necessary to have cordial understanding in one's prayer in all things and places. And to this fruit persons of the church must exert themselves, and, even more, religious, for to achieve this and devote oneself to this or be well habituated to this in the middle of the world is most arduous. Thus it is not a necessity for all.') Fo. 140v.

74 Gerson, *De orationis conditione atque sanctitate*, in his *Oeuvres complètes*, vol. 1, p. 187.

75 Wessel Gansfort, *Opera* (Groningen, 1614), pp. 13-17.

76 Guillaume Fillastre in his *Thoison d'or*, prepared for Charles the Bold, declared: 'En oultre, sapience n'est pas acquise seullement par oyr, mais aussi se acquient

et augmente par estude, par lire et par subtillement penser et mediter a ce qu'on a leu et estudie, pour ce dit le saige: Estudie mon filz, estudie sapience car par ce tu me resjoyera le cuer [Proverbs 23. 12–15]. Les livres ne sont pas en vain ou pour folie donnez aux hommes, mais pour pure neccessite, car ils sont faiz pour suppleer et secourir a la foiblesse de la memoire qui flue et coule comme l'eau par le russel. Parquoy peu proffiteroit oyr ou interroguer pour apprendre se memoire ne le retient. Or ne souffist memoire pour tout retenir pour ce qu'elle est habille comme dit est. Parquoy a est necessaire l'etude des livres pour retenir ce qu'on apris par encquerir et par oyr. Es livres aussi sont trouvees souvent doctrines non oyes parquoy l'homme peut apprendre et retenir par livre et estudier science et sapience sans docteur ou instructeur. Car trop plus ferme est le sens de la veue que le oye et rent l'homme trop plus certain pour ce que la parolle est transitoire, mais la lettre escript demeure et plus se imprime en l'endendement du lisant.' ('Furthermore, knowledge is not acquired by hearing alone, but also is acquired and increases by study, by reading and by subtly thinking and meditating on what one has read and studied, for as the sage says: Apply your mind to instruction and your ear to words of knowledge. . . . because in this manner you will make my heart rejoice. Books are not given to men in vain or for amusement, but out of pure necessity, for they are made to supplement and come to the aid of the weakness of memory, which flows away and runs like water in the stream. By which it would profit little to hear or to ask question to learn if memory does not retain it. Thus, for all its skill, as it is said, memory does not suffice for retention. This is why the study of books is necessary in order to retain what one has learned by inquiry and by hearing. In books there are also often found doctrines not heard by which man may learn and retain by reading and studying knowledge and wisdom without a teacher or instructor. For the sense of sight is much firmer than hearing and makes man much more certain, because the spoken word is transitory, but the written letter remains and impresses itself more in the understanding of the reader.') Paris, BN, MS fr. 140, fo. 98v.

77 'Car il n'y a rien en quelque bonne science qui ne puisse servir en temps et en lieu. Et si tu ne peux tout lire (car les livres sont preque infinis) a tout le moyns estudies toy a lire les plus utiles et qui plus fait a la bonte des meurs. Et te doibs plus employer et estudier a savoir les choses qui plus sont proffitables a ton salut. A cestuy enseignement doibvent prandre garde ceulx qui sont en l'escole de predicacion et qui frequentent les sermons ou predications. Ilz ne doubvent jamais despriser la doctrine ou predication quelle que ce soit tant soit simple ou nue ou petitement fondee, mais quelle ne soit scandaleuse ou exornee. Car en une predication, y a gens de tous estatz: sciens et ignorans, grans et petis, simples et aultres, et chacun doibt faire son proffit de tout. Example avons de nostre maistre Ihesu Crist qui en presences des apostres qui estoient plus illuminez prechoit souvent en simple langage et par paraboles de choses comunes.' ('For there is nothing in some good knowledge that cannot serve in its time and place. And if you cannot read everything (for books are nearly infinite), at least study to read the most useful and [the ones] that do most for goodness of comportment. And you should employ yourself more and study to know the things that are most profitable to your salvation. Those who are in the

school for preaching and who frequent sermons or predications should pay attention to this teaching. They should never scorn doctrine or preaching whatever it may be, no matter how simple or elementary, as long as it is not scandalous or exaggerated. For at sermons there are people of all conditions: learned and ignorant, great and humble, simple and others, and each has to take profit from everything. We have the example of our master Jesus Christ who, in the presence of the apostles, who were more learned, often preached in simple language and by parables about common things.') Ciboule, *Le Livre de sainct meditation*, fo. 4v.

78 'Le ix[e] degre [de humilite] est tenir silence jusques a ce que on soit interroge qui est contre ceulx qui habondent en langaige et delutent estre ouys. Certes est plus seure choses ouyr parler que parler et n'y a si saige qui ne doye avoir crainte a parler longement, soit en exhortacion, sermon, predication, lecon ou narration. Celui aussi ou celle qui prent plaisir a parler et quaqueter impertinentement sans edification n'est humble quelque apparence que il en monstre par dehors' ('The ninth degree [of humility] is to keep silence until one is interrogated, which is against those who abound in speech and delight in being heard. Certainly, it is surer to hear things said than to speak, and there is no wise person who must not have fear of speaking at length, either in exhortation, sermon, predication, reading, or narration. Also the man or the woman who takes pleasure in speaking and chattering impertinently without edification is not humble, no matter what appearance he shows externally.') Robert Ciboule, *Le traictie de perfection*, Paris, BN, MS fr. 1841, fo. 75.

79 Gerson, *Moralité de la passion*, in his *Oeuvres complètes*, vol. 7, p. 143.

80 Such scenes were also included in early printed books of hours; see for example Wordsworth (ed.), *Horae Eboracenses*, p. 80.

81 Giacomo Lercaro, *Metodi di orazione mentale* (2nd edition, Bevilacqua e Solari, Genoa, 1957), p. 31; Merton, *Contemplative Prayer*, p. 75; Barrault (ed.), *Cisneros, obras completas*, vol. 2, *passim*.

82 'Oraison est une conversation et familiere collocution de l'ame avec Dieu qui ne se peut proprement difinir ne humainement comprendre, car c'est une elevation que l'ame fait d'avec les choses crees et temporelles pour adherer aux eternelles dont elle est capable par son immortel estre. Quand donc elle s'adresse a son facteur elle oblie les faultez et fonctions du corps ou elle est associée a raison de leur corruption que est cause que les afections du corps ne la peuvent suyvr.' ('Prayer is a conversation and familiar dialogue of the soul with God that cannot properly be defined nor humanly understood, for it is an elevation that the soul makes from created and temporal things to adhere to those eternal, of which it is capable by its immortal being. When it thus addresses its maker, it forgets the faults and functions of the body to which it is associated because of their corruption, which is the reason why the affections of the body cannot follow it.') Paris, BN, MS nouv. acq. Lat. 82, fo. 3.

83 Lercaro, *Metodi*, pp. 177–94.

84 Alphonsus Rodríguez, *A Treatise of Mental Prayer* (Douay [?], 1630, p. 153).

85 Calvin, *In lib. Sam.*, cap. 1, col. 270. The Protestant prejudice against the repetition of set formulas in prayer is exemplified during the Reformation by the deletion from books of hours of rubrics offering papal indulgences; see, for

example, Newberry MS 35, fos 94 and 98, and Newberry MS 82, fos 152v and 166v.

86 'nous commandant que quant nous vouldrons pryer, nous pryons ainsy, Nostre pere qui es, *etc*. Et pour ce nous devons dire ceste oraison avec une tres grande reverence et humilite de cueur et tres grand ferveur d'esprit, en pensant tous les mots qui sont en la dicte oraison pour l'honneur decellui qui nous a baille la forme de ainsy pryer. En quoy jusques a maintenant les brebis ont este tres mal instruictes par la grande negligence des pasteurs qui les debvoient instruire de pryer en languaige qu'on entendist, et non pas ainsy seulement barboter des levres sans riens entendre. Car comme dit saint Pol, Si je prie de langue, mon endendement est sans fruist [I Corinthians 14. 19]. Et pour tant il commende que tout ce que on dist en la congregation des fideles qui est l'eglise qu'on le dist en languaige que tous l'entendent autrement qu'on se taise. Et se on eust bien observe ce commandement jamais si grandes tenebres ne nous feussent advenues. Car on prieroit encore le pere celeste en soy, es cieulx, en esperit et verite et non pas es creatures en la terre charnellement et en vanite.' ('commanding us that when we wish to pray, we pray thus: Our Father, who art, etc. And for this reason we must say this prayer with a very great reverence and humility of heart and very great fervour of spirit, thinking all the words that are in the said prayer for the honour of Him who gave us the form of praying thus. In which up to now the sheep have been very badly instructed by the great negligence of the pastors who must instruct them to pray in language that one understands, and not simply mumbling with the lips without understanding anything. For as St Paul says, "For if I pray in a tongue, my spirit prays but my mind is unfruitful." And yet, he recommends that all that one says in the congregation of the faithful that is the Church be said in a tongue that all understand, or else keep silent. And if one had observed this commandment well, such great misfortunes would never have happened to us. For one would still pray to the celestial Father in the heavens, in spirit and truth, and not to creatures on earth, carnally and in vanity.') Paris, BN, MS fr. 19246, fo. 4–4v.

87 ibid., fo. 10.

88 'Moraliter ex isto capitulo edocemur exemplo Annae. ... In omni quoque angustia ad Deum confugere confidenter, et coram eo cor nostrum effundere orando instanter, et taliter exorare cum disciplina atque silentio, ne sibilo oris et gestibus indisciplinatis, aliorum deuotionem impediamus, aut eorum orationes interrumpamus; imo qui taliter orat iram colligit magis quam ueniam, secondum Chrysostomum [see PG 57: 276–7]; imo talis Deum offendit et proximam, et sua inquietudine ac indescritione grauiter laedid se ipsum.' *Enarratio in cap. 1 libri primi regum*, art. 2, in *Opera omnia*, 3:258. See also his *De oratione*, art. 27, in *Opera omnia*, 41:53.

89 See R. Cresswell, 'Le geste manuel associé au langage', *Langages*, 10 (1968), pp. 119–27.

90 *De oratione*, art. 25, in *Opera omnia*, 41:51.

91 Ludovico Barbo, *Metodo di pregare e di meditare*, Scritti monastici editi dai Monaci Benedettini di Praglia (Padua, 1924), p. 58.

92 On these iconographic changes, see François Garnier, *Le Langage de l'image au moyen-âge: Signification et symbolique* (Léopard d'or, Paris, 1982), pp. 128 and 212–

13; Louis Gougaud, *Devotional and Ascetic Practices in the Middle Ages*, tr. G. C. Bateman (Burns, Oates and Washbourne, London, 1929), ch. 1, 'Attitudes of Prayer'; G. B. Ladner, 'The Gestures of Prayer in Papal Iconography of the Thirteenth and Early Fourteenth Centuries', in *Didascaliae: Studies in Honor of Anselm M. Albareda* (B. M. Rosenthal, NY, 1961).

93 For example, Moulins, BM, MS 89, fo. 16 (Paris, BN, Collection Porcher, boîte no. 35), and Newberry MS 43, fo. 27.

94 For example, Paris, Arsenal, MS 434, fo. 53, and Lyons, BM, MS 431, fo. 583 (Paris, BN, Collection Porcher, boîte no. 34).

95 *Vita Sanctae Coletae* with commentary of Charles Van Corstanje, Yves Cazau, Johan Decavele, and Albert Derolex (E. J. Brill, Leiden, 1982), p. 201.

96 This posture was known since the mid-thirteenth century; see, for example, Erlande-Brandenburg, *Le Roi est mort*, figures 115-16, and Ladner, *Gestures of Prayer*, fig 14, but its popularity increased markedly in the period after 1350.

97 Hubert Jedin and John Doland, *Handbook of Church History* (4 vols, Herder and Herder, NY, 1965), pp. 570-2.

98 *Statuta legenda in concilio Oxoniensi edita per D. Stephanum Langton A.D. 1221*; Giovanni Domenico Mansi, *Sacrorum conciliorum nova et amplissima collectio . . .* (52 vols, H. Welter, Paris, 1901-27), XXII, col. 1175.

99 See S. J. P. Van Dijk, *Sources of the Modern Roman Liturgy* (2 vols, E. J. Brill, Leiden, 1963), vol. 1, p. 40.

100 See A. K. Pugh, 'The Development of Silent Reading', in W. Latham (ed.), *The Road to Effective Reading: Proceedings of the Tenth Annual Study Conference of the United Kingdom Reading Association* (Totley, Thornbridge, 1973), pp. 111-14. The reading habits traditionally associated with books of prayer brought to the Mass are exemplified by the instruction contained in an early eighteenth-century example of such a text as the *Heures, prières et offices à l'usage et devotion particulièrement des damoiselles de la maison de Saint Louis à Saint Cyr* (chez Jacques Collombat, Paris, 1714), p. 16: 'L'on ne pretend pas obliger de faire a toutes les messes toutes ces considerations, mais seulement d'en fournir quelque idée; ni d'en reciter toutes les prières, mais d'en donner seulement quelque modèle, chacune suivant sa devotion particulière.' ('One does not claim to obligate carrying out all these considerations at all masses, but only to furnish some idea of them; nor to recite all the prayers, but to give merely a few models for them, each [*damoiselle*] following her own devotion.')

101 *Bibliotheca franciscana ascetica medii aevi*, IV (Quaracchi, 1904).

102 'Or pouvons nous bien veoir par raison que c'est plus noble chose de penser devotement a nostre seigneur tandis qu'on dit la messe que de dire heures ou autres choses.' ('Thus we can see well by reason that it is a nobler thing to think devoutly on Our Lord while one says the Mass than to say hours or other things.') Fo. 107v.

103 Paris, BN, MS fr. 2439. The 'Oroison contemplative pour la messe' falls on fos 22-41.

104 Fo. 170.

105 The full rubric reads: 'Quiconques vouldra oyr la messe en contemplant les misteres qui y sont representez sans ren dire de bouche pourra tenir tel ordre', fo. 237.

106 Fo. 108.

107 Fo. 16. The treatise begins: 'Premierement quant on oyt sonner la messe, on doit penser . . .' ('First, when one hears the bell ring for mass, one should think'). Specific vernacular prayers are provided for the principal divisions of the ceremony.

108 This tract is contained in Paris, BN, MS fr. 24439, fols 38–41v. The rubric preceding it reads: 'S'ensuivent aucunes devotes meditacions que la personne doit mediter en oyant la sainte messe.' ('There follow some devout meditations that the person must meditate on hearing the holy Mass.') The incipit is: 'Pour oyr messe devotement au profit de son ame est a noter selon ce que dit le docteur devot en son quart livre de sentences, Que celuy ot la messe prouffitablement, non pas celuy qui escoute les motz de la messe, mais qui met son cueur et sa memoire a remembrer la benefice de la passion nostre seigneur [see Peter Lombard, *Sententiae*, bk 4, dist. 9].' ('In order to hear Mass devoutly to the profit of one's soul is to note what the devout doctor says in his fourth book of sentences, that "He hears the Mass profitably, not who listens to the words of the mass, but who employs his heart and his memory to remember the benefice of the Passion of our Lord." '). The treatise ends: 'Je vous supply que veuillez prier pour le povre pecher qui ce pitit . . . dicte . . . a escript. C'est le povre et indevot frere Olivier.' ('I beg you to pray for the poor sinner who wrote this small text. He is the poor and unworthy brother Olivier.') The following text in this *recueil*, also attributed to 'frere Olivier', has been identified as a work of Olivier Maillard, for whom a comprehensive bibliographical study is sorely needed.

109 'Et notez, que tout plus devotement et en plus grant ferveur l'on dit ces choses ycy en son cueur que l'en peult plaire a Dieu. Et se pourroiet faire alcuneffoiz si devotement que celuy qui escoute la messe profiteroit plus que celuy qui la dit.' ('And note that the more devoutly and fervently one says these things in one's heart, the more one can please God. And it could at times be done so devoutly that the person listening to the Mass would profit more than he who says it.)) Paris, BN, MS fr. 24439, fo. 41.

110 Van Dijk, *Sources of the Modern Roman Liturgy*, vol. 1, pp. 68–94.

111 For a general overview of the kinds of prayer frequently contained in private books of devotion, see Leroquais, *Les Livres d'heures*, vol. 1, pp. xxix-xxxii, and Sonet, *Répertoire d'incipit*.

112 Edward Waite Miller, *Wessel Gansfort, Life and Writings*, principal works tr. Jared Waterbury Scudder (2 vols, G. P. Putnam's Sons, NY and London, 1917), vol. 2, pp. 3–4.

113 W. A. Pantin, 'Instructions for a Devout and Literate Layman', in J. J. G. Alexander (ed.), *Medieval Learning and Literature: Essays Presented to Richard William Hunt* (Clarendon Press, Oxford, 1976), pp. 398–420.

114 Paul Lacombe, *Livres d'heures imprimés au XVe et au XVIe siècle conservés dans les bibliothèques puibliques de Paris* (Imprimerie nationale, Paris, 1907), p. lxxvii.

115 See, for example, Brisson, *A Critical Edition*, p. 236.

116 See Leroquais, *Les Livres d'heures*, vol. 1, pp. xxxiv-xxxv.

117 See, for example, Paris, BN, MS fr. 2439, fo. 77.

118 For example, 'Et scias quod istud breue est utile super feminam parientem; liga

illud breue super eam et statim pariet sine periculo; et scias quod qui istud secum deportauerit non dubitabit suum inimicum nec aliquod periculum nec in uia, nec in aqua, nec in bello uictus, nec in turnamento, nec coram iudice, sed semper superiorem manum habebit.' (Newberry MS 104.5, fo. 28v.; 'Sciendum est quod quicumque hos uersos quotdidie dixerit et eos secum porauerit nullo modo morietur inconfessus.' (Newberry MS 104.5, fo. 41v). For similar examples, see Leroquais, *Les Livres d'heures*, vol. 1, pp. 3 and 233; vol. 2, p. 208. For another example from the end of the fifteenth century, see Rouen Bibliothèque Municipale, MS 3027, fols 156v-157v.

119   Erotic scenes developed particularly as frontispieces to Book IX of the French translation of Valerius Maximus, *Facta et dicta memorabilia*; see, for example, Paris, BN, MS fr. 20320, fo. 318v, and especially MS fr. 289, fo. 414v. Equally erotic hand-painted scenes decorate some of the surviving copies of the first incunable edition of this text, for example, Paris, BN, Rés. Z 210 and Rés. Z 203.

120   See, for example, Rouen, BM, MS 3027, fo. 86. In other manuscripts, a transparent towel revealed intimate details of Bathsheba's figure; see Paris, Arsenal, MS 428, fo. 93, and Walter's Art Gallery MS W 449, fo. 76.

121   In the wall fresco of the Last Judgement in the cathedral of Albi, prayer books discreetly cover the private parts of the naked dead, which were often explicitly depicted in the actual prayer books of that period.

122   The austerely decorated books printed by Christopher Plantin may be contrasted to the sensuality displayed in the illuminations and woodcuts of earlier manuscripts and printed editions.

5

From Ritual to the Hearth: Marriage Charters in Seventeenth-Century Lyons

ROGER CHARTIER

Pierre Bénier wedded Martine Palatin in Lyons on 11 May 1631. The ritual governing the ceremony dated from 1542, but the better part of it was older and based on texts printed in 1526 and even before, in 1498. Under the porch outside the church the priest, with the man at his right and the woman at his left, proclaimed the banns a fourth time, asked the couple if there was any impediment to their union, then blessed *annulum et chartam totum simul* – 'the ring and the charter together'.

The ritual mentions this *charte* several times. It was blessed and held by the marrying couple, along with the ring, during the rite of the joining of their hands and the priest's recitation of the blessing of Tobias: *Deus Abraam, Deus Ysaac et Deus Jacob: ipse vos conjugat*. The document is mentioned explicitly in the formula in French recited in unison by the priest and the husband (before 1542, the bride had recited it with them):

In the name of the Holy Trinity, of the Father and of the Son and of the blessed Holy Ghost, I, ——, wed you, ——, with this ring and with this charter as God has said and St Paul has written and the law of Rome confirms; and from this day I recommend to you my benefactions and my alms.

The priest then pronounced the words, *Quod Deus conjuxit homo non separet. In nomine Patris et Filii et Spiritus Sanctus*. The husband then gave the charter to his bride, just before placing the ring onto the fourth finger of her right hand. A final prayer was said, and the bride and bridegroom were asperged with holy water before they entered the church.

The charter that Pierre, master hat-maker, gave to Martine when he took her to wife in 1631 is conserved in the Archives des Hospices civils in Lyons. It contains both text and images, and at the centre of the sheet, used widthways, there is a formula printed in black and red letters, leaving space to add the names of the couple and the date of their marriage. The text reads:

In the name of the Holy and Individuate Trinity, Father, & Son, & Holy Ghost, I [Pierre Bénier] take you for my wife, whose name is [Martine Palatin] And I recommend to you my alms, as God has said, St Paul has written, and Roman Law confirms. May man not separate what God has conjoined. Given in [Lyons] in the year of our Lord one thousand six hundred [thirty-one] the [eleventh] of the month of [May 1631].

Beneath this formula is the signature of one Aubert, the priest who celebrated the marriage and filled in the blanks.

The tablet containing this text, which repeats the wording of the rite, is surrounded by a hand-coloured decorative woodcut divided into sections. The four evangelists occupy the four corners of the sheet; in the centre at the very top, God the Father looks down from his clouds at Adam and Eve in Paradise; the marriage of the Virgin is placed symmetrically at the bottom of the sheet. The Angel of the Annunciation decorates one side of the written portion, the Virgin Mary the other. There is no doubt that for Pierre and Martine this object must have borne a high emotional charge. It recalled a decisive moment in their lives, and it carried the memory of the rite that had sanctioned their union into the intimacy of their home. It brought into the domestic sphere images of some of the major figures of Christian iconography familiar from the stained glass windows, altar-pieces, sculptures, and many paintings in the city's churches. A piece of printed matter not destined for 'reading' and of plural intentions and uses, the marriage charter, like confraternity images or pilgrimage certificates, was an object closely associated with the life of its possessor. It was both a lasting witness to an essential engagement (here marriage; elsewhere confraternity membership or a devotional voyage) and a support to everyday piety (plates X to XIV).

Painted Images and Printed Images

The archives of the Hospices civils in Lyons contain a collection of 174 wedding charters.[1] This study is founded on the idea that materials of this sort, which belonged fully to print culture, also had an essential place in the ways in which Christianity was once lived and conceived of. For several years now, historians have shown an interest in religious imagery between the sixteenth and the eighteenth centuries, but their attention has focused on one particular form, the painted picture, in its various genres, altar pictures, chapel paintings, ex-votos, or the reredos.[2] This has led historians to favour a specific class of images governed by pictorial aesthetics and best viewed from a distance. In general they bear no written text, or, like the ex-voto, give only an extremely concise formula and a few summary identifying elements (the year, a name), and they were thought by the people who commissioned them and those who made them to be a means of teaching Christian history and Christian truth to the illiterate.

The printed religious image by no means shared the same set of constraints and intentions. It was made to be manipulated, to be given to one person by another, to be used in ritual or devotion, to be kept in a secure place or exhibited on the wall. It was thus a familiar object, close to hand and manipulable. Furthermore, it combined writing and images. Some paintings were based on texts, but the text remained outside the object and was not necessarily familiar to all who viewed the picture. Unlike them, the printed broadsheet combined on the same page a text (ritual, devotional, commemorative, etc.) and the iconographic representations linked with it. The relation between writing and the image was thus profoundly changed, facilitating transfer from the one to the other and encouraging entry into written culture even for people who were not truly literate. Such texts were encountered daily; they were known because they were heard when they were repeated and deciphered by those who could read for those who could not. Not only did they undoubtedly constitute one of the major routes to Christian acculturation and familiarization with Scripture, but this learning took place outside the schools and without reliance on their tightly regulated methods.

The marriage charters, like confraternity images, undeniably transmitted the teachings of the Church and, by that token, may be considered religious pedagogical materials. It is also certain, however, that their quite special link to the individual and his or her life choices, as well as their form, which permitted the individual to leave his or her mark on the object, created a familiarity with printed broadsheets that was impossible with painted church furnishings. The restricted number of fragile and ephemeral objects of this sort that remain to us and their low commercial value should not lead us astray. They were printed in large quantity; they were owned by common people everywhere; they were installed in the home. Thus they lay at the heart of *ancien régime* culture, which depended upon the products of the printing presses as repositories for memory, preserves for the spoken word, and supports for pious acts.

The Charter and Ritual

In Lyons, the existence of the nuptial charter was closely tied to the modalities of diocesan marriage ritual, which stipulated that the wedded couple have it in hand and that the priest bless it. The charter as an object, however, seems to have appeared before it became an obligatory part of ritual, since the oldest attested mention is in 1453 and the oldest extant example is dated 1465, whereas documented ritual before 1498 mentions only the giving of the ring. Furthermore, although the marriage charter figures in Lyons rituals in 1667, 1692, and 1724, the latest extant charter is dated 1691.

Marriage liturgy thus did not wholly control the use of the nuptial charter, which had its own reasons for being. Historically speaking, it gave spiritual form to the ancient practice of the *charte de dotation* (charter of endowment) by which the husband constituted his wife as mistress of his possessions and guaranteed her marriage portion. The 1667 ritual reflected this shift when it introduced into the diocese (where it did not previously exist) the custom of the husband's gift to the wife of 13 *deniers*, a clear symbol for the actual endowment of the wife by her spouse. Looking forward in time, the marriage charters disappeared sooner in actual social use than in the text of the rites, because their practical utility declined at the end of the seventeenth century. Until that time, they had served as legal proof of marriage and could attest to the legitimacy of a union – which is what explains their presence in the archives of the Hôpital général de la Charité.

In the seventeenth century, by the rules of the institution, children were taken in only if they were true orphans with neither father nor mother, had reached seven years of age but were not yet fourteen, and had been born of the legitimized marriage of parents who were citizens of Lyons or had resided in the city (excluding its *faubourgs*) for at least seven years. Along with other papers – the child's baptismal certificate, attestation to the death of his or her parents, and a certificate of residence delivered by the captain of the quarter (called *pennonage* in Lyons) – the nuptial charter could be presented in lieu of a marriage contract to authenticate the legitimate union of Lyonnais parents who had passed to their reward.[3] Thus it could prove to be a necessary document to get an orphaned child accepted into the Hôpital général. (Another institution, the Hôtel-Dieu, opened its doors to illegitimate children and to children under seven years of age.) Families thus kept the document carefully. But when the ordinance of 1667 provided for deposit of a copy of parish registers with the secretariat of the royal courts, it became easier to obtain extracts (on payment of a fee to the parish priest or the official who delivered the copy), and from that moment on the marriage charter became less necessary. Furthermore, it infringed upon the rights of the keepers of the registers, the clergy first among them. This doubtless explains the slow disappearance of a custom (and an object) still recalled in ritual but no longer part of social practice.

At its height, was the marriage charter unique to the diocese of Lyons? First, it should be noted that the French wording of the formula, required by ritual and inscribed on the charters themselves, is frequently encountered in southern France in regions in which the exchange of vows was active (the couple pronouncing the formula themselves) and mutual (pronouncing it together). In Lyons the ceremonial underwent a two-fold evolution. First, at the end of the fifteenth century, when unison recitation of the formula by which husband and wife promised themselves to each other first appeared, it replaced an older usage in which only the husband responded to the priest's question, 'Is it your pleasure to take to wife and spouse ——, here present?'. Later, in 1542, the husband alone recited the French text, which meant that the words of the rite conformed with the formula written on the charters, in which only the man was portrayed as taking an active part. Although this brought Lyonnais ritual back to a uniquely male declaration, it remained, up to the early eighteenth century, a ritual of active donation in which the priest intervened only to pronounce the blessing of Tobias at the rite of the joining of hands, and the Biblical *Quod Deus conjuxit homo non separet*, the formula that concluded the union of the couple.

Although the formula of active donation was widespread in southern France in the dioceses of Provence, Languedoc, and the Massif central, and although it even persisted after the decree *Tametsi* of the Council of Trent had shifted sacramental authority to the priest's words (*Ego vos in matrimonium conjugo in nomine Patris et Filiis et Spiritus Sanctus*), the husband giving a nuptial charter to the wife is mentioned specifically and with some consistency in only thirteen diocesan rituals during the sixteenth century. Three of these dioceses were in the Rhône valley (Lyons, Vienne, Valence), one in Provence (Aix), six in the province of Bourges (Limoges, Clermont, Saint-Flour, Mende, Rodez, Vabres), to which the dioceses of Périgueux, Angoulême and Bazas should be added (the latter was an exception in Aquitaine).[4] Perhaps effective use of a marriage charter – called, according to the region, *charte*, *carta*, *cartula*, *pagina*, or *littera* – was observed unequally. The fact remains that over a large part of southern France a written and pictorial object, at first a woodcut and later a copperplate, was invested with a highly charged ceremonial significance. It stood at the heart of the matrimonial rite and, at least for Lyons, it was valid for the certification of marriages contracted in the city.

The Marriage Formula

The formula used in marriage charters in Lyons, given first in Latin, then in French, keeps closely to the text of the ritual itself, but with significant differences. The written version combines the words said by the wedded couple (later by the husband alone) and the priest, arranging them, however, in an order of its own. Whether the formula was printed, engraved, or in calligraphy, it contained the same elements:

1 The invocation, which showed two changes from the spoken formula. The Trinity is called *Sainct et Individuée*, perhaps for apologetic and anti-Protestant reasons, and the Holy Ghost looses its qualification as *benoît*, perhaps considered too archaic.

2 The formula of the vow, *Moy* — *te prends pour femme qui te nommes* —, which is closer to the ritual when the words *Je* — *espouse toi* — *avec cestuy annel et avec ceste chartre* are said only by the husband. I might note that certain charters offer a variant of the initial formula, which becomes *Au Nom de la Saincte Trinité, du Père, du Fils et du Sainct Esprit Amen. Je* —*prends pour ma femme et loyalle épouse, vous* —, which substitutes the more formal form of direct address, introduces the notion of loyalty (mentioned in several rituals of active donation), and suggests, by ending the invocation with *Amen*, the close tie between the written text and ritual recitation.

3 The statement, *Et je te recommande mes aumônes*, which gives spiritual and charitable form to the endowment of the wife by the husband. The charters add the three-part formula, *comme Dieu a dit, saint Paul a écrit et la loi romaine confirme*, which ritual placed with the declaration of mutual (later only the husband's) engagement.

4 The Biblical words, given here in French, that were pronounced by the priest: *Que l'homme ne sépare ce que Dieu a conjoint.*

There is variation in the way the two formulas, separated in the actual rite, are combined. According to the charters, for example, the three-part reference to God, St Paul, and Roman law is at times attached to what precedes – the matrimonial vows and the promise of alms (or only the latter) – and in other instances to what follows – the Biblical phrase pronounced by the priest. Variations in punctuation and the use of capital letters express these choices by placing a period between *confirme* and *Que* (as in our 1691 marriage charter), or breaking the sentence between *aumônes* and the phrase *Ainsi que Dieu l'a dit, Saint Paul l'a escrit et la loy de Rome le confirme ce que Dieu a conjoinct l'homme ne peut séparer*. The accent shifts somewhat from one formulation to the other, since the reference to the authority of Scripture and to the Church, in the one case, recalls that marriage is founded on the engagement and donation of the couple, whereas in the other, it underscores the indissoluble nature of marriage conferred by the Biblical obligation, as stated by the Church.

The Popularization of an Object

A survey of the sociological aspects of the use of marriage charters in the diocese of Lyons has been made possible by P. B. Berlioz's identification of the social status and the occupation of husbands mentioned in them.

Working with the archives of the Hôpital de la Charité and the parish registers of the city, and having consulted the charters of the Hospices civils of Lyons, but also those of other public and private collections, Berlioz has found the distribution of occupations among 296 documentable husbands in the seventeenth century to be as follows:[5]

notables (nobles, officials, liberal professions):	11.1%
merchants:	13.1%
leather, foods, building, wood, and metal trades:	32.0%
textile, clothing, hat trades:	19.5%
silk trades:	12.5%
unspecialized workers (men-of-all-work, carters, boatmen, day workers)	11.4%

In the seventeenth century, which was the period of greatest use of the nuptial charter, the object was thus present in all strata of the population of Lyons, and its obligatory presence in the marriage ritual seems in fact to have been respected. In spite of the special bias of the documentation that has been preserved (predominantly charters from couples whose orphaned children were adopted by the Hôpital général, thus who might be supposed not to belong to the highest social spheres), the social identity of the possessors of charters can be taken as a fairly faithful reflection of the distribution of the population of Lyons as a whole, as indicated in marriage contracts in the following century, where we find 9 per cent notables, 7 per cent merchants, 71 per cent artisans (of whom 27 per cent were in the silk trades) and 12 per cent unspecialized workers.[6] We can then conclude that whatever their social status, all married men and women in Lyons in the seventeenth century – before 1670, in any event – owned, preserved, and handled an object of the sort. In some of its forms, print culture was thus by no means the exclusive privilege of the elites, nor of the literate, since a number of illiterates were thus made familiar with texts they could not read but had heard and spoken.

If marriage charters were universally present in Lyons in the seventeenth century it was because of the 'popularization' of an object. When it first appeared in the late fifteenth or the early sixteenth century, the marriage charter seems to have been used primarily among the families of nobles or rich notables. This is attested not only by social distribution in the remaining sixteenth-century charters, two-thirds of which bear the names of notable citizens, but even more by their form. On the one hand, there is often space left for the crests of the families allied by marriage, a custom that did not disappear in the seventeenth century when the majority of couples had no arms to put on the blank shield. On the other hand, the object itself was at first undoubtedly more costly, since it was an illuminated parchment written by a master calligrapher. Although ritual consistently mentioned marriage charters, this does not seem, in the majority of cases, to have signified that all possessed a charter, at least not in the form that was to become prevalent in the seventeenth century, when even the humblest citizens of Lyons could acquire mass-produced *tableaux de mariage* decorated with woodcuts or copperplate engravings, and printed in large quantity by several *marchands imagiers* or *maîtres graveurs en taille douce* of the city.

The Eclipse of the Secular

Nearly all the charters in the archives of the Hospices civils of Lyons date from the seventeenth century, and more than half of them (55 per cent) were used for marriages celebrated between 1640 and 1670. If we want to understand how they were used, we need to pursue two lines of investigation: Christian imagery aiming at the inculcation of the teachings that the Catholic Reformation considered essential, and the unique appropriations that Lyonnais couples might have made of an object that they handled, possessed, and safeguarded in the privacy of their homes. In the seventeenth century the charter was almost never totally secular in its decoration. In point of fact, charters that devoted the entire border portion surrounding the written text to motifs that have nothing to do with religion are extremely rare. One charter dated February 1639 shows caricatures, grotesques, fruits, and infant musicians; two others from 1639 and 1646 depict fauns and dragons, serpents and winged snails, surrounding a naked man and woman facing each other. These may be survivals of older models, which often bore only foliated scrollwork, garlands, or a frieze of flowers and leaves bordering the central text.

Some secular motifs nonetheless remained along with religious themes in charters of a decorative and iconographic style dating from the late sixteenth or early seventeenth centuries. This is the case in one series, the first extant example of which comes from 1595, which places the names and the symbols of the evangelists in the four corners of the sheet and leaves a good deal of space for flowers, leaves, and butterflies surrounding the tablet that contains the text. The same disposition is found after 1599 in another series of charters, smaller in size, that differ only in the presence of shields for family crests centred at the top and the bottom of the sheet. This can also be seen in one charter of which only one example exists, used for a marriage in 1615, in which the creation of Eve is pictured in the upper register and her temptation in the lower. Animals, insects, and horns of abundance decorate the border, and on either side of the text there is a vignette of a faun and a woman, both nude, whose intimate parts are being licked by dogs. The same is true of another charter extant in one copy, probably from 1636, on which a

TABLE 5.1 Iconography in marriage charters conserved in the archives of the Hospices Civils, Lyons

References:

A = The four evangelists	G = The creation of Adam and Eve
B = St Peter and St Paul	H = The temptation of Eve
C = God the Father	I = Adam and Eve driven from the Garden
D = The Infant Jesus	J = Cain and Abel
E = The Holy Ghost	K = The Marriage of the Virgin
F = The Trinity	L = The Annunciation

M = The Nativity of Christ	
N = The Marriage of Cana	
O = Man and Woman with joined hands	
P = Man and Woman at table	
Q = Louis XIV and Marie-Thérèse	

Type	Dates of first and last example extant	Number of charters	A	B	C	D	E	F	G	H	I	J	K	L	M	N	O	P	Q
1	1589–1654	6	X	X															
2	1595–1653	15	X	X															
3	1601–1667	31	X	X	X														
4	1608–1666	5	X	X															
5	1615	1					X												
6	1615–1677	25	X	X										X					
7	1619	1	X	X															
8	1629–1671	11	X	X					X	X									
9	1634–1654	4	X	X	X														
10	1636	1	X		X														
11	1640–1679	10	X	X															
12	1650–1673	33	X	X				X	X	X			X	X					
13	1665–1687	3	X	X	X				X	X		X	X	X	X		X		
14	1667–1668	2	X	X									X	X					
15	1668	2	X	X	X														
16	1668–1675	3	X	X	X				X	X			X	X					
17	1669–1671	2	X	X														X	
18	1671	1	X	X				X											
19	1672–1690	2	X	X				X					X	X					
20	1677	1	X																
21	1680–1682	2																	X

1. The date of the oldest known example of a charter cannot be taken as its date of publication.
2. This survey takes into account only major figures on the charters, excluding purely decorative elements.

great number of cupids carrying wreaths and palm fronds, fruits, flowers, and pierced flaming hearts leave little room for God the Father on his cloud (at the top) and the joined hands of a man and a woman (at the bottom). Such images were the exception, however. It is as if the Christianization of the marriage charter had turned back the tide of secular marriage emblems – cupids and hearts, horns of abundance and magnificent fruits – in favour of Biblical and Catholic imagery. The marriage charter thus found a place among the many objects – church paintings and mission banners, pictorial flysheets and confraternity broadsheets – that were used by the Catholic Reformation in its campaign for using images to further Christianization.

A Christian Iconography

The 161 charters of the Charité – all of which bear a figurative iconography and were mass-produced from wood blocks or copperplates, coloured and uncoloured – can also be categorized by the twenty-one different iconographic motifs they contain (twenty-two if the two variant formats of the initial series of 1595 are counted separately). Two motifs appear most frequently and occur in more than one layout out of two: the four evangelists (sixteen series) and saints Peter and Paul (twelve series). All these figures are represented in conformity with tradition. Each evangelist is either evoked by his symbol and his name, written on a scroll, or he is figured in human form and accompanied by his symbol. Peter carries keys and Paul a sword. The text of the charters refers to them as well, invoking the authority of Peter and Paul ('as St Paul has written and Roman law confirms'), and placing the matrimonial rite and the couple's vows within the context of respect of the Word of God, as transcribed in the Gospel. The formula 'as God has said' leads to the representation of the Father in five series of charters, alone at the top of the sheet, installed in a cloud, holding the globe or giving his blessing, and, twice, dominating the depiction of the earthly paradise and its creatures.[7]

The other persons of the 'Holy and Individuate Trinity' are more rarely shown alone. There is one instance of Christ carrying the globe and the cross in the oldest series and one of the Holy Ghost in a wording attested after 1608. After the mid-seventeenth century, representation of the Trinity dominates, appearing in seven of the ten series of engravings in use at that time in Lyons, grouped together as in the invocation in the text to 'Father, Son and Holy Ghost'. A decline in theocentric piety at the beginning of the century, the need to invoke the tripartite God against the Protestants (who formed an important minority of the population of the city), and the search for better conformity between the formula recited and the image depicted all played a part in the portrayal of the holy Trinity in nuptial charters in the diocese of Lyons. The Trinity had ample precedent in church paintings, but it was rare in other widely distributed forms of printed matter such as confraternity images or pilgrimage certificates, which were more interested in the saints and the Virgin.

The three representations most frequently encountered in the charters – the evangelists, Peter and Paul, and the Trinity – together or separately, were thus the ones that the text suggested with the greatest immediacy, either in the invocation or the matrimonial formula. The image showed what had been said and was preserved in writing but, exceptions apart, it did not show the pronouncing of vows and the act of donation that lay at the centre of the rite. The man and the woman united in matrimony appear only in the charter with the cupids (discussed above). Only one example of this series exists; it is of late date (attested in 1669 and 1671), is signed 'I. Philip fec. & ed.', and shows the couple installed on either side of a long table. The couple figures in two other series, but it is hardly an ordinary couple. In the border of a charter of 1665 we read, 'This charter was made in the year of the Marriage of King Louis 14, year 1660, and is sold *chez* Louis Pinchar' (upper part of plate XII). The motif seems to have had a certain success, since the king and Marie-Thérèse can be found on another charter, copies of which exist dated 1667 and 1668. It had a certain longevity as well, as it was used throughout the decade of the 1660s and even for a marriage in 1687. Although accompanied by the Trinity and the evangelists, the royal couple introduces a secular dimension into the image, underscored in the first series by a host of cupids. These cases aside, however, the Lyons wedding charters put their entire iconography to the service of acculturation in Christian teachings, neglecting wordly and secular figures, and only on one exceptional occasion serving the cause of the glory of the monarch.

Eve and Mary

Lyons image-makers took inspiration from another source besides the words pronounced during the marriage rite and written on the charters: the two contrasting motifs of Eve, tempted and punished after the fall, and the life of the Virgin, represented in the two episodes of Mary's marriage and the Annunciation. In four versions the two unions – the one of the first man and woman (born of him) and the one of Mary and Joseph – are placed facing each other, either at the top and bottom of the sheet (as in the two series whose oldest examples date from 1629 and 1640) or at the two sides (as in a charter used in 1668, in which the depiction of the terrestrial paradise, on the left, is surmounted by the ten commandments of God, and the marriage of the Virgin, on the right, is placed under the commandments of the Church). The pedagogical and moralizing intent is clear here, contrasting

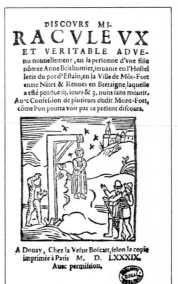

PLATE I Hagiographic pamphlet on St Louis of Anjou published in L'Aquila in the late sixteenth century (first and last pages).

PLATE II Title page of an *occasionnel* on the story of the hanged woman of Montfort miraculously saved by the intervention of the Virgin. The pamphlet is given as published in Rennes in 1588.

PLATE III Title page of the *occasionnel* from Douai on the same story published in 1589. It differs from the Rennes publication in its longer title (better adapted to sale by a street crier), and in the illustration showing the discovery of the miracle.

PLATE IV Title page from one of the earliest editions of Perrault's *Contes* in the *Bibliothèque bleue*, Troyes. The permission accorded to Pierre Garnier is dated 1723.

PLATE V (*Below*): Illustrations of key moments in Perrault's tales.

(a) Reference to classical antiquity is both aesthetic and moral (*Blue Beard*).

(b) Marriage: the choice of a symbolic synthesis of the tale (*Cinderella*).

(c) A re-used woodcut from the Pierre Garnier edition of *Sleeping Beauty*, 1737.

PLATE VI Two motifs in the tale presented simultaneously: the meeting of Little Red Riding Hood and the wolf, and the wolf devouring the grandmother. The process of making the end more decorous begins (*Histoires ou Contes du Temps passé, Avec des Moralités. Par M. Perrault*... A La Haye, et se trouve à Liège, chez Bassompierre, 1777).

PLATE VII (*Left*): Persistence of a correspondence between image and text from the earliest editions from Troyes (*Les Contes des Fées, Avec des Moralités. Par M. Perrault*. A Troyes, chez Madame Garnier). (*Right*): The illustration betrays the wording of the text: 'The poor lady was almost as dead as her husband, and had not strength enough to rise and embrace her brothers' (*Grisélidis, Peau d'Ane et les Souhaits Ridicules. Contes Par Perrault, de l'Académie Française*. A Paris, chez Lamy, 1781).

LE PETIT

CHAPERON ROUGE.

CONTE.

PLATE VIII (*Above*): scene of butchering suggesting a happy ending contrary to Perrault's ending (*Les Contes des Fées, Avec des Moralités. Par M. Perrault.* A Troyes, chez Madame Garnier). (*Below*): elimination of the violent ending, and a move towards stylization and suspense – Little Red Riding Hood in sight of her grandmother's house (*Le Petit Chaperon rouge, Orphise ou l'enfant du malheur, Les enfants égarés...* A Troyes, chez Veuve André).

LE PETIT

CHAPERON ROUGE,

CONTE.

PLATE IX Book of hours in the style of Rouen, second half of the fifteenth century. Secular motifs find a place in the margins of a book of devotions.

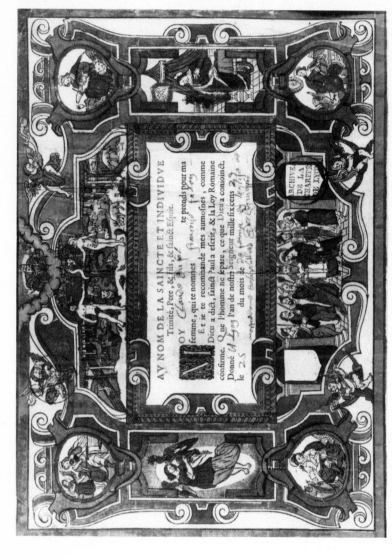

PLATE X A marriage charter used for a wedding in 1629. It carries a printed ritual formula filled in by the priest, and hand-coloured woodcuts showing the Evangelists, God the Father, the Garden of Eden, the Annunciation, and the Marriage of the Virgin.

PLATE XI These two illustrations were executed from the same design, using two different techniques. (*Above*): Woodcut and typography. (*Below*): coloured copperplate engraving with engraved text.

PLATE XII (*Above*): a marriage charter and monarchical representation: the royal couple of 1660. (*Below*): marriage charters and Christian catechism: the Ten Commandments of God and those of the Holy Church.

PLATE XIII (*Above*): an example of re-use showing three marriages on one charter. (*Below*): an amateur artist has added a frieze of cherubs bordering a traditional charter.

PLATE XIV An unusual charter, not from the shop of an image-maker: hand-coloured bouquets of flowers, an engraving (the marriage of the Virgin) cut out from another source and pasted on, and a painted landscape.

PLATE XV Interrogation of Václav Slavík, 16 July 1748. 'Examination of Václav Slavík, prisoner at Říčany, for suspicion of heresy, which was conducted at the rectory of the same city, 16 July 1748 by us, commissioners established for this purpose by the reverend vicarial office of Chocerady.'

PLATE XVI (*Left*): title page of the postil of Antonín Koniáš, 3rd edition, Prague, 1756. (*Below*): lists of heretical books confiscated and sent to the archdiocese of Prague by the vicar of Libeznice in 1749.

PLATE XVII Three page layouts of the same emblem from Alciati (1531, 1534 and 1549).

PLATE XVIII Emblem LXIII ('L'Occasion') from the collection of Guillaume de La Perrière, *Le Théâtre des bons engins*, 1539.

PLATE XIX Plate of devices commemorating the birth of Louis XIV, from C. F. Ménestrier, *L'Histoire du Roy Louis le Grand par les médailles...*, 1689.

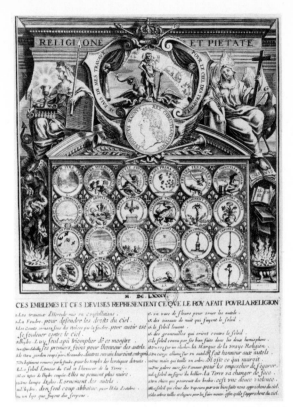

PLATE XXII (*Above*): before the entry – the *prévôt des marchands* harangues the King.
(*Below*): rue Saint-Jacques – the arch devoted to love of the people.

RELATION
VERITABLE
ET IOVRNALIERE,

De tout ce qui s'est fait & passé en la
reduction de la Ville de la Rochelle,
à l'obeïssance du Roy.

A PARIS,
De l'Imprimerie de IEAN BARBOTE,
en l'Isle du Palais, ruë de Harlay,
à la fleur de Lys Couronnée.
M. D C. XXVIII.
AVEC PERMISSION.

20

RELATION
VERITABLE DE
tout ce qui s'est passé, dans
la Rochelle, tant deuant
qu'apres que le Roy y a
fait son entrée le iour de
la Toussaincts.

*La Harangue & les submissions des
Maire & Habitans de ladite
ville, auec la response que leur fit
le Roy. L'Ordre qui fut gardé pour
les conduire à sa Maiesté, &
autres particularitez.*

A PARIS,
Chez Anthoine Vitray.
M. DC. XXVIII.
Auec Permission.

PLATE XXIII (*Above*): two accounts, published in Paris, of the end of the siege of La
Rochelle. (*Below*): two *mémoires* listing the prices of foodstuffs in the famine-straitened
city.

MEMOIRE
VERITABLE DV
prix excessif des viures
de la Rochelle pendant
le Siege.

Enuoyé à la Royne Mere:

A PARIS,
Par Nicolas Callemont, de-
meurant ruë Quiquetonne.
M. DC. XXVIII

MEMOIRE
TRES-PARTICVLIER
de la despence qui a esté faicte
dans la ville de la Rochelle,

*Auec le prix & qualité des viandes qui
ont esté excessiuement venduës
en ladite Ville,*

Depuis le commencement du mois
d'Octobre, iusque à sa Reduction,

A PARIS,
Chez CHARLES HVLPEAV, sur le
Pont S. Michel, à l'Ancre double, & en
sa Boutique dans la grand' Salle
du Palais. 1628
Auec Permission.

PLATE XXIV (*Above*): a *placard* – the allegory of the King in triumph over La Rochelle. The dike has been given the form of two keys. (*Below*): by the same author, the same allegory, but without illustration and in the form of a pamphlet.

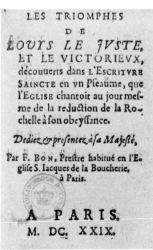

term by term sinful insubordination and respectful alliance, the disobedience of the first couple and the later couple's acceptance of the divine decree, the bad couple and the good.

Whether linked with the marriage of the Virgin or associated with other motifs, the depiction of the Garden of Eden keeps closely to the Biblical text. On occasion it is represented by one scene, with God in his cloud looking on from above while the animals form a background for Eve, who offers the forbidden fruit to Adam at the urging of the serpent tempter. Or, reaching further back in Genesis, two scenes show the creation of man and of woman (on four series of charters between 1634 and 1654). One charter, extant in only one copy dated 1615, depicts the creation of Eve and her temptation; a 1677 charter (in one copy) bears four corner medallions showing the earthly paradise before the fall, the temptation, the expulsion from the Garden of Eden, and, to complete the series, Cain and Abel, the sons of Eve. A minority of charters thus bears a representation of the primordial scene, the original foundation of marriage ('A man, therefore, will leave his father and mother and will cling to his wife, and the two will become one flesh'), and the sign of his misery when he disobeys the law of God.

Scriptural authority for the image of good marriage is weaker, since the theme of the marriage of Mary with Joseph (designated for this union by the appearance of the divine dove and by the flowering of the staff he carries) appears late and is given in apocryphal Gospels or legendary narratives recast several times between the ninth and the thirteenth centuries. The basis for the ten occurrences of the scene comes rather from the iconographic tradition established by painters (Italian for the most part) of the late fifteenth and the sixteenth centuries, before, but even more with, the Counter-Reformation. The new tradition eclipsed the previous one, which often presented the episode as a charivari aimed at an older bridegroom. It made it the model for Christian marriage between a couple of not too marked an age difference who promise themselves to each other in the presence of a priest, who blesses them, by the joining of their hands and the giving of a ring.[8]

The ten representations of this scene in Lyons echo the major traits of this iconography. Few people are present at the marriage: in two series there is no one, in three there are one or two people, and in two others there are, respectively, seven and nine people present. They are portrayed as peaceable witnesses to the exemplary union of a young woman and a man who is never shown as a truly old man. In only four images is there any hint of the motif of the miraculous staff that distinguishes Joseph from the other men of Judah (whose staffs fail to burst into blossom), and even these depictions are discreet. Only a few in the company carry a staff and only one charter shows Joseph's staff in flower.

In the seventeenth century, after the Church had established its authority over matrimonial ritual, the marriage of the Virgin was not a frequent subject in altar paintings or chapel decoration. The theme persisted without interruption in the charters, however, as its late re-elaboration portrayed an exemplary marriage in full conformity with diocesan dispositions for the ritual. The motif can also be seen as a representation on the charter of the ceremony that had prompted its purchase and its use, and as a projection, in pictorial form and in the guise of the ideal model couple, of even the humblest bride and bridegroom in the parishes of Lyons.

In five series of charters the marriage of the Virgin is accompanied by another scene that is its sequel, the Annunciation (present also in four instances without the marriage of the Virgin). The episode is treated in the traditional manner, with the Virgin and the Angel either appearing together in one space or, in a solution suggested by the form of the sheet, situated on opposing sides of the central tablet containing the written text. As secular symbolism of abundance and fecundity declined, it was the iconography of the Annunciation that was charged with recalling the first and fundamental reason for alliance between man and woman.

Marriage charters, in regions where they were in use and along with other widely distributed materials, contributed to bringing into the intimacy of the home representations designed to offer dogmatic, moral, spiritual, and worldly teachings. Usually given in images alone (to be 'read' separately or together), this pedagogy also attempted to use the charters to aid in the penetration – even among illiterates[9] – of the fundamental texts of the Christian religion. Thus a charter sold by Robert Pigaut and known by two marriages in 1668 enumerates 'the ten commandments of God', 'the commandments of the Holy Church' and 'the Sacraments of the Church' (lower part of plate XII). The addition of texts other than the one in the central portion of the sheet remains exceptional, however, as if a broader catechetic mission could not easily be reconciled with the strictly ritual function of the charter, or as if religious printed matter was distributed in carefully differentiated and mutually exclusive genres.

Similarly, a 'Jesuit' charter used for a marriage in October 1636 shows Ignatius of Loyola, Francis Xavier, Louis Gonzaga and Stanislas Kostka in the four corners, Peter and Paul at the right and left hands, the initials IHS formed by the instruments of the Passion at the top, and MRA in the form of a calligram at the bottom. This charter is a hapax, however, in a corpus in which invention was held back by the charter's rules of composition, obligatory text, and ceremonial use. The nuptial charter was thus fixed in a series of limiting constraints stronger than the ones that governed the production of devotional flysheets, and even stronger than the ones imposed on the confraternity broadsheets. By that token, it stands as the most 'standardized' of the print objects that offered a fixed text and a closed repertory of images.

Techniques and Hands

Marriage charters in Lyons underwent three major changes in form during the course of the seventeenth century: paper won out over parchment, copperplate engraving over woodcuts, printed or engraved texts over hand calligraphy. Some of these innovations were linked, as when an engraved text was substituted for typographic composition because decorative copperplate engravings (which demanded a special press) were substituted for woodcuts (which could be inserted into the same print form and printed by the same press as the text in movable characters).

These changes did not take place in linear fashion, however. Until the end of the century there were in each series charters on vellum – more precious, more solemn, less fragile – charters made from wood blocks (even for types of charters that also used copperplates after a certain date), and charters written and painted by hand. Thus in a series of thirty-one charters, the oldest of which is dated 1601, the use of copperplate engraving after 1624 fails to eliminate the use of woodcuts, which continued to be used up to 1645. In another series attested as early as 1615, copperplates, which appeared in 1645 and were printed on paper or vellum, also failed to do away with the use of wood-block illustrations. There were differences in the application of colour to the sheets, as well as in the bearing material, and the techniques used for the illustrations and the text. Each series, in fact, provides black and white copies in ink alone and others that were illuminated, painted with gouache, or printed in colours (plate XI).

This means that charters of the same type show an extreme diversity. For example, of extant copies of a type used between 1615 and 1677 and that show the four evangelists, Peter and Paul, the marriage of the Virgin, and the Annunciation, the oldest five copies (1615, 1620, 1623, 1626, 1642) are on parchment, bear a text in calligraphy in blue and gold, are illustrated with woodcuts, and are painted. Seventeen others, used for marriages between 1645 and 1677 and signed 'Cl. Audr.' (Claude Audran), are decorated by copperplate engravings with an engraved text. Four of them are on vellum and thirteen on paper; three are uncoloured, but the other fourteen are all coloured differently. Finally, two charters on paper, dated 1677, return to hand-tinted woodcuts framing a printed text. Thus, by variation in the materials and the techniques used, an object that was universally possessed could find at least a relative individuality to associate its particular aspect with the unique circumstance that prompted its acquisition and its use. By this interplay of possible differences within an obligatory uniformity, the Lyonnais charters unite the two major properties of printed matter, the imposition of like forms on all buyers, and the ability to assure unique appropriation of these common print pieces.

Traces of this interplay can be seen in the different hands through which charters passed. First, the printer-*imagier* occasionally signed his piece. The charters in the Charité collection show such signatures and addresses as 'A Lion chez Louis Puisant rue Ferrandière A l'enseigne du Purgatoire' (series 3); 'Claude Audran fecit' (series 6 and 12); 'Blanchin fecit, Savari et Gaultier excud.' (series 11); 'Se vend chez Louis Pinchar' (series 14); 'A Lyon. Chez Robert Pigaut demeurant en la rue Thomassin' (series 15); 'F. Demasso ex.' (series 16); 'I. Philip ex.' (series 17). All these men were master engravers and print merchants in the central city; they were the ones who worked out the iconographic and decorative layouts (from a repertory of obligatory motifs) and who printed both pictures and texts. The charter as an object owed its form to them, and this was increasingly true as others involved in their production – the master writers, for example, who wrote in calligraphy the text of the ritual formula, and the illuminators who painted the images – played a diminishing role. The priest who celebrated the marriage put his signature under the text and filled in the names of the wedded couple, the place and date of their union, and the parish in which the marriage was celebrated, in the spaces remaining in the calligraphic, printed, or engraved text. It was also the priest who crossed out a name when a charter was re-used for a next alliance (Claude Goy used his charter for his first marriage, 4 February 1646 and again for his remarriage, 23 August 1667), or even for three ceremonies (as in one charter of 1632 with cancellations in 1649 and again in 1670; see plate XIII).[10]

Unique Appropriations

Often, however, the wedded couple wanted to take more personal possession of this object imposed by ritual and sold in quantity, so they added touches of their own. Signatures were placed outside the framing border, at times reduced simply to an X, or there might be a hand-drawn border around the printed border. For example, on a charter engraved by Claude Audran and used for a marriage in July 1657 someone, perhaps the husband or the wife, added a note from another tradition to the religious imagery of the charter by drawing a frieze of winged cherubs, part angel, part cupid, around the edges (lower part of plate XIII). Backgrounds and costumes, architectural and decorative elements, were hand-coloured, freely and at times unskilfully, or blanks in the pictures were filled in, in particular the escutcheons designed to hold nobles' or notables' heraldic insignia.

This effort to make a ritual object more fully one's own at times went as far as the creation of unique charters drawn by unprofessional hands and bearing a handwritten text. Certain of these repeat elements of religious iconography from the printed charters (the four evangelists – or rather the three, since the charter in question is unfinished – for a marriage in February 1649; saints Peter and Paul for another of 1661). Others kept to a simple floral decoration at a time when secular ornamentation had disappeared from the mass-produced charters (one example, June 1662). One of the charters conserved by the Charité, used in May 1660, attests to this desire for personal creation (and perhaps also for economy) in a unique example that reaches the limit of the genre. On either side of the handwritten text a bouquet of flowers is drawn and hand-coloured. Between them at the top of the sheet a hand-painted engraving representing the marriage of the Virgin, cut out of another piece, is pasted onto the parchment, framed by hand-drawn sprays of leaves. Beneath the text at the centre, a small painted landscape set into an oval scrolled frame shows a tree at the edge of the lake (plate XIV). Like Pierre Ignace Chavatte, the contemporary worker in Lille who liked to cut out and paste into the chronicle that he kept devotional images or the engravings he found in *occasionnels*,[11] this amateur painter in Lyons (who may have been the husband himself) composed an object that was both universal in its use and unique of its kind. He put into it all the care that the circumstances required for the major event in his life.

The marriage charter was thus a print genre of complex uses. It participated in the church rite of which it was, along with the ring and the gift of 13 *deniers*, one of the symbolic objects, but it also accompanied a lifetime in the privacy of the home. It aided in the Christianization of an entire people by its reminder of authority and its obligatory imagery. But it also authorized free, unique, and creative appropriations. Until it disappeared, it repeated the forms and the motifs of its elite origin, stabilized by pastoral teaching; it also became the most 'popular' of print objects in its ubiquitous presence in society. In their particular use in Lyons and by the good fortune of their conservation in number in a hospital archive, nuptial charters permit us a glimpse into a form of print culture in the *ancien régime* that was not limited to typographical products alone, did not involve only the literate and was the bearer of practices that went beyond simple reading.

Notes

1 Archives des Hospices civils de Lyon, musée des Hospices civils de Lyon. My thanks to Mlle J. Roubert, curator of this collection, for the aid she offered me. On marriage charters, see P. B. Berlioz, *Les Chartes de mariage en pays lyonnais* (Imprimerie Audinienne, Lyons, 1941).

2 To cite a few examples: on altar-pieces for main altars or chapel altars, Gaby and Michel Vovelle, *Vision de la mort et de l'au-delà en Provence d'après les autels des âmes du Purgatoire, XVe-XXe siècles* (A. Colin, Paris, 1970); Marie-Hélène Froeschlé-Chopard, *La Religion populaire de la Provence orientale au XVIIIe siècle* (Beauchesne, Paris, 1980). On the reredos, see Victor L. Tapié, J.-Paul Le Flem,

and A. Parhailhé-Galabrun, *Retables baroques de Bretagne* (Presses Universitaires Françaises, Paris, 1972); M. Ménard, *L'Histoire des mentalités religieuses des XVIIe et XVIIIe siècles. Mille retables de l'ancien diocèse du Mans* (Beauchesne, Paris, 1980). On ex-votos, see Bernard Cousin, *Le Miracle et le quotidien. Les ex-voto provençaux, images d'une société* (Aix-en-Provence, 1982).

3 Paul Gonnet, *L'Adoption lyonnaise des orphelins légitimes (1536-1793)* (2 vols, Libraire générale de droit et de jurisprudence, Paris, 1935), pp. 420–60. On the Hôpital général de la Charité, see Jean-Pierre Gutton, *La Société et les pauvres. L'exemple de la Généralité de Lyon, 1534-1789* (Les Belles-Lettres, Paris, 1971), pp. 298–303 and 326–42.

4 On all these points, and in particular on the comparison between ritual in Lyons and other dioceses, see Jean-Baptiste Molin and Protais Mutembé, *Le Rituel du mariage en France du XIIe au XVIe siècle* (Beauchesne, Paris, 1974), pp. 186–93 (for marriage charters). See also André Burguière, 'Le Rituel du mariage en France: pratiques ecclésiastiques et pratiques populaires (XVIe-XVIIIe siècles)', *Annales ESC*, 33(1978), pp. 637–49.

5 Berlioz, *Les Chartes de mariage*, alphabetical table of charters, pp. 45–102.

6 Maurice Garden, *Lyon et les Lyonnais au XVIIIe siècle* (Les Belles-Lettres, Paris, 1970), p. 224, table 3.

7 François Boespflug, *Dieu dans l'art. Sollicitudini Vostrae de Benoît XIV (1745) et l'affaire Crescence de Kaufbeuren* (Editions du Cerf, Paris, 1984).

8 Christiane Klapisch-Zuber, 'Zacharie ou le Père évincé. Les Rites nuptiaux toscans entre Giotto et le Concile de Trente', *Annales ESC*, 34(1979), pp. 1216–43 ['Zacharius, or the Ousted Father: Nuptial Rites in Tuscany between Giotto and the Council of Trent', in Christiane Klapisch-Zuber, *Women, Family, and Ritual in Renaissance Italy*, tr. Lydia G. Cochrane (University of Chicago Press, Chicago, 1985), pp. 178–212].

9 In Lyons in the beginning of the thirteenth century, the percentage of male illiterates in the popular classes was from 75 to 80 per cent among gardeners, men-of-all-work, and masons; 50 per cent among cabinet-makers, carpenters, and domestic servants; 35 per cent among cobblers and bakers; 30 per cent among silk-workers. See Roger Chartier, Dominique Julia, and Marie-Madeleine Compère, *L'Education en France du XVIe au XVIIIe siècle (Société d'édition de l'enseignement supérieur*, Paris, 1976), table, p. 102, after Garden, *Lyon et les Lyonnais*.

10 The number of cases of re-use encountered in the corpus studied is insufficient to tell whether the practice was habitual or exceptional.

11 Roger Chartier and Daniel Roche, 'Les Pratiques urbaines de l'imprimé', in *Histoire de l'édition française*, gen. eds Henri-Jean Martin and Roger Chartier (3 vols, Promodis, Paris, 1982-), vol. 2, *Le Livre conquérant, 1660-1830*, pp. 402–29, in particular p. 424 ['Urban reading practices, 1660-1780', in Roger Chartier, *The Cultural Uses of Print in Early Modern France*, tr. Lydia G. Cochrane (Princeton University Press, Princeton, 1987), pp. 183–239].

6

Reading unto Death: Books and Readers in Eighteenth-Century Bohemia

MARIE-ELISABETH DUCREUX

In eighteenth-century Bohemia, in both towns and countryside, books seemed to play a central role in the life of peasants and modest artisans. It fed their thoughts, forged their identity and, in certain cases, encouraged their faith. Retracing that culture of the book is not an easy task, however. The dossiers that accumulated under a 'mild' Austrian Inquisition portray the suspects who were interrogated in a rudimentary and relatively unchanging role that from the outset reflected pre-established views of an improvised and composite Protestantism. Thus the motivations of people whose love of books led them into an accusation of heresy are concealed behind the judges' description of their actions, by their own attempts at camouflage to escape the opprobrium of public excommunication or, worse, a penal sentence, and by a contemporary perception of the non-Catholic traditions of the Czech population, credited by legend and history with a great weight that still remains to be evaluated.

Who, then, were these suspects guilty of loving books too much? To begin with, most of them were rural folk. All strata of peasant society, rich and poor, seem to be among them, from the landless day labourer who worked as a farmhand for a *sedlák* (yeoman) or a better-off relative, to the *rychtář* (judge, baillif), the head of the village community appointed by the seignory and its intermediary with the villagers. Shepherds, tavern-keepers, millers and millworkers, artisan-farmers who exercised their trade as tailors or cobblers to round out their budgets – all these could be named as readers of prohibited books in a denunciation or during a missioner's visit. They cannot be defined by profession, although landholders – yeoman farmers (*sedláci*), peasant farm-holders (*chalupníci*), and gardeners (*zahradníci*)[1] – appear more often in this rural world.

In contrast, the cantors who also functioned as schoolmasters (*kantoři* or *cantores*) rarely appear: ten at the most are mentioned and only four were interrogated, among whom only one was a true rural *cantor*. Seignorial administration is almost totally absent: there was one important affair in 1749 involving Václav Trubáč, *revidens* and *official* (administrative agent) of the seignory of Rychenburk in the circle of Chrudim in Eastern Bohemia,[2] who, moreover, had been a *cantor* in his youth. This near absence is hardly surprising, since as a rule schoolmasters and seignorial administrators were obliged to swear to their Catholic orthodoxy before assuming their posts.[3] The rest of the seignorial bureaucracy – intendants, regents, secretaries, and copy clerks in the châteaux – belonged to other social strata than the peasantry. As for the profession of *cantor*, it included quite different sorts of men, from a simple peasant who taught the rudiments as best he could to a parish cantor with a certain degree of competence, who was usually a good musician.

Bohemia boasted a number of cities but they were nearly all fairly rural market-place towns dependent in varying degrees on a seignory. They furnished their quota of artisans, who appear in good number in our sources. The royal free cities, which had been converted in the early seventeenth century, provided few suspects and, except for citizens of Prague and Kutná Hora, burghers hardly ever figure in the interrogations. Whether cities were royal or seignorial, their population was not large: the largest of them reached perhaps 3,000 to 4,000 inhabitants at the beginning of the eighteenth century, and some counted their population only in the hundreds.

With around 40,000 inhabitants (75,000 at the end of the century), Prague stood out in marked contrast to the rest of the urban scene.[4] After 1730, however, not many citizens of the capital are to be found in the archdiocesan investigations. In the preceding decade, investigations had uncovered a veritable network of gatherings to discuss the Scriptures and other texts, sometimes even in the presence of a preacher from abroad. Curiously, when this commerce was dismantled, 'heresy' disappeared in the capital, or at least it no longer gave rise to systematic inquisition. In 1728 Vojtěch Blaha, citizen of Prague's New Town (Nové Město) and considered, along with the innkeeper and great book reader, Jakub Vorlíček, one of the instigators of the meetings, was sent to the galleys for three years. He paid for the others, for the most part humble folk. The Blaha-Vorlíček affair reflects an urban universe very different from that of the rest of Bohemia, a world of domestic servants, street trades, and prostitutes, and an atypical world of migrants freed from their juridical ties to their seignory.[5]

Thus as the depositions unfold we begin to see a variegated range of social and cultural spheres. Digging deeper, we see life itself in behaviours and habits, relations with the seignory (omnipresent in country areas), family structure, the practice of trades, and more. Finally, these examinations for heresy teach us much about cultural transmission and the complexity of relations with the written word. They also oblige us to confront the question of the survival or the resurgence of a particular form of Protestantism.

The keys to this history lie in a brief survey of the context that produced these readers who seemed so dangerous to their eighteen-century censors. Books became an extremely important issue for the Church and for the state in the years following the Battle of the White Mountain. The attitudes towards reading displayed in the interrogations cannot be understood fully without the dimension of a centuries-long religious conflict. Czechs of the seventeenth and eighteenth centuries lived in a climate of militant conversion to Catholicism; nonetheless, they came from a past affected, from the early fifteenth century on, by the Reformation – Hussite first, then Protestant. This background had left its mark. But the Counter-Reformation had done much in a century and a half to transform both society and individuals, whether they liked it or not. Their world was now mixed. The old Protestant environment could still serve as a conscious or unconscious reference, but its dismemberment no longer furnished most people with more than the pieces of a puzzle that, after 1620, had been completed by a resolutely Catholic present.

Catholic and Austrian Bohemia

Around 1700, Czech lands had nearly recovered from the upheavals and the ruptures of the first half of the seventeenth century.[6] The Peace of Westphalia in 1648 had definitively sealed the fate of Bohemia. The kingdom now had to keep pace with a new world, one that saw the formation of the absolutist Austrian state. The crisis inaugurated in 1618–20 by the rebellion of the estates of Bohemia and the victory of Emperor Ferdinand II at the White Mountain had resulted in a recasting of institutions, politics. and culture. Socially, it rescrambled the elites. A new nobility of foreign origin and loyal to the Habsburgs partially supplanted the old aristocracy, who became impoverished or migrated abroad for religious reasons. The old kingdom had lost its independence. The emperor, who was also king of Bohemia, was declared hereditary sovereign in the 'renewed' Constitution of 1627.

With a stroke of the pen, the elective monarchy that had resulted from the Hussite wars of the end of the fifteenth century thus became a divine-right monarchy. Reinforced exercise of power by an increasingly centralized monarchy tended to reduce the role of the diet to registering imperial decrees. Prague was the capital of the country, but its head was in Vienna. After the death of Emperor Rudolph II in 1611, no Habsburg elected to reside in the Hradčany palace. The king of Bohemia spoke German and no longer understood the language of his Czech subjects. Although edicts and patents were written in the two languages, German gradually replaced Czech in the conduct of affairs of state. The organizational shift to Vienna encouraged cosmopolitanism among the governing strata. When Maria Theresa combined the offices of chancellor of Bohemia and of Austria into one in 1749, she eliminated the Chancery of Bohemia, the last remaining formal vestige of Czech autonomy, and integrated the kingdom once and for all into the 'hereditary German lands' – that is, into Austria. 'The Catholic religion has much increased, but the Czechs are no longer what they once were [and] their glory and their liberties have been swallowed up in eternal ruin.'[7]

Thus after 1627, in a continuing process begun in the sixteenth century, the Czechs lost their independence. Protestants to 80 or 90 per cent,[8] they were forced to change faith. After the White Mountain, the nobility and the bourgeoisie of royal towns had to convert or leave the country. Between 1620 and 1628 evangelical pastors – Czech Brethren, Lutherans, 'Neo-Utraquists',[9] a few Calvinists – were expelled; nobles and bourgeois had to submit or face exile. After 1648 all the remaining aristocracy in Bohemia had joined the Habsburgs. As subjects attached to their seignory, peasants and rural artisans did not have the option of departure and had to embrace the Roman and Apostolic religion. But if up to the end of the Thirty Years War country people refused to hear Mass or make confession, the last sweeps of the dragoons soon persuaded them. After 1651 only an infinitely small handful of recalcitrants still refused to practise Catholicism and receive the sacraments.

Conversion was, in Bohemia, one of the key words of the seventeenth and eighteenth centuries. It was an obsession for the government and the archdiocese. Judged accomplished the first time in 1651, thanks to the efforts of the army, it was nevertheless proclaimed incessantly until nearly the end of the eighteenth century, when a renewal of heresy was suspected in country areas. In spite of all, around 1700 eight decades of militant Counter-Reformation seemed to have transformed the land of the Hussites into a Catholic land. For some time Bohemia had been celebrated as 'renascent', 'Christian' (that is, Catholic) 'devout' and 'Austrian'.[10] From 1620 on, the split between the old and the new struck the very heart of families and individuals. In the cities, the sons of the rebels of 1618 now swelled the ranks of the Society of Jesus, the religious group most actively promoting conversion. In a few generations the transformation seemed complete and, at the turn of the eighteenth century, the state that the émigré jurist Pavel Stránský had described in 1634 in his *Res Publica Bojema*, published by the Elzévirs in Leiden, had disappeared.

Conversion, a Snare and a Reality

In Vienna, the state thought of the Czechs as rebels, but in the eighteenth century they came to be viewed as heretics. These two connotations already existed in the seventeenth century, however, and they reflect fairly well the ambiguous situation of the kingdom within the Habsburg conglomerate. In point of fact, the state functioned thanks to the Bohemian aristocracy, which furnished it with counsellors, ministers and major office-holders – and with the money that the tax office collected from their lands. As we have seen, Catholicism had been the one religion of the kingdom since 1627. A few incidents aside – such as the election in 1741 of Charles VII ('of Bavaria') to the throne of Bohemia – the body politic had been brought to heel, and the government turned its suspicion towards the lower strata of society, suspected of crypto-Protestantism.

In the eighteenth century, the image of heresy held (for different reasons) by the Church and the Viennese state offices encouraged and justified the pursuit of a policy of re-Catholicization, a policy that remained unmodified until Joseph II and the Patent of Toleration of 1781. Until that time, it appeared vital to the Austrian state that its Czech subjects showed themselves to be true Catholics.

On the morrow of the Thirty Years War, then, a converted people lived in Bohemia, a people converted more often by force than of its own will, occasionally by calculation. How are we to evaluate the depth and the authenticity of this conversion that continued over a century and a half? Historically, the problem remains open. There are too many overlapping and contradictory elements in a process that was both personal and social, and that was played out on the scale of the individual but also in the long term. A change in religion over an entire land is, at some point, the sum of thousands of individual conversions. That this was the case in Bohemia remains doubtful and, in any event, unmeasurable.

Intimate belief cannot be measured in the same ways as frequentation of the sacraments or attendance at Mass, its supposed signs. Without the slightest doubt, the first generations after 1620 practised a dual language that could be carried as far as Nicodemism or as concealing secret practice of the ancestral religion. More simply, they refused novelties. Thus, for exmple, people's attachment to the chalice can also be seen as a direct link to traditional religion. Little by little, however, and more rapidly in the cities, Catholicism gained a firmer hold.

Unlike nearby Germany, seventeenth- and eighteenth-century Bohemia was not a country of confessional coexistence. It was a land of symbiosis in which a Protestant mentality (in the larger sense) was gradually covered over with Catholic practices, often in the more global context of 'magic' thought and relying on a pietistic sentivity common to Catholics and recalcitrants alike. Hapsburg reasons of state had required that the Czechs become Catholic; statistically, the goal had been attained. The pomp but also the spirit of a Baroque religion seduced them bit by bit, by winning them over to the new devotions and by offering new forms of sociability in lay sodalities. Certain customs from their non-Catholic past were safeguarded, however, such as the reading of postils, the singing of hymns and the so-called 'literary' confraternities (*literátska bratrstva*). Thus, over a century and a half, religious impregnation came primarily through ceremonies and pious practices. The cult (even to excess) of miraculous images proved attractive in the long run with its thaumaturgic saints and protectors, its crying Virgins, its bleeding crucifixes, and its innumerable pilgrimages.

Still, what was proposed to the people of Bohemia was more a way of life than a faith. In other words, in the reconversion that constituted the quite literal extirpation of Czech forms of Protestantism and the re-Catholiciza-tion of an entire people, deculturation in some respects played a larger role than acculturation. As Philippe Joutard has written in connection with Cévenol Camisards, 'in many cases, the anti-Protestant struggle ended up only in disbelief without accomplishing the integration of the prevailing religion.'[11]

The oft-repeated constraint for outward adherence to Catholicism left room for inner free choice. This perhaps explains the tepidity and the detachment historically characteristic of Czech Catholicism. It may also explain the lost unity, into the nineteenth and the first half of the twentieth centuries, of a Protestantism fractured into several churches, all of which claimed to represent the Reformation locally, and a proliferation of sects and self-appointed prophets typical of religious wanderings in Eastern Bohemia. Similarly, when the Patent of Toleration was proclaimed in 1781, the Czechs saw little of interest to them in the Lutheranism and the Calvinism that they were henceforth permitted to exercise but that had few familiar connota-tions. Only 75,000 people in Bohemia and Moravia together – barely 2 per cent of the population – declared themselves Lutheran or Calvinist. In a final paradox, this reversal that seemed such an achievement was compared somewhat later to the extraordinary success of the model of national renascence that contrasted the autochthonous Protestantism of the Hussites, and the tradition of the Czech or Moravian Brethren, to 'German' Catholicism, a model promoted by Masaryk to the status of national identity, in terms that have continued to prompt debate.

In this country that accomplished the first Reformation in Europe a century before Luther, an obligatory conversion to Catholicism thus probably contributed to the laicization of people's consciences.

The Defeat of Heresy

The massive and forced conversion to Catholicism of a population estimated to have numbered 1,700,000 inhabitants in 1627[12] long preoccupied the king-emperor, the diet, and the archdiocese. The last of these took on the task of tallying Easter communion by keeping statistics (systematically after 1671) on confession. In 1651, by order of Ferdinand III, a count of subjects according to their faith listed the new Catholics and those, already in the minority, whose conversion appeared hopeless. The parish priests of the diocese of Prague, in response to a questionnaire addressed to them in 1677,[13] declared that in general their parishes contained no manifest heretics.

Nonetheless, at the end of the seventeenth century, the archdiocesan consistory, which had never believed conversion to be universal, was concerned by a recrudescence of heresy among these apparent Catholics. The organization of parishes had never completely recovered from the secularization of benefices that had been launched under Hussitism. The parish network was in a piteous state despite real efforts on the part of the archbishops and, impossible to staff properly, it was even less able to assure all parishioners a religious instruction in accordance with Tridentine norms. Almost every parish in Bohemia grouped several churches, annexed churches or commendatory parishes, each of which served a varying number of villages and isolated hamlets. A parish thus included a group of often rather widely scattered localities.

In this situation, the parish priests, even when they were assisted by a chaplain, rarely gave catechism instruction elsewhere than where they resided. In the best of cases, when bad weather or the state of the roads did not preclude it, Mass on Sunday was said in turn in the principal churches. In the absence of the parish priest, the *cantor* – that is, the schoolmaster and choirmaster – would sometimes substitute for him in an associated church, using a hymnbook or a postil, books often of doubtful Catholic authenticity.[14] In an attempt to remedy the shortcomings of the parish network (which had no easy solution), seasonal Jesuit missions were backed up, between 1725 and 1733, by permanent missions operating for several months at a time in one area. These missions reported directly to the archdiocesan consistory and were charged with instruction, preaching, explaining the new devotional practices, hearing confessions, and celebrating Mass, but also with conducting investigations of persons and books and keeping an eye on the suspects.

In 1726, a rescript of Emperor Charles VI codified penalties for heresy, which had become a crime against the state in 1627. Such sanctions ranged from death for the seller of books (a 'seducer' of the conscience) to forced labour, most commonly on the lands of the local lord or in the city holding the prisoner, or exile, or service in the galleys. Two patents preceded this ordinance, in 1717 and 1721, ordering parish and missionary clergy to seek out books in pedlars' bundles and in people's houses. After 1721, the sovereign attempted a permanent removal from the ecclesiastical courts of jurisdiction for heresy trials, a move firmly opposed by the archbishop. In practice, however, missioners, parish priests, and vicars forane[15] continued to carry out the first interrogations, but once the suspect was formally charged with heresy, they had to transfer the case to the 'Royal Council of Appeal of the Castle of Prague', the only court empowered to render sentences. Empress Maria Theresa renewed her father's legislation in this respect, first in 1749, then yearly from 1764 to 1780.

When it came down to cases, however, the severe punishments stipulated by Charles VI and his daughter were not applied in their full rigour. Between 1704 and July 1781, 729 dossiers concerning heresy in the three dioceses of Bohemia[16] were transmitted to the Court of Appeals in Prague.[17] The court released 181 prisoners and sentenced the rest, almost systematically, to forced labour (74.7 per cent) or, less often and after 1748, to the penitentiary (11.5 per cent), in eight cases to prison, twice to exile, and, in 1747-8, five times to army service. The death sentence was pronounced 44 times, or in about 8 per cent of cases, but it is less than sure that all the executions actually took place.[18] The sentences fall into three overall groups, 1727-37 (53.4 per cent of sentences), 1748-52 (23.6 per cent) and 1759-64 (9.7 per cent). After 1765 it seems, for the most part, as if repressive propaganda and the existence of the laws were enough to hold heresy at bay. I might note, though, that three months before the publication of the Patent of Toleration, the Court of Appeals sentenced another ten people to forced labour.

In reality, the state lacked the financial and political means for radical, massive, and continued persecution. It was the state, through the diet, that paid the permanent missioners who served as inquisitioners at the lowest level, along with the parish priests and the vicars forane. There were twelve such missioners at the most, their number varying yearly according to need, but also, and above all, according to what was available in the kingdom's treasury and to the attention that Vienna periodically paid to the problem. The landed nobility, which had powers of justice over its subjects, and on which the first arrest of the suspects generally depended, feared that its peasants would flee to Saxony, Prussia or Hungary. Thus when it could it demonstrated less than total zeal in the pursuit of heresy.[19]

A quibbling and constant local surveillance was preferred to pitiless repression. The 729 cases that the Court of Appeals did judge served above all as examples. They corresponded to copious lists of names and statistics on suspects in the archives of the archdiocesan consistory in Prague,[20] only a small proportion of whom were transmitted to civil justice. Thousands of people were interrogated in this manner. After, they were typically excommunicated, then solemnly received back into the bosom of the Church after making an act of faith and swearing, in a simple recantation or a sworn *reversales juratae*, not to fall into their past errors again. Thus, between 1725 and 1728, more than 400 deviants were noted down in the city of Prague alone. A few were burghers, but most were simple folk, used goods sellers, domestic servants, and vintners (involved in a complicated affair of communion in the two kinds). In the diocese as a whole we find 458 suspects in 1735, 359 (22 of whom were turned over to the civil authorities) in 1750, and 105 in 1751. Each year brought its harvest of names.[21] The interrogations of these people, not all of which have been conserved, testify to a certain ambiguity in the form of inquisition followed. Although it was backed up by the 'muscle' of local, city, and seignorial authorities, it did not lead to penal sanctions in the majority of cases.[22]

The Book as a Sign of Heresy

The men and women sentenced by the Court of Appeals and the suspects who were only interrogated usually shared a trait: they had read, listened to a reading of, possessed, sold, bought, exchanged, lent, or even simply praised books that their parish priest had not expressly permitted them. Their relation to books was often a determinant factor in the pursuits and the surveillance to which they were subjected. In this sense, the book was a sign of heresy. The equation functioned clearly in the mind of the clergy, who, both in Prague and locally, launched and conducted the interrogations. It explains the widespread hunt for books and their readers that took place in Bohemia until 1781, the systematic searches of suspects' houses, and the pressures for denunciation of neighbours or for bringing prohibited works to confession. It also explains the reiterated condemnations from the pulpit of any reading matter not specifically countersigned by the parish priest. Finally, it explains a Catholic publishing strategy inspired by a desire to imitate both the form and the uses of illicit reading matter that the people had quite apparently found to their liking.

Obviously, for the missioner or the parish priest who acted as inquisitor, heresy had other characteristics than the presence in the home of a book that was or was judged to be Protestant. Still, locating heretics was organized around the presence of books, since the book made the heretic. When a book was found, the rest of the procedures were automatic. A blasphemer without books remained just that in the eyes of the judges, but if he or she read or possessed forbidden volumes, the same crime became heresy.

The very order of the interrogations, which became codified soon after

1733, speaks to the power that the Catholic clergy saw in the book. The first thing that suspects were asked, after answering the usual questions involving civil identity and religion in the family, involved their reading and their library. Questions on doctrine and beliefs only came after. Furthermore, the questions themselves (exceptional cases aside) traced the stereotypical portrait of a formal heretic fallen into 'the Lutheran error': they concentrated on the cult of the saints and the Virgin, on the sacraments, on communion in the two kinds, occasionally on the real presence, on Purgatory, and on the respect due the pope and the clergy. The 'Lutheran error' was, as we have seen, far from being the majority belief in the Czech countryside.

The book identified the heretic. For one thing, its use in the course of traditional family religious practices encouraged heterodoxy: 'Heretical books, old and new, read and sung in private, are the cause of the persistence of heresy', wrote the missioner Jakub Firmus.[23] In the same period (around 1725), the Jesuit František Mateřovský gave clear expression to a state of mind animating a good number of other missioners' reports and that can be found somewhat later in the annual reports 'on the state of religion' written by the archbishop for the Diet and for Rome. Father Mateřovský says,

The perquisitions and examinations that I have conducted recently make me aware of the stealthy advance of the vice of heresy, which surreptitiously insinuates itself into Prague and various places in the diocese. . . . We must grasp clearly that above all else, the true root of all this evil is the heretical books that teem in Bohemia and are hidden in the most varied places. They are the heritage of the ancestors, and in ever greater number are peddled here from heretical countries. The most efficacious remedy is to eliminate them.[24]

Thus the book contaminated by contact, and if all people in Bohemia were not heretics, all could contract the disease. As the chief judge of the ecclesiastical court and Grand Vicar of Prague recalled to the diocesan clergy in 1735, 'Often the books which the Czechs, in particular the simple people, notoriously love unto death to read furnish them the occasion to doubt faith, or even turn them away completely from the salutary Catholic and Roman religion.'[25] Monsignor Martini was explicit: it was reading that produced a heretic. This is also the lesson that could on occasion be drawn from the interrogations and the oaths of reconversion of those who promised to mend their ways and do penance: 'As long as I had no evangelical books I was Catholic; then when I had some, I was evangelical, and now that I no longer have any I am Catholic', declared Jiří Wolf around 1780, expressing in exemplary fashion the irresistible pull of the frequentation of books, which by themselves conferred a religious identity.[26]

In 1753, Anna Němečková, the wife of a tailor in Vestec and the daughter of a yeoman farmer, twenty-nine years of age, said something slightly different: the ability to read made one seek out the forbidden and the heretical. Conversely, if a person was illiterate and unable to understand books, it was a proof of Catholic orthodoxy: 'And I would really think that he [her father] was Catholic, because he did not know how to read very well, not enough to go seeking something heretical. But my mother, she looked at heretical books.'[27]

Conversion by the Book

Such dangerous books could thus be eradicated by burning them,[28] by correcting them (when possible), or by confiscating them after noting their titles carefully, often along with the name and social status of their proprietors. Like the ancient Hydra, however, they always sprang up again. They could even be found in the hands of people whom the archdiocese had sentenced, even after they had solemnly sworn never again to have such a thing in their possession. The attraction or the need for books remained too strong: 'This people loves books so', one missioner sighed, 'that it is impossible ever to see them disappear.'[29]

If heresy passed by means of the book, conversion logically would take the same route. Since there was little hope of curing the common people of its habit of reading, its books would be replaced by others. For the missioner Father Třebický, writing in 1717, conversion was unthinkable without substituting safe books for dangerous ones .[30] Catholic works must take the place of the sequestered volumes, works imitating as closely as possible the form and structure of the ones that circulated in Bohemia from Zittau, Leipzig, or Halle. The idea was not new. As early as the late sixteenth century, the first collections of hymns of the Counter-Reformation, counting on the seduction of song, borrowed the internal organization and a good number of hymns of the Brethren and the Utraquists. In the seventeenth century, the same intent guided the publications of the Prague Jesuits, who went as far as to translate the Bible so they could distribute it in an authorized version.[31] In the eighteenth century, however, this intent was embodied in a veritable politics of Catholic publishing in the vernacular, fostered above all by the work of the missioners, some of whom went beyond simple imitation of genres and titles to recommend Protestant ways of reading.

This was the case of one zealous missioner, Antonín Koniáš, whose reputation (in point of fact exaggerated) as an ardent book-burner has come down through the centuries (see plate XVI). This Jesuit played an important role in both the organization of missions and in official Catholic publishing in the eighteenth century. The compiler of an Index in Czech that served as a manual for book confiscators,[32] he was also the author of a number of

works written for distribution to the people. Among them was a postil
published in two versions, Czech and German, and printed three times
between 1740 and 1756. The work seems to have been received favourably
by the masses, thanks to the stories and the *exempla* that it contained.[33] This
sort of book, typically Protestant after the sixteenth century, consisted of a
collection of Epistles and Gospels for the day in the vernacular, with
commentaries and occasionally a hymn following each reading. Thus it
progressed with the liturgical year, replacing the sermon that usually followed the Epistle and the Gospel reading for the day. Koniáš wrote in his
preface,

On Sundays and feast days it is not enough to hear the Word of God proclaimed in
public in church. One must, at the hour of rest, read oneself or hear the reading of
salutary books countersigned by the spiritual authority [the parish priest]. . . . God
gives us Sundays and feast days . . . in order that the reading or the hearing of His
divine discourse may sow in our hearts the seed of His Word which, in its time, will
procure for us the profit of eternal salvation.

He concludes,

Surely, if we take to heart the doctrine that we read or hear [someone] read, if we
remember it often during the course of our daily occupations, the eternal Word of the
Father - His Son Jesus - will not abandon us.[34]

Reading at home, aloud, for an audience of the family and the domestic
circle - reading texts so well memorized that they became an integral part
of the individual - these are the traits of what is known as intensive reading
as it was practised in European societies before the nineteenth century.[35]
Even though this sort of reading was common to both Catholics and Protestants in both its procedures and its nature, as has recently been demonstrated,[36] not only was it much more frequent and socially more generalized
among Protestants, but its very status was different. For Reformed
worshippers, the reading of religious books (the model for which remained
Holy Writ), repeated hundreds of times until their contents were assimilated
and incorporated into people's thought, constituted one of the essential acts
of faith. It was the translation of *Sola Scriptura*, which transmitted the
efficacy of salvation. This is exactly what our Jesuit promises. Pretending to
address his remarks to good Catholics, he was really aiming at those who
still used the written word as their heterodox ancestors had done. In
adopting Koniáš's postil, they had no need to change either their habits or
their reasons for reading.

One might ask, however, whether this substitution met with the success
expected. The ecclesiastical authorities pursued an ambiguous policy,
divided between the extermination of volumes judged to be heretical and
the distribution of a literature necessarily more restrained in its themes.

Indeed, Koniáš's Index, which missioners and parish priests possessed, ensured orthodoxy by rejecting all that had been written and printed in Czech lands from 1436 to 1620. The archbishop's pastoral instructions and letters, on the other hand, reflected the circumspection habitual in the rest of Europe at the time concerning secular use of liturgical or scriptural texts in the vernacular. In spite of all, prohibition and destruction failed to uproot a taste for reading. There was no way to get around the Hussite and Protestant liturgical practices reflected in widely distributed printed matter like hymnals, so the Church adopted them, adapting them to a varying extent. Above all, it attempted, as we have seen, to strike all possible malcontents of the faith at the very heart of their relations with books.

Thus, although there existed modes of reading specifically connected with the Reformation, we can to some extent speak of a Protestant acculturation of the Catholic Counter-Reformation in Bohemia. Conversely, the repression that struck Czech bookselling for 160 years set up a process of deculturation. I might also note that Catholic reading reached its full meaning only within a radically different system of signs and hierarchical practices. Reading came after attending Mass in importance, and receiving the sacraments, participating in processions and pilgrimages, belonging to one or more confraternal organizations, wearing a scapular, reciting the Rosary, making the sign of the cross, making fasts and abstinences, possessing images, and using particular forms of salutation.[37] In its relatively long course, did re-Catholicization manage to extirpate (or at least shift) many Czechs' veneration of the book? Perhaps, but it is certain that, around 1800, reference to the Scriptures could still evoke personal affirmations of identity and philosophy that, for some people, were situated explicitly outside any confessional context.

To give a single example, in the first year of the nineteenth century at least four villagers in the seignory of Rychenburk, in the circle of Chrudim, met at one another's houses to take turns reading the Apocalypse, the Bible, and the Gospels.[38] All 'dwelt in the living peace' and were, according to their own statements, 'of no religion'. Josef Veselý, 46 years of age, a tailor who farmed a field owned by his brother and sold pearl barley on the side, declared that this peace was 'to the letter' the one Jesus Christ had proclaimed and that he had found it by himself through the spirit. It was true that he had declared himself 'Helvetian'(Calvinist after the Edict of Toleration), but now he was 'non-Catholic in the living peace'. His companion, Josef Suchý, 37 years of age with four children, a yeoman farmer who worked a 'half landholding' in the village of Svratouch,[39] had been known until then as a Catholic. Suddenly, however, he declared himself Calvinist. Interrogated on this sudden change on 5 July 1800 in the offices of the castle, he explained that he had only done so 'for people', for in reality he dwelt 'in the peace', a peace 'holy and golden' found in the Gospel. He owned only one book, inherited from his father, containing the Gospels and the Epistles. Its small format and place of publication, the city of Litomyšl, indicate that it was probably a pedlar's book comparable to those of the French *Bibliothèque bleue*.[40] The work was in any event Catholic. Its literal reading had delivered Josef from existing religious structures and revealed to him his capacity to think for himself. He expressed his change of heart with citations that served him as guidelines, so thoroughly assimilated that they made up his deepest sense of identity.

19 – How long ago did you leave the Catholic religion?
    – Almost two years ago.
20 – By what route?
    – By love: 'Love thy neighbor as thyself.'
21 – But that Gospel, you must already have read it while your father was alive.
    – Yes, but it is particularly since his death six years ago that I have read it.
22 – But since you were reading it before, why didn't you abandon Catholicism sooner?
    – Because I did not understand as well, and one fine day, I understood better. For in one epistle there (I don't remember any more if it is in the Gospels or the Epistles) it is written, 'I was displeased in the temple built by the hands of men, but you, you are the temples of the Holy Spirit, and the spirit resides in you.'[41]

Here it is the written text that contains absolute truth and produces truth in its reader, legitimizing his spiritual and individual freedom of choice.[42] The spirit illuminating these marginal figures emancipated from confessional ties wafts through the words of the book. It would be difficult not to see the distorted but living shadow of a tradition of Protestant reading behind this nonconformity.

Protestant Roots and Individual Identity

The connection between heresy and reading or the presence of unapproved books seems clear. Still, a problem remains (in spite of Monsignor Martini's unidirectional solution): did people have books because they were heretics? Or was it the book that led to heresy?

It is of course a delicate matter to judge people's true character fairly on the basis of interrogations in which it was fair game to attempt to mask what might get one in trouble. Still, and taking all the precautions required by a source that pre-defined the 'heresy' of an arrested suspect, we can say that the population we see in these interrogations on the whole draws its attitudes towards the written word and the book from the habits of an already remote non-Catholic past. Such habits quite certainly played an important role in the stubborn attachment to writing (to print, but also to the manuscript) of these humble and anonymous survivors of two centuries of Hussitism and Protestantism.

Among some of them – a minority – the will to remain faithful to the religion of their fathers still lay behind their refusal of assimilation into the now-dominant Catholicism. But with time, even among these recalcitrants, the contents of their ancestral faith had often been reduced to symbolic reference points, when it was not subjected to radical reinterpretation. Furthermore, the meaning of their stance fluctuated with the generations and with individuals, and later exchanges with émigrés from neighbouring Protestant countries both reinforced and transformed their behaviour. More frequently, it was the memory – vivid or latent – of the family or the group that perpetuated or awakened a feeling of difference, channelled through attitudes towards reading in particular. The influence of kin and neighbours and the attraction that the more audacious of the village autodidacts exerted over their entourage kept these resurgences alive. More simply, the pull of tradition could lead individuals until then in conformity with the Catholic norm to an affirmation of self. This could occur thanks to words repeatedly heard, or through participation in clandestine assemblies in which religion and perhaps other topics were debated, and in which there was singing and occasionally communion in the two kinds, just as well as it could by a frequentation of illicit books.

Furthermore, purely religious motivation was always founded in other impulses inherent in the lively anticlericalism of rural folk and in their syncretistic and materialistic mentality. In 1729, for example, an old farmhand, 75 years of age, Jan Černý, who owned four books and enjoyed preaching to his neighbours, argued against the utility of prayers for the souls in Purgatory: 'In the old days,' he declared, 'Masses for the dead were never said in the morning, and that is why crops came up much better.' Does this enigmatic statement mean that corpses fertilized the soil? For 'where one falls after death one remains after' and 'there is no Purgatory, for we make it here on earth.' This was indeed a Protestant belief, and one shared by the Czech Brethren.[43]

Thus if in the eighteenth century an attachment to the book could go beyond a strictly confessional context, it is clear that it was well rooted in a Hussite, then in a Protestant, *habitus*. The responses that one peasant woman of 28 years of age made on 14 December 1778, in the parish church of Heřmanice, to Matthias Střelský, her parish priest, summarize elements of doctrine found – in part or in full – among the suspects most thoroughly rooted in this religious tradition. Kateřina Koldová was particularly conscious of that tradition. She came from a family that in 1781 had declared itself of the 'faith of the Lamb' (*víra beránkova*). She later commented on her sweeping rejection of Catholicism, Calvinism and Lutheranism,[44] 'From my childhood we have been like this, and our relatives on both sides. I think that this faith will go well with the Helvetic confession. My parents used to tell me that it came to us from Master Hus.' Three years earlier she had testified against her parents in the following terms:[45]

I often heard them say this, which is contrary to the Catholic articles of faith:
  1 There are only two sacraments, Baptism and the Lord's Supper. It is written in the New Testament;
  2 The saints of God are in heaven, but they cannot in any way aid us on earth;
  3 Holy images are idols;
  4 The Catholic faith is not good; the evangelic faith is;
  5 The sacrament of the altar should not be received in the one kind but in the two. One must drink from the chalice, but the priests do not want this for us and refuse it to us;
  6 Indulgences serve no purpose, nor do confraternities;
  7 Going on pilgrimages is something for good-for-nothings;
  8 There is no Purgatory after death, but only two places, heaven and hell and nothing more.
Yes, I truly believed all that, but when I listened to the Word of God in Church it was you I believed.[46]

A small number of detainees undoubtedly saw themselves more or less consciously as the heirs of the Czech Brethren. The interrogations do not tell us so directly, but other somewhat rare writings do, such as those of Jan Šlerka. A harness-maker in the small city of Polička in Eastern Bohemia, in 1762 Šlerka led a small company of companions first to Silesia, then to Berlin, where his nonconformist sectarianism made him unwelcome. We find him later in the Slovak provinces of what was then Hungary. His *History of Bohemia*, which remained in manuscript and has only recently been rediscovered, tells us that he considered himself, like his grandfather, an 'ancient' of the Brethren. He relates the tribulations of the Czech nation, which had held to the true faith and 'for a thousand years' had opposed 'the Antichrist of Rome'.

Šlerka's case was an exception, however. There were few who expressly declared their connection with the Czech Brethren when given an opportunity to do so. After 1781, they declared themselves of the 'faith of the Lamb' or they chose Calvinism, as if even the memory of their heritage had dimmed. Moreover, Eastern Bohemia at the end of the eighteenth and in the nineteenth century offered propitious terrain for sects and for what was called 'the errance' (*blouznivectví*). These groups combined various elements derived from Hussitism and Protestantism with the conviction of direct inspiration by the Spirit within, which was not always the Holy Ghost. Thus, to the horror of the authorities, groups such as the 'Moroccans', the 'Brothers of Paradise', the 'Abrahamites' and many others, deists and social utopians, flourished after 1781 as the direct descendents of the suspects of the eighteenth century.

Until 1779, not one suspect interrogated admits to being non-Catholic, and all agree in general terms to mend their ways and undergo the penalties stipulated by the Church in the hope of avoiding the more severe penalties that the state held in reserve. The few exceptions concern émigrés who returned from foreign lands to visit family and friends, also carrying books and passing along missives. The emigration of peasants and artisans for reasons of faith, particularly from East Bohemia and western Moravia, remained a fact of life in the eighteenth century.

Motives could of course be mixed. The activities of one Count Zinzendorf who reconstituted a *Unitas Fratrum* at Herrnhut that differed widely from the original, and the existence in Rixdorf, near Berlin, of colonies of Czech émigrés, both reinforced these small refractory groups from the outside and influenced them. Emigration had its confessional constraints. In the frontier city of Zittau in Saxon Lusatia, the arriving émigré had to take instruction in Lutheranism and undergo an examination before hoping to become a citizen of the city. People left Bohemia for other reasons than religion. There were many couples whom their lord refused the right to marry as they pleased and who fled by night to a nearby land to be wed by a pastor. They often returned to their place of origin in spite of the pursuits that awaited them there. Similarly, artisans who made their journeyman's tour or weavers often went to Zittau. Their case is less clear, since economic motives might have coincided with more suspect reasons.

In any event, if the new inhabitant remained awhile in Zittau, he or she became Lutheran. Translations of authors widely distributed in Germanic lands – Arndt, Moller, Kegel, Francke, or Milde – popularized a Lutheranism of marked spiritualist tendencies. All this literature provided a pietism that awakened all the more echoes in Bohemia for appealing to the prevalent sensibilities. The works printed in Zittau at the beginning of the century by a Bohemian, Václav Kleych, were distributed by the thousand in Bohemia, and they rekindled the Czech evangelical tradition.

Those who manifested their particular identity by their reading and their attachment to the book, their language and their rejection of Catholic practices such as fasting, abstinence, or the veneration of the saints, found themselves in fact hemmed in by two visions of themselves imposed from the outside. In their own land they were seen as the incarnation of the 'vice' or the 'plague' of heresy, but Protestant Europe, with Prussia at its head, perceived them as witnesses to constancy of faith under persecution. Censors and protectors called them Protestant according to their own definitions. In Bohemia, however, the situation went beyond a simple confessional cleavage. The suspects' positions were often not what the foreigners or the Catholic authorities expected them to be. The absence of pastors, the scarcity of preachers, the clandestinity of book circulation, and laws limiting freedom of action favoured a multitude of ways of being that were totally

individual, just as it encouraged the formation of relational networks, often
limited to the area surrounding one seignory, but on occasion following the
perigrinations of one migrant throughout Bohemia.

The Circulation of Written Texts

How could anyone make up a library in rural Czech lands in the eighteenth
century? More accurately, the 'library' was a book collection varying in size
from one volume to twenty or more, and much of the time it was made up of
two or three books, to which tracts or manuscript papers might be added.

Of course, the only books whose presence we can grasp clearly were the
ones forbidden by the Church. Sporadically, our documents note that a
suspect possessed a book that had passed the tests of censorship. How
legitimate Catholic publications were received, or the majority of the
pedlars' songsheets,[47] for the most part escapes our knowledge. There is
probably no extant source that gives information on all books owned in rural
areas comparable to the wealth of material in heresy interrogations.[48]

We know by their reports and by the archdiocesan *agenda* that missioners
and parish priests distributed hymns, catechisms, and postils as well as
rosaries and images of the saints. Thus people might easily have had in their
homes both Catholic books and prohibited books.[49] Several of those
interrogated had neglected to get their parish priest to countersign perfectly
acceptable books, such as the *Rose of Paradise*, the Jesuit Bible (also called the
Bible of St Wenceslas), or the Psalter of Paroubek. Moreover, as they
conducted their house searches, the missioners regularly encountered
manuscript or printed tracts of a Catholic acculturation ('superstitious'
though they might be), such as the *Imprecations of Christ*, the *Missive of Pope
Leo*, the *Prayers to St Christopher* (which help to find buried treasure), the
Golden Letter of Bethany, and so forth. Thus, the literature consumed was of a
mixed nature, but unfortunately it is difficult to ascertain exactly how mixed
people's reading was.

These reservations expressed, let us return to the forbidden books. At the
time of their interrogations in 1755, Lidmila Horynová and her husband,
farm-holders in the seignory of Pardubice, had hidden twenty-one
prohibited books and several manuscripts in three different hiding places in
their house. Karel Hrubeš, the *rychtář* of Lozice, owned a dozen such works
in 1729. Among these was Václav Hájek's *Chronicle*.[50] He simply kept them
in a chest or left them out on a bench, which was what led to his
denunciation.

Some of the books were inherited from kin. But family inheritance did not
necessarily follow in the direct line: an aunt, a sister-in-law, a cousin could

also bequeath a book. Did personal preferences play a role? Was the choice determined by reading skill? Did the luck of book circulation enter in?

Book distribution was, in point of fact, a fairly complex matter. The oldest works had been published in Bohemia or Moravia before 1620. In the seventeenth century, other works came from German presses at the instigation of émigré Czech clergy. The newest ones arrived en masse from Zittau, in Lusatia, then from other German cities such as Leipzig, Berlin, and Halle, to mention only the most frequent ones. Thus books that were transmitted from generation to generation or among collaterals formed a moving whole. At any moment a link in the chain might change the overall picture by retiring a work or introducing something new. Books were lent, exchanged, sold, and resold continually. When Lidmila Horynová's books were sequestered only fifteen or so belonged to her, which was most commonly the case. A book was movable goods, and although certain books were forbidden, they did not stay hidden away.

There were two sorts of market for such publications. There was an internal market that included kin and the circles of rural sociability. The other market lay outside the framework of daily life. A good many people travelled Bohemian roads. It even seems that they travelled a lot in this country in which, in principle, a dual subjection, personal and economic, kept the population immobile. Carters, the *formané* charged with the *robota* of transport, never appear in our sources. Pedlars of cloth and stuffs, soldiers, discharged or on campaign, journeyman 'companions' doing their grand tour, peasants moving from one seignory to another, beggars, vagabonds, 'errants' of all sorts were all itinerants who might carry books. A messenger from abroad – someone like Václav Slavík, Mátěj Čoudil, or Josef Einsidler, who died in prison – or a preacher making his way through the countryside might lurk under one of these identities. With the police forces of the kingdom at their heels, these book 'passers' arrived from Germany and the Slovak provinces of Hungary. Some established themselves outside the frontiers, others lived inside the country and went back and forth. Buyers more often than not had to be satisfied with the titles they proposed, although some, like Jiří Samec in 1779, took direct orders. Less often, a book could be bought during one's own travels. A *cantor*, for example, who was visiting his son (who had risen in society as a *Regenschori* in Prague) took advantage of the occasion to buy a Bible.

The suspects' responses to questions often give the price they paid for the book. In 1729, for example, Karel Hrubeš paid 1 Rhenish florin 40 kreutzers to a 'knave who buys linen in Zittau', and another Rhenish florin for a second book to a man from Chroustovice, his parish seat.[51] At the same date near Prague, Jiří Štastný, a farmhand who was always between two masters, bought the *Špalíček* of Václav Kleych from a peasant's son from Roudnice for 1 florin 24 kreutzers.[52] In 1745, Rosina Čepková spent only 17 kreutzers for the same title. An independent farmer, Pavel Čačák, still in 1729, borrowed 4 florins to buy himself a commentary by Martin Philadelphus. A great many more examples could be given.[53]

Transactions of the sort were not paid for only in coin. Between neighbours a book could be exchanged for another that appealed more, and there were even more surprising trades. Books could repay a loan when cash was lacking, on occasion even with interest. The blacksmith of Turov, to whom Karel Hrubeš had lent 7 florins, repaid his debt in three instalments by getting books for Hrubeš for a value of 3 florins on each occasion. The book was a means of payment. Another person we have seen, Lidmila Horynová (who incidentally sold books herself), received from the tailor of a nearby village a copy of *Hymns on the Gospels* in payment for a pair of trousers that a thief had stolen from his shop. In 1729, a yeoman farmer, 36 years of age, Matěj Bina, gave a large Gospel to Jakub Hrubeš, the tavernkeeper of Telčice and a relation of Karel Hrubeš, in exchange for a *vērtel* of oats.[54] "As he talked to me no more about it, I thought that it was because of the book', he comments. Matěj Hlaváček remembered in 1753 that his father had acquired a New Testament eight years earlier in exchange for a calf. Václav Šultz from Ovenec, near Prague, thought that a book would cover the funeral expenses for his farmhand, who had died of the plague in 1729. Čačák offered peas in payment for a second book. In country areas, where coin was cruelly lacking, the book entered into the circuit of substitute coin. Above all, it had a market value independent of anyone's personal, affective, or religious investment in it.

Manuscripts

The interrogations offer perhaps fewer indications concerning the circulation of manuscript works. Local exchange networks existed, however, and there were quasi-professional writers, schoolmasters, cantors and rectors, on occasion *rychtáři*, people like one Vavák, a good Catholic and a wealthy yeoman farmer (*sedlák*) of Milčice who wrote his memoirs at the turn of the nineteenth century. He was not the only one to do so.

The museums of Bohemia and Moravia contain a good many example of manuscript volumes from the eighteenth century – hymnals, for example, which might be illustrated and were usually made for use in the parish choirschool. They are always copies, integral or fragmentary, of printed models, to which other texts were added at the whim of the copyist or the copyist's clients. Thus it was not only the illegality of a book that led to its being copied, since this happened with perfectly Catholic works. Other factors came into play, such as the difficulty of obtaining the work and perhaps its price. Many of our readers owned manuscript works penned by others, but they also produced them themselves. In 1775, Jan Černík, to cite one example, copied extracts from two books he himself owned. The missioner who interrogated him did not ask him for what purpose he patiently compiled objects that he already owned in printed form, but what Černík told him sheds some light on the question:

16 You have admitted to owning the *Špalíček*, the New Testament and the meditations.[55] But here is another Bible, a small book, written and bound, another, also written, but unbound, *item* three manuscript sheets. From whom did you get all this?

I found the Bible in a hayrick on my meadow four years ago. (*Sunt Biblia Hallensis*.) I do not know whose it was. This one is mine; I wrote it from the *Špalíček* and the meditations. That other one I also copied out of the *Špalíček*; it is hymns. The three sheets belong to Frantz Hron, called Pánek. He lent them to me.

Inquiretur, the protocol reads. Copying the same book several times made perfect sense. For one thing, the choice of extracts made a new text. Thus writing constantly created new objects based on material that had already been printed. Furthermore, making a book for oneself, doubtless a painstaking process, reinforced a sense of possession. 'This one is mine; I made it', Černík says. Copying a book was a way to appropriate it more completely.

Access to Literacy

These peasants or artisans were typically people of little schooling, but they read (or at least listened to others read). How did they learn to read? The interrogations seldom say. Family, school, and workplace are mentioned, but it is impossible to weigh their respective roles. Transmission of reading skills in private by close kin must have played a large part, but usually we can only guess that it occurred by reading between the lines when inherited books used by preceding generations are mentioned. Occasionally, though, familial initiation in reading is explicit. Magdalena Židličká, who was 49 years of age in 1748, declares that her parents taught her to read a bit. Several suspects, on the other hand, state that they attended school as children, but this statement has no obligatory relation to their real ability to decipher texts. Anna Němečková, for example (whom we have already met) states that in 1753 her parents sent her, her brothers, and her sisters to the parish school. Her father, a yeoman farmer (*sedlák*)in Vestec, was nearly illiterate, which for Anna was a sign of Catholic orthodoxy. In contrast, her mother was an assiduous reader of heretical books who must already have made her profession of faith. Anna states, 'She forced us to read them, but as for us, we amused ourselves more willingly with other things, like all young people.'

In one of the many interrogations to which he was subjected before 1748, Václav Slavík, a smuggler and *disseminator* of books, explains his cultural itinerary. We are in July 1748, when he was 37 years of age. He was born in

Ředice, a village in the royal demesne of Pardubice that lay in the parish of Holice. His parents, illiterate peasants, sent him to learn 'the art of letters' with the *cantor* Pleskota in the village of Bělá. There was a school with a schoolmaster and an organist in Holice, a town of perhaps 2,500 inhabitants when Slavík was a child.[56] The family doubtless had its reasons for sending him to a village *cantor* who lived more than twenty kilometers from their home. Slavík's later career might make one suspect a common tradition of crypto-Protestantism. He later went to another village to apprentice with the tailor Valenti, who furnished him with books.

It is easier to understand how one came by books. The professional environment could, as it did here, play a part in consolidating and reinforcing earlier acquisitions. In certain cases it promoted the use of books. This is, at any event, the impression one gets from several interrogations. Jakub Dvořák, a vintner from the immediate suburbs of Prague, discovered books because of a colleague who 'much praised' them to him. Other cases seem more obscure. Josef Brabeneč claimed that when business matters took him to Gersdorf in Lusatia he was offered two books by the tailor who lodged him, a Czech émigré. On his return to Bohemia, our traveller began to preach, proud of his new knowledge, but also confident of being protected by the thumb of a hanged man that he had bought for a hefty price from a hangman.'There are seventy-seven religions, all salutary', he declares. 'The Saxons and the Silesians do not hold the Mother of God as holy. . . . Death is nothing but a dream; hell, nothing but a sort of obscurity.' He was arrested and sentenced on 9 January 1767 to one year of forced labour, chained hand and foot, for scandalous speech and suspicion of heresy.[57]

In the villages and towns our readers might encounter Jews, who reappear several times in the sources. They sold old clothes and used books. Above all, they attracted people, since they were a bookish lot if ever there was one. Lidmila Baurová, Václav Janeček, Jiří Chaloupka and Jiří Psenička, Jan Pita, Jan Vacek and Lukáš Volný paid with their lives in 1748 for frequenting them, and were burned at the stake in five towns in the circles of Hradec and Bydžov. Following the example of the tailor Pita, disciple of the rabbi Mendl, had they sought to interpret the Bible too closely?[58] At least sixty-one people were implicated in their 'Hebraic sect', as the missives from Vienna called it, and they were all sentenced to a variety of punishments by the Court of Appeals. Twenty years later, the movement revived in the same region.[59] These exchanges between Christians and Jews also had fainter echos, perceptible since at the end of the seventeenth century, and in June 1752, when the state was concerned over what it considered apostasies, the piarist missioner, Father Victorinus a Matre Gratiarum, noted two peasants in the village of Oseček 'who discussed the Scriptures among the Jews of Poděbrady, proposing their interpretations and reciting the psalms of David from memory'.[60]

What can we know about the schools that our suspects might have known? They varied enormously between a humble village and the royal city. In the early eighteenth century, cities like Chrudim or Čáslav maintained a *cantor*, a *subcantor*, a *ludimagister*, and several assistants. There is little information on education in the village before the reforms of Maria Theresa and the application of the law of 6 December 1774, which stipulated obligatory schooling for all children over six years of age. Still, we can reconstruct the building of schools and the presence of schoolmasters in each parish of the diocese of Prague in 1677, 1700, and 1713. In 1700, according to my own calculations, we can estimate that there were approximately 1,000 schools in the diocese.[61] In 1779, there were apparently 1,272 elementary schools in Bohemia and, if the statistics are reliable, 1,906 in 1781.[62] In any event, the school population rose rapidly. Estimated at 30,000 children in 1775 (perhaps 15 to 16 per cent of children from six to twelve years of age), it more than doubled by 1779, and in 1784 had already reached 119,000 children (or 59.4 per cent of the age levels concerned). In 1828, 91.1 per cent of children frequented primary ('trivial') schools.[63] Before the last twenty years of the eighteenth century, then, a minority of all children went to school for part of the year, in general during the winter. In principal, they learned the catechism, then reading, writing, and some arithmetic, along with musical instruction. Often, however, the instruction provided did not cover all of these headings.

Bohemian artisans and farm folk suspected of heresy did not often sign their *reversales*. The ability to sign one's name is today no longer considered a sure index to the mastery of writing, so it is out of the question to calculate percentages of those who could read or write on the basis of these documents. Nevertheless, one formula returns constantly at the foot of professions of faith: 'Because I do not know how to write, I have signed with three crosses by my own hand.' Does the lack of a signature by and large correspond to an inability to write? It is with this hypothesis that, in spite of all, I offer a few figures here. We have the *reversales* of 192 individuals, all rural folk from the north-east quarter of Bohemia, who swore their oaths between 1713 and 1771.[64] Two-thirds are men. Only 42 per cent of them sign, as does only one woman. But 90 per cent of these men and women are accused of having read and kept books in their homes. One of the women who does not sign had also lent and sold books. Thus in Czech country areas, the ability to read seems to have exceeded the mastery of writing.

Moreover, even when it is established that they do know how to read, our suspects were not necessarily able to handle all types of written texts. Their interrogations from time to time show proof of only relative ability to decipher a text. 'I know how to read printed letters, but not written letters', explains Jan Stěna, a twenty-nine-year old shoemaker in Veletov in 1729. He could write, however, but probably not in cursive. Fifty years later Kateřina

Koldová was incapable of reading the hymns that her husband had had copied because she could not decipher manuscript writing. Could we go further and suggest that discrimination between printed characters and manuscript letters defines a norm? This seems to be implied by Tomáš Vobocký in 1729 when he found it difficult to clarify the level of reading competence of his friend Čačák:

28 –Didn't Čačák have a manuscript hymn in his house?
   –Yes.
30 –Did he sing it?
   –Yes, and I with him.
31 –So he could read manuscript letters?
   –If he can read *other things* [other than print], I don't know. In any event, he sang.[65]

Did not these 'other things' designate what was not ordinarily mastered in the blacksmith's social circle?

Reading a text did not necessarily mean that one could follow it word for word from beginning to end. There were more economical ways that did for a portion of our readers. Here the level of formal recognition joined that of comprehension. Grasping the tenor of a book through its structure was yet another way of knowing. Vobocký prompted disbelief in his examiner when he described the contents of one of Čačák's books in minute detail, yet claimed to have had it in his hands only twice. Did this mean that in truth he had read it much more than twice? 'No,' he countered, 'it is just that every time I looked at what chapters there were in it.' He then recited, in order, the titles of the various parts of the volume: the Acts of the Apostles, the Epistles of St Peter and of St Paul, the Apocalypse, and a commentary on the Old Testament.

Testimony of this sort raises the vast problem of the real comprehension of the texts' message. What real knowledge did these semi-illiterate rural people have of the books they frequented? Their acquaintance rarely seems direct and total. Many say that they have not understood their reading or what was read to them (which might be a ruse to put off the inquisitor). All understood the text basically in their own manner. Very often the book was simply a support, an indispensable reference. Whether the reader was truly imbued with the substance of a text read and reread step by step, or whether he or she had appropriated it by means of mnemonic reference points, free course was always given to the faculties of interpretation.

Doctrinal Illiterates

Not only were the persons interrogated of restricted literacy;[66] their knowledge of doctrine was also very incomplete. Jiří Černík, who was 33 years of age in 1775, thought that confirmation served to pardon sins. In Prague in 1725, a poor woman, Anna Vojtěchovská, confused the sacraments and the Trinity. 'How many sacraments are there?' the missioner, František Mateřovský asked her. Three: God the Father, God the Son, and God the Holy Ghost. 'What, aren't you ashamed? You can't even tell the number of the sacraments!' She answered, 'Your lordship, I am old, I am more than eighty years old, and when I was young, they did not teach like that, and now, I cannot learn any more.' In like fashion, some of the examinees with more clearly perceptible post-Hussite roots placed themselves in a dogmatic interconfessionality that can in part be explained by ignorance. Again in 1725 and still in Prague, a servant, Anna Uhlířová, saw no difference between the Lord's Supper in the two kinds and the Catholic Eucharist.

This woman was not moved by a desire to follow the faith of her fathers, nor, conversely, by an interest in showing fervent adherence to Catholic beliefs and rites. Like many others, she asked above all an immediate efficacy from gestures that were felt, basically, as magical. Frequenting a secret assembly, listening to the readings, communicating in the two kinds also (and primarily) meant hoping for aid and protection. Uhlířová states that since she had communicated with bread and wine in a garden of the capital, as much impelled by curiosity as at her brother's urging, she had 'received no more blessings at all, nor found one single piece of bread'. The practice even turned out to have harmful effects: far from enriching her, it had brought the charge of heresy on her. Thus we see her arguing the need to honour the Virgin and the saints. Was this a reflection of her true belief or was it camouflage? That was not what counted. But how had she learned that the cult of the saints was obligatory? 'I read it in the books', she answers, thinking to give her inquisitors the greatest possible proof of her sincerity. It never crossed her mind that by saying so she aroused their suspicions even more: would not a good Catholic have invoked her parish priest's preaching on this point? For her, the book constituted an authority superior to that of her examiners, who professed to teach the one licit and salutary faith. The testimony of Jan Fiala, another person implicated in the Blaha-Vorlíček affair, also attests to the inherent power of the book. This itinerant poultry-pedlar told his judges, in effect, 'I do not know how to read, but since what you are telling me is in a book, I will accept and believe it.'

For the book had an immediate power to instill belief. 'Give one look at this book and you too will believe it', the farmhand Matěj Švančara objected in 1778 to his brother-in-law Vojtěch Kohout. Kohout was so persuaded that he was arrested. That they were arguing the uselessness of the cult of the saints is secondary. 'One single glance at the Holy Scripture', Švančara insisted, 'and you will immediately know that what is forbidden to you is good, and what is permitted to you is bad.' The book was knowledge and truth to the absolute degree. Little wonder, then, if priests did not want to share the source of their power – here, the possession and use of religious books. 'Old books are better', this same Švančara , a sort of local 'guru', told Kohout, 'and this is why the priests keep them to themselves and give the bad ones to people.'

The truth of the book went beyond the truth of its contents. It was truth itself, concretely, in its full materiality. Anna Notovná was so persuaded of this that she rejected belief in Purgatory without hesitation: 'They claim that there is one, but it is not to be found anywhere. It is not in the *Syrach*,[67] it is not in the Gospel, and I told my son, it is not in that other little Gospel there either. So I added, I don't believe in it at all.'

The object of all fascinations, the book, mysterious and forbidden, under-standably attracted even the best Catholics. In Bezno in 1779, Veronika Niznerová and František Šakr lived near this Anna Notovná, who was known to harbour inappropriate thoughts. Veronika ceded to the seductions of the hymns that Notovná sang to her, perhaps to help her bear her unhappiness at her brother's illness. She bought a hymnal from Kleych. As his inquisitor remarked, Šakr, a worker mason and a farmholder, had only good books in his house. He was none the less intimately persuaded that there was no resisting the charm of a 'Lutheran' book. 'Fool!' he shouted at his servant woman. 'Don't you know that if their Bible fell into your hands you would throw away your books on the spot, so sweet it is and so suave!' This was how the magic of prohibited texts operated on the eve of the Patent of Toleration.

'The Catholic religion is evil, for it comes from the executioner; the Evangelical is good, for it comes from the Gospel.' These juxtaposed statements, which operate by exactly the same process of analogical deriva-tion, are often encountered. It hardly matters whether they are correct etym-ologically. If *kat*, the executioner, led to *katolická*, thus proving the constitutive decline of the Roman Church and its doctrine, the Gospels – *evangelium* – fully legitimized the true faith, which drew its origin from the sacred Book. What it transmitted was revealed truth. Underlying the attachment to the book manifested by our suspects we find the model of Scripture, at once the materialization and the mediation of the divine Word and the real presence of God. It may be coincidence, but in Czech the noun designating the ability to read – *čtení* – also signifies the Gospel, reading *par excellence* proclaimed aloud.

'From the Gospel all that is good and all that is evil can be known, and all sermons must be drawn from it.'[68]

The New Testament first, then.[69] On this base, the foundation of behaviour, other books were also psychologically invested with a transfer of the sacred function. In most cases such works were religious: hymns, prayers, postils and commentaries, summaries, 'manuals'. Before 1620 the authors and editions of such works vary enormously. In 1753, the prize possession of

Václav Čepek from Velenice, a workman, 35 years of age, was a commentary on the Gospel of St Matthew written by Erasmus.[70] Čepek had inherited it from his father. He was not alone: titles that today we would class under other categories than devotion enjoyed a similar veneration. These were for the most part sixteenth-century editions, such as the *Czech Chronicle* of Václav Hájek or the laws of the royal cities. Such secular books were indeed rarer.[71]

Furthermore, several interrogations clearly show that the term 'Scripture' was very broadly defined. The word in Czech – *písmo* – was also applied to other sources of the Word than the Bible, in whole or in part. God expressed himself in hymns and postils as well, and his elect – readers, auditors, and singers – sure of possessing His spirit, did not doubt of their clairvoyance in literal interpretation of the texts. Often, then, Scripture was mentioned when there was explicit reference to another type of printed matter. It is probably from a postil (perhaps even a Catholic work) that Jan Nahlovský or Jan Jandík memorized the Gospel around 1729, for they placed the passage in the liturgical year: 'He spoke of the Gospel and said that there is something like that written at the Monday of Pentecost, that he who enters into the sheepcote, if it is not me, is a criminal and a brigand.'[72] Or again, 'Priests are the false prophets spoken of in the last Gospel after Pentecost, those who show you the Christ: There is the Christ; here is the Christ.'[73]

Hymns joined postils as mediators and perhaps played an even more important role. The Czech *Gesangbücher* or *kancionály* often included sung paraphrases of the Gospels, placed back into the liturgical calendar. It is not impossible that our two suspects also drew their knowledge from a publication of the sort.

Sung Reading

In general, hymns and canticles occupied a very special place in the statements of the people interrogated. By itself, the hymn constituted a mode of acquisition of knowledge, and thanks to its form – sung, versified, and rhymed – it was a favoured medium. A Czech hymn-book, Protestant or Catholic, was the lay equivalent of a ritual, a missal, a prayer book, or a psalter. It contained the entire liturgical year, the temporale and the sanctorale, and as a supplement it offered spiritual munitions for all the occasions of daily life. Thanks to it, for example, one could struggle against anxiety, ward off a bad reputation, leave for a voyage with a lighter heart, bury one's dead in a dignified manner, remember that humankind is but dust and many things besides – and all in song. Such books were thus nearly indispensable and much used, and they were published in an enormous number of editions and in all formats.

The hymnbook also offered the catechism, doctrine, and scriptural texts in a form only slightly adapted to the requirements of rhyme. A good many were found in the possession of our suspects. The figures that follow give a general but highly relative notion of how many, since they concern only the year 1753.[74] Out of 614 volumes sequestered by the ten of the missioners sent out at the time by the archdiocese, 492 titles were actually turned in. Collections of hymns represented 21 per cent of these copies, even 36.5 per cent if one counts psalters. In comparison, postils accounted for 28 per cent of copies and prayer books 26 per cent. These two types of printed work often included hymns, which increases even further the true proportion of the latter. The New Testament accounted for 15 per cent, and complete Bibles only 5 per cent of the volumes seized that year. In their monthly reports, the missioners constantly noted the presence of flysheets or pamphlets containing *cantilenae*, printed and manuscript, that might be *haereticae*, *scandalosae*, and *venerae*.

In light of the examinations for suspicion of heresy, the hymn appears as a purveyor of formulas and of evidence that offered a framework to thought or took its place. If, as Pavel Čačák claimed in 1729, the mother of God 'cannot aid us in anything, being herself quite content to be in heaven', 'In whom will we find refuge?', Jan Pekař responded, repeating word for word the *incipit* of a pilgrimage hymn to the Virgin of the early eighteenth century, much in fashion at the time.[75] While he was working to repair a wooden part in the millworks, the carpenter Jan Nahlovský, around the same date, struck up two hymns with his fellow-workers. One contained the Hussite version of the Lord's Prayer and the other, Utraquist, went back at least to 1559, but neither would have been out of place in most of the Catholic hymnals that bore the *imprimatur*. Nahlovský, who seems always to have been ready to talk, took advantage of the situation to instruct the miller and his helper in the unworthiness of the cult of saints. The first of these, Matěj Škrěta, reported:

Then we sang the hymn that begins 'Let us pray "Our Father", in silence and in humility'. When we arrived at the verse: 'Oh desperate man, you are so hardened that you give yourself over to the cruellest torments. Take thought that there will be for you no guarantor'.[76] He explained to me that this guarantor means that it would be vain to implore St Wenceslas, 'St Guy, pray for us', etc. Then we sang, 'Let us prepare, oh ye faithful, for prayer.' Later on in this hymn it goes, 'Not in the name of another'. 'You can plainly see', he said to me then, 'the Scripture proves it to you: we must not invoke anyone other than God.[77]

It was also because she accorded this absolute faith to the message of the hymn, considered the divine Word, that a servant woman of Prague, Kateřina Černá, communicated once with both bread and wine in 1709 or 1710. Her brother, of whom she was thoroughly afraid, sang her a Hussite refrain that she knew by heart: 'You who do not take in the two kinds, take great fear of the fires of hell!'[78]

Obviously, it is essential to include singing among the practices studied. By means of the hymn or the song, the text took flight from the pages of the book to reach the illiterate or the semiliterate even when, as was the case here, they had daily acquaintance with reading.

Oral Practice and Memory

In order to make the holy Word one's own, it was thus not necessary to know how to read. The first transmission of texts often passed by oral means and by memorization, and this was true even in the fifteenth century, when the papal nuncio Eneas Silvius Piccolomini, the future pope Pius II, expressed his astonishment that in Bohemia a simple old woman should be capable of reciting the Bible by heart. Books – which one hid and prized, but also exchanged, sold, and lent to others – offered an equivalent to the sermon. Their preaching was intimate and immediately internalized; they were a voice that every reader and every listener could make his of her own and interpret freely. All the examinations for heresy show proof of this. The link between the written text and oral practice was permanent. Did our readers also read with their eyes alone and for their inner selves? We cannot know, for it is their voices that betrayed them: it was often when they read aloud, for themselves and for others, that they were taken. Generally, their reading was collective. The master of the house read for his household or attracted a neighbour's visit by promising to read a page or two. This was the case of the yeoman farmer Pavel Čačák in 1729, who invited Tomáš Vobocký, the village blacksmith, to come to his house to see a work of Martin Philadelphus, an Utraquist priest and humanist of the later sixteenth century. Once again, the work was a postil.[79]

This affair is exemplary, for it reflects a reading situation that was, in the last analysis, very complex, elements of which can be found dispersed in many of the interrogations. The host first read the passages concerning the Sybil, then the Marriage of Cana. Then Vobocký took the book and read the sermon for St Steven's Day and 'the commentary on the sheepcote'.[80] From that day on, he formed the habit of going to Čačák's house, where the two men soon began a second book, some sort of abridged version of the New Testament. The whole household assembled to listen to them, but 'from the village there was no one'. From this semi-private reading, they quite naturally passed to singing. There too, the oral was based on the written and the written on the oral. Čačák owned a manuscript hymn – a 'little song' (písnička), since Czech does not distinguish between hymn and song. But we will never know if this sheet was really read or served simply to jog their memories. It might also have played a third role of a symbolic nature. These handwritten pages that people copied and at times composed and that were discovered in their houses were not necessarily used directly, since many people – Čačák perhaps among them – only knew how to read print.

What we see developing is a quite different set of relations with writing, no longer founded on the decipherment or the hearing of a text, but on the possession or the simple presence of the book as an object. It was certainly a talisman, although that aspect is little documented. At the same time, it was the site of a transference on the part of its owners. The book or the paper that one locks up carefully, at some risk, is a figure of oneself, and by that token it becomes one's most prized possession. It becomes a part of the individual in what makes him or her most irreducibly unique, projected to the external world. There is no need to know how to read to respect writing when it is invested with such force. This, in my opinion, surpasses a simple affirmation of secret Protestantism. Dorota Kopečná procured four books in 1776 so that at the proper moment she could show the *hejtman* of the circle of Chrudim that she too was 'of that religion'.[81] Eighteen years earlier in the same region, Magdalena Teplá, a landless farm woman of 63 years of age who did domestic service, declared herself ready to die rather than to leave her books in the hands of the missioners.[82]

The Ultimate Appropriation

Pushed to its logical term, the assimilation of the written word rendered the book useless. The last phase of this process of internalization returned to oral practice, closing the circle. Our suspects show a surprising ability to retain what they have read and heard others read all through their lives. There again, words known 'by heart' are deeply graven in the individual who makes them his or her own. This explains the insistent tendency to quote that we see in all the people interrogated, witnesses and accused alike. Their thought found expression in references and formulas, but also it revealed the sincerity of the speakers. A reading internalized to this point also furnished a way to decipher and understand the meaning of existence and of the surrounding world. The false prophets of the Scriptures quite banally came to mean the priests that these people lived next to and saw daily. Jiří Janda, a peasant from Mečeřice, no longer distinguished real events from predictions in the holy books. His neighbours declared that it was impossible to repeat all that he said, and he backed up his every word with Scripture, showing a predilection for Revelations, which he cited from memory, verse after verse. He declares:

Moreover, the priests are very clever, and they will fool the whole world. Here, this is what is noted in this chapter: they will go wallow with women and will seek them harm. Watch out, that may have already happened in your village; in ours, that is what has already happened.

When asked to say where he got his information, Janda continues:

I said: in the book of a Saxon soldier. He [the parish priest] answered me: That isn't possible, one cannot understand those books immediately! And me, I proposed to him to give me whatever book he wanted. I would read it to him one time only, and then I would recite it right away from memory!

No need to learn in order to understand and to interpret. God had made all men worthy of penetrating his secrets:

His Word is not for some only, but for all. Me, I possess the holy Spirit, and I speak by the holy Spirit. I have read more than a hundred Bibles and many books. Someone came to look for them in my house, in the well, everywhere. They didn't find anything: me, I have it all in my head. No one can take anything from me.

They took his life, though. The Court of Appeals sentenced him to death, 'the sword at his throat', according to the traditional expression, on 18 September 1761.[83]

References to Interrogations Cited

With the exception of the interrogations of Suchý and Veselý, the catalogue numbers for which are indicated in the notes, all references to interrogations are taken from sources of the archdiocese of Prague (APA I) housed in the Central State Archives (Státní ústřední archiv, here abbreviated SÚA) of Prague. In order to avoid unmanageable repetition, the files used are listed here, with the names of the suspects or the witnesses interrogated.

H 2/1–2[4287]	Matěj Bina, Blaha-Vorlíček affair, Pavel Čačák, Jakub Dvořák, Jakub Hrubeš, Karel Hrubeš, Jan Pekař, Jan Petr, Jan Stěna, Václav Šultz, Tomáš Vobocký
H 2/4[4288]	Václav Trubáč
H 2/5[4290]	Václav Čepek, Rosina Čepková, Matěj Hlaváček, Anna Němečková
H 2/5[4292]	Josef Brabeneč, Kateřina Černá, Jan Černý, Blaha-Vorlíček affair, Jan Fiala, Jan Jandík and Jan Vaněšovský, Jan Nahlovský and Matěj Škřeta, Anna Uhlířová, Anna Vojtěchovská
H 2/5[4293]	Jan Černík, Jiří Černík, Lidmila Horynová
H 2/6[4294]	Jiří Janda
H 2/9[4439]	Vojtěch Kohout and Matěj Švančara, Kateřina Koldová, Veronika Niznerová, Anna Novotná, František Šakr
H 5/2–3[4314]	Václav Slavík

Abbreviations of References Cited in Notes

SÚA Prague; SÚA Zámrsk Central State Archives of Czechoslovakia, Prague or
Zámrsk
APA I   Archives of the Archdiocese of Prague
AS   Archives of the Court of Appeals of Prague
BA   Archives of the Diocese of Hradec Králové

Notes

1 These terms are obviously approximate. In Czech countries, what I am calling a yeoman farmer was a *sedlák*; a peasant farm-holder was a *chalupník*; a gardener, *zahradník*. In Czech these terms refer to the fiscal category of the peasants as a function of the quality of their land and the acreage of their holdings. In the eighteenth century one encounters 'demi-yeoman-farmers' or 'quarter-yeoman-farmers' (*půl sedlák*; *čtvrtláník*, etc.).

2 A 'circle' was an administrative and geographic division that corresponded to a province, perhaps to a county or a shire in early modern England.

3 It seems that this oath was not always required, at least until the last third of the seventeenth century, no matter what the law said. Several testimonies of parish priests of the diocese of Prague state as much in 1677. In the eighteenth century, pastoral visits indicated that, on the contrary, such swearing was universally practised. This did not prevent certain schoolmasters from making use of Protestant books, however. In 1752 a schoolmaster in the parish of Skalsko, in the circle of Boleslav, used Luther's catechism in class, brought to him by a pupil, the daughter of the shoemaker Pecháček (SÚA, APA I, H 2/5^{4291}).

4 On the population of Prague, see Ludmila Karníková, *Vývoj obyvatelstva v Českých zemích, 1754–1914* (Nakl. Československá akademie věd, Prague 1965), p. 59. I have relied my own research for provincial cities.

5 SÚA, APA I, H 2/5^{4292} and H 2/1–2^{4287}; AS 152, 41/350 and 159, 41/351.

6 Bohemia proper, Moravia, and Silesia. I am concerned here only with Bohemia.

7 Václav Jílenský, *Líčení dnešních pohrom...* ('An account of the catastrophes that now overwhelm the Czech nation'), MS in the Library of the National Museum in Prague, catalogue number VD52 (written after 1627 by a Catholic).

8 František Šmahel, *La Révolution hussite, une anomalie historique* (Presses Universitaires Françaises, Paris, 1985), p. 112.

9 The Hussites were called 'Utraquists' as early as the fifteenth century. The term is formed from the expression *sub utraque specia*, 'in the two species'. It thus designated those who communicated with both bread and wine. At the end of the fifteenth century, Bohemia was the only biconfessional country in Europe. The Utraquist church possessed its own institutions even though it had not resolved with Rome the problem of ordination. The structures of the Catholic Church were much affected by the Hussite crisis. Between 1471 and 1561 the archiepiscopal seat of Prague remained vacant. In the sixteenth century, the Utraquists were separated into 'old Utraquists', more and more in the minority,

and 'neo-Utraquists', who were in reality crypto-Lutherans. In fact, the two official religions were still Catholicism and Utraquism, except between 1609 and 1620. In 1609, Rudolph II accorded religious liberty to Czech evangelicals. The term grouped the Utraquists, who professed all tendencies of European Lutheranism, Calvinists in small number, and the Czech Brethren (created in 1457), more often called Moravian Brethren outside their own country. The latter group issued from radical Hussitism in its most profound and popular form. From the beginning the Brethren were organized as a separate church, and its illegality earned it repeated persecutions. As a community of predestined 'saints' divided into congregations with their priests (called 'servitors') and their 'ancients', male and female, it remained numerically in a very small minority. But it was well implanted throughout Bohemia and in eastern Moravia, and its cultural influence far outstripped the small number of its members.

10 These modifiers, highly representative of triumphalist phraseology and the rhetorical formulas to celebrate the victory of the Habsburgs, have been furnished to me by S. Sidecius, *Bohemia exoriens, christiana, bellicose, imperans, gentilis, pia, regnans, austriaca . . . concinnata a Collegii Societatis Jesu Pragensi* (Prague, 1627). The Jesuits of Clementinum had had the work printed on the occasion of the crowning of Ferdinand III as king of Bohemia.

11 Philippe Joutard, *La Légende des Camisards, une sensibilité au passé* (Gallimard, Paris, 1977), p. 41.

12 There are several different estimates of the population of Bohemia before the Thirty Years War. They oscillate between 1,200,000 and 1,700,000. Moravia is supposed to have had a population of from 800,000 to 900,000. See Eduard Maur, *Československé dějiny, 1648-1781* (Prague, 1976), p. 6.

13 APA I, registers B 11/7 to B 13/4a.

14 This suspicion comes from the inventories of parish books annexed to the responses to the 1677 questionnaire.

15 From around 1634, *vicarii foranei* in each circle were charged with supervision of twelve to fifteen parish priests.

16 The dioceses of Prague, Litoměřice (formed in 1655), and Hradec Králové (formed in 1660).

17 The verdicts of this Court of Appeals can be found in the Central State Archives in Prague, collection AS, *Ortelní knihy. Urtheil-Bücher*, catalogue numbers (from 1704 to 1781) AS 146, 41/344 to AS 182, 43/378. All the figures given here have been elaborated on the basis of these documents. In reality, the first cases of heresy judged by the Court of Appeals go back to 1689. The series of verdicts presents two lacunae (for the years 1749-50 and 1756) and one register is unusable (1757-8). The number of 'heretics' sentenced must thus be somewhat greater than calculated here.

18 At least four of those sentenced to death had their sentence commuted to forced labour in perpetuity.

19 For its part, the officiality of the diocese of Prague seems to have been also less than zealous in transmitting to the Court of Appeals well-documented cases of individuals who had relapsed several times, like Václav Polák, who fell into 'heresy' seven times before the Church turned him over to the court.

20 I will use here only the investigations from the diocese of Prague, which covered

all Bohemia with the exceptions of the circles of Litoměřice, Hradec Králové (and Nový Bydžov), which made up the dioceses of the same name. Interrogations in the diocese of Hradec Králové have not yet been made public.

21 APA I, H 2/5[4293] (De statu religionis autem, 1735, 1750, 1751).

22 The distinction between the charged prisoners transmitted or not to the state's justice recalls the separation between reconciliati and relaxati of the medieval Inquisition. The analogy is not total, however, and the system – if indeed there was one – functioned differently. The officiality at times absolved suspects without transferring them to the Court of Appeals when, according to its own criteria, they were heretics. In many cases, the raising of the excommunication was accompanied by various penalties inflicted by the lord of the penitent: fines, the construction of a chapel, funding pious works, etc.

23 Augustin Neumann, Prostonárodní náboženské hnutí dle dokladů konsistoře Králové hradecké (Hradec Králové, 1931), p. 161, appendix 27.

24 APA I, H 2/5[4293.]

25 APA I, Patentes vernales, 31 March 1735.

26 Karel V. Adámek, Listiny k dějinám lidového hnutí náboženského na ceském východě v XVIII a XIX věku (2 vols, Nákl. České akademie věd a uměň, Prague, 1911), p. 62.

27 APA I, H 2/5[4290.]

28 Pyres of books began to burn in the seventeenth century. After 1719 the ecclesiastical authorities counselled confiscation instead. Despite this, books were still burned.

29 APA I, H 2/9[4296.]

30 'Sine hac substitione parum produit omnis labor, atque conatus conversioni haereticorum impensus', cited in Neumann, Prostonárodní náboženské hnutí, p. 108.

31 The Bible known as 'of St Wenceslas', printed in three volumes between 1677 and 1715.

32 Clavis haeresim claudens et aperiens. Klíč kacířské bludy rozvírající a zavírající... (1st edition, Hradec Králové, 1729; 2nd edition revised and augmented, 1749; 3rd edition 1770).

33 APA I, H 2/9[4296], Circa libros. Pro memoria.

34 Antonin Koníaš, Postilla, aneb celo-roční vejkladové... (3rd edition, Prague 1756), Preface, 2nd, 3rd, and last leaves (unpaginated).

35 On intensive reading, see Rolf Engelsing, Der Bürger als Leser. Lesergeschichte in Deutschland 1500–1800 (Metzler, Stuttgart, 1974).

36 Roger Chartier, 'Du livre au lire', in Roger Chartier (ed.), Pratiques de la lecture (Rivages, Marseille, 1985), pp. 62–88, esp. pp. 70–1.

37 During the eighteenth century, missioners did their best to impose a purely Catholic way of greeting one another. In the place of the habitual Pozdrav Pán Bůh (God greets you), they introduced the translation of Laudetur Jesus Christus, Pochválen bud Pán Ježíž. Koníaš propagandizes for this salutation in a dialogue in his book, The Infallible and Golden Dawn (Zlatá neomylná dennice, [Prague, 1733]). Furthermore, the second edition of 1738 of his hymnbook, The Cythare of the New Testament (eight editions before 1800) includes a hymn that praises the advantages of the use of this formula.

38 Central State Archives of Eastern Bohemia, Zámrsk, velkostatek Rychenburk, C.36, fos 298–301.

39 Meaning that he was a 'half-farmer' (*půlsedlák*).

40 The national bibliography of works printed in Czech and in Slovak up to 1800 (*Knihopis*) cites three different editions of a collection of Gospels and Epistles published in Litomyšl, all by the printer Tureček, who specialized in broadsheets and books sold at markets and fairs. The book is certainly one of these. (*Evangelia i[těž] epištoly na nedeli a svátky...*, 1785, 16°, 238pp.; 1791, 16°, 224pp.; 1800, 16°, 224pp.).

41 The first citation is obscure, but it is perhaps a reminiscence of the Acts of the Apostles 17:24. The second refers to the first letter of Paul to the Corinthians, 3.16, with modifications.

42 Suchý, Vesely and the many 'errants' located during the same period cannot be classified in either of the two categories of religious dissidence proposed by Troeltsch since they respond neither to the *Kirchentypus* nor to the *Sektentypus*. See Henri Desroche, *Les Religions de contrebande* (Mame, Paris, 1974); in particular, his reflections on cultural nonconformity, pp. 128–33.

43 Interrogation of Jan Černý, from Žeřcice, parish Dobrovice, responses nos 9 and 12; interrogation of Jiří Hlaváč, responses nos 10 and 12. APA I, H2/5[4292.]

44 Václav Schultz, *Listář náboženského hnutí poddaného lidu na panství litomýslském v století XVIII* (Prague, 1915), p. 132. In the village of Horky, where Jan Kolda and his wife Kateřina lived, 80 persons (17 households) preferred to declare themselves 'evangelical first', rather than Lutheran or Calvinist, following the Patent of Toleration in December 1781.

45 APA I, H 2/9[4439.]

46 Kateřina Kodová and her husband possessed prohibited books. It is difficult to untangle what in her confession may have come from her reading and what from family religious tradition. Whatever the answer, Protestant impregnation in the larger sense is obvious. Many of the ideas she expresses are already found around 1420 among the radical Hussites, the Taborites. The *Sola Scriptura* is already in Hus, and the New Testament long constituted the sacred book of the Brethren and the foundation of their theology. After around 1532, the Czech Brethren retained only two sacraments, the Lord's Supper and Baptism. Like the Taborites, it recognized neither transubstantiation nor consubstantiation. For the Unity, the Lord's Supper was both commemoration and the spiritual presence of Christ among the assembly. In the absence of ministers, laymen could quite well administer communion in the two kinds to one another. Faith alone guaranteed salvation, but it should find expression in charity. Like Luther and Calvin and like the Taborites, the Brethren refused the intercession of the saints, the cult of images, confraternal organizations, pilgrimages, prayers for the dead, and Purgatory. One of the essential contributions of Hussitism was to have abolished the distinction between priests and the laity. The Four Articles of Prague, the minimal programme around which moderates and radicals joined after 1420, demanded liberty to preach the word of God without shackles, communion for all in bread and wine, the poverty of the Church and the clergy, and the punishment of mortal sins. Among the people subjected to interrogation

226 MARIE-ELISABETH DUCREUX

on suspicion of heresy, iconoclasm and anticlericalism are frequent. These traits and others, such as a tendency towards blasphemy, are not automatically the signs of a Protestant heritage. This set of elements poses two interrelated problems, however: that of the conditions of transmission, and that of the possibility of recreation of certain ideas or attitudes outside of a constructed system.

47 At the end of the seventeenth and into the eighteenth century, most of the flysheets and the chapbooks sold by pedlars and in the fairs and markets were in the form of hymns and songs. The national bibliography of printed works in Czech and in Slovak up to 1800 (*Knihopis*), which does not include flysheets, lists 6,694 tracts and pamphlets, nearly all during the eighteenth century. This figure is minimal. The subject matter of these publications was 75 per cent religious.

48 Jiří Pokorný in Prague has studied the inventories after death of Prague burghers of the eighteenth century: Dr Nový's students have analysed similar parish priests' inventories in the diocese of Prague after the White Mountain. These two population groups are not comparable to the one studied here. The interrogations do not include all the books owned at a given moment by an individual, but they show them in context, which is their great advantage.

49 Among the suspects cited in these pages, this was true of František Šakr, Matěj Švancara, and Jan Černík, among others.

50 Václav Hájek Z Libočan, *Kronyka česká* ... (1st edition, Prague, 1541), quarto, 527pp. These annals of Czech history met with great success.

51 The Rhenish florin was an accounting denomination; the kreutzer, both a coin and an accounting denomination. One florin was worth 60 kreutzers. Prices should be compared with those in practice on the official book market and with the purchasing power of the sums involved.

52 *Špalíček* (little log) was the usual designation for one of the books of the émigré publisher, Václav Kleych, so called because they were thick books higher than they were wide. *Nábožních křestanů ruční knížka* ... (Brief manual for pious Christians ...) (Zittau, 1709), 16°; eight editions before 1782.

53 Interrogations for heresy are rich enough in detail to suggest that they might make a contribution to the history of book prices. In the case of these suspects, it goes without saying that a sentimental value might have interfered with the books' commercial worth.

54 A *vértel* was a fourth of a *strých*, or about 13.3 kg, an equivalent measure proposed for a *vértel* of rye by Emanuel Janoušek, *Historický vývoj produktivity práce v zemědělství v období pobelohorskem* (Prague, 1967), p. 108.

55 These meditations are probably the translation of Philip Kegel's *Zwölff geistliche Andachten*, a work often encountered among the suspects who owned books. Its title in Czech is *Dvanáctero duchovné nabožné přemyšlování* (1st edition, Zilina, 1669, followed by seven more editions before 1783, eight in Slovakia, three in Lusatia, one in Prague, one with no mention of place).

56 This is indicated in a manuscript inquiry of 1713 on parish revenues (APA I, register B 15/1, circle of Chrudim, vicariat of Pardubice, parish of Holice).

57 SÚA Prague, AS 178, 43/374.

58 SÚA Prague, AS 170, 41/368, APA I, H 2/5⁴²⁹⁰, H 2/5⁴²⁹², H 5/2-3⁴³¹⁵/SÚA

Zámrsk, BA IVi/21, box 33. A report dated 12 April 1747, and signed by the vicar of Chrudim, presents Mendl as a pedlar who went from village to village and dissuaded people from reading their Catholic books, which were, according to him, 'false and deceiving'.

59   SÚA Zámrsk, BA IV 1/4, box 74.

60   APA, I, H 2/5⁴²⁹¹. The peasants were named Francouz and Chadima.

61   Survey of 1677: APA I, registers B 11/7 to B 13/4a; survey of 1700: B 13/4 to B 14/17; survey of 1713: B 14/18 to B 14/29, and B 15/1 to B 15/24.

62   By extrapolation from the figures for 1779 to 1789. To give a comparison, only 4 per cent of Silesian children attended school in 1770. See Otakar Kadner, *Vývoj a dnešní soustava školství I* (4 vols, Nákl. České adademie věd a uměń, Prague, 1929–38), vol. 1, p. 59.

63   Michail Kuzmin, *Vývoj školství a vzdělaní v Československu* (Prague, 1981), p. 64. Kuzmin elaborated these data on the basis of official Austrian statistics and from several other scholars' figures.

64   APA I, C 105/5²⁰¹⁸, H 5/2-3⁴³¹⁴. H 2/6⁴²⁹⁴.

65   My italics.

66   A term suggested by Jack Goody: John Rankine Goody, *Literacy in Traditional Societies* (Cambridge University Press, Cambridge, 1968).

67   *Syrach*, or Sirach, is the name consistently used for Ecclesiasticus in these sources.

68   Interrogation of Jan Vaněškovský, Jan Jandík affair, around 1729, response 13, APA I, H 2/5⁴²⁹².

69   With the exception of Ecclesiasticus and the Psalms, it is very unusual to find references to the Old Testament in the suspects' testimony. Their scriptural baggage came from the New Testament, in particular the Gospels and the Epistles of the liturgical year and the Book of Revelations (the Apocalypse). Quite often these texts were transmitted to them by other books than the Scriptures themselves. Furthermore, among the confiscated books, complete Bibles were relatively rare. They might be editions of the sixteenth and the beginning of the seventeenth century, in general those of the great Prague printers Melantrich and Veleslavín, or, less often, the Bible of the Czech Brethren of the Unity, called the Králice Bible, in its one-volume version. In the eighteenth century, pietist Bibles printed in Halle in Czech were introduced illegally into Bohemia. The editions of the New Testament are always in greater number than those of complete Bibles (fifteenth century: 2 Bibles, 2 New Testaments; sixteenth century up to 1620: 14 Bibles, 25 New Testaments, of which 3 Catholic; eighteenth century: 6 Bibles, 2 of them Catholic, 18 New Testaments, 3 of them Catholic). None the less, in 1643, a printer in Litomýšl republished, in the middle of the Counter-Reformation, the Bible of the Czech Brethren of 1525. In contrast, the Old Testament had only one separate edition, in 1541 in Nuremberg. This predominance of the New Testament has also been noted in certain Lutheran, German-language regions. See, for example, Miriam Usher Chrisman, *Lay Culture, Learned Culture: Books and Social Change in Strasbourg, 1480–1559* (Yale University Press, New Haven, 1982).

70   Desiderius Erasmus, *Paraphrasis in evangelium Matthaei per D. Erasmum . . . nunc*

primum nata et edita . . . (Mayence, 1522). It was translated in Czech and printed in Litoměřice in 1542 with the title, *Evangelium Ježíše Krista Syna Božího podle sepsáni sv. Mattouše, kterěž Erazym Roterodamskaj v širších slovích a iako s vajkladem . . . vydal* . . . .

71   In 1753, for example, out of 492 titles listed by ten missioners, the proportion of religious to secular printed works was 94 per cent to 6 per cent.

72   An allusion to the Gospel according to St John, 10.1–21 (here verses 9–10). Vaněšovský and Jandík were incorrect in placing this extract at Pentecost Monday, since at this time in figured under Pentecost Tuesday. Interrogation of Jan Vaněšovský, question 12, Jan Jandík affair.

73   An allusion to the Gospel according to St Matthew, 24.24. It was at the time the reading for the twenty-fourth and last Sunday after Pentecost (today the twenty-third in the Catholic Church). Jan Nahlovský affair, interrogation of Matěj Škréta, question 4.

74   This number is minimal, since it was established on the basis of monthly reports, quarterly catalogues, or annual lists of confiscated books sent by ten missionaries the archdiocese. The entire diocese is not represented equally; certain documents may have been lost. Furthermore, one should add to these 614 volumes those sequestered by the Jesuits during their 'penitential' missions, but I have been unable to find the figures for these for the year 1753. The overall number of books confiscated each year varied. If we can believe the archdiocesan protocols, there were 333 in 1752; 2,667 in 1754; 3,068 in 1775, for example.

75   Interrogation of the cowherd Jan Petr, Pavel Čačák affair. To my knowledge, the hymn in question (*Komuž se utéci máme, než k Panné Marii*) was printed for the first time in the collection of Božan in 1719.

76   The hymn ends thus: 'Take thought that there will be for you no guarantor if your life is placed on the eternal brazier.' (*Modleme se Otci svému v pokoře a tichosti*, the first printing in a collection, Olomouc, 1559, Jan Gunther; taken up by all the Catholic hymnals in the seventeenth and eighteenth centuries).

77   This was the hymn, *Připravmež se věrni k modlení*, a first version of which figures in the Hussite hymnal called of Jistebnice. The Utraquists, the Lutherans, and the Brethren reproduced a slightly different text. This is the one picked up by most of the Catholic hymnals of the seventeenth century and the one sung by Škrěta and Nahlovský. Koníaš picked a variant for his *Cithare of the New Testament*. Nahlovský confirmed Škrěta's deposition but he referred to a different verse of the same hymn to prove the uselessness of the cult of the saints: 'In the second song there is this: "Raise our eyes towards the Lord our God in person".' The interpretation is the same.

78   Hymn *Kdo by nebral z obojího, boj se pekla horoucího.*

79   There were two people named Philadelphus. The author of Čačák's book seems to have been Martin Philadelphus Zámrský (1550–92), the author of an *Evangelical Postil or Commentary* . . . *that one reads each Year in Christianity, that one Meditates and that one Explains in the Assemblies of the Church of the People of God* (Jezdkovice u Opavy, 1592, 2°, 1310 pp. This book also contained hymns.

80   A possible allusion to the Gospel according to St John, 10:1–18, already cited by Jandík and Vaněšovský.

81 Václav Oliva, 'Z minulosti Chlumku u Luže a jeho okolí', *Sborník historického kroužku*, 6(1905), p. 137.

82 *Reversales* of Magdalena Teplá, APA I, H 2/6[4294].

83 SÚA Prague, AS 175, 43/373.

PART III

Political Representation and Persuasion

INTRODUCTION

In the sixteenth and the seventeenth centuries power was illustrated: ritual and pomp, ranks and honours, decisions and acts all passed massively into print in pieces combining image and text. The profusion of these objects – which ranged from the most rough-hewn to the carefully elaborated – poses several problems.

Was illustration passive, involving multiple agents, or active? Did its multiplicity and diversity reflect 'opinions' or varying points of view about public affairs? Or is that diversity instead a manifestation of the universal and supple hold on the public of the wielders of power?

What role did printing play, or did people want it to play? Was it simply charged with proliferating and perpetuating the passing event or the current institution – in a word, was popularizing its chief task? Or was it to explain and persuade? Or, indeed, was it to prescribe, impose a message, and fascinate?

The massive presence of the image incites us to return to the old medieval problem of the utility of images for popularization. Did they exist, as Gregory the Great and an entire Christian tradition insisted, because they were more accessible to the unlettered than text? In reality, the situation had become more complex. The image was joined with the text in a mobile relationship of implication, proximity, and hierarchy; hence we need to try to define the direction and the scope of this relationship. Moreover, the image no longer referred to one text alone (the Scriptures), but to a variety of writings, previous and contemporary, and to immediately available scenes of decorum and rite. The question of the inherent force of the figure – of representation and of the reproduction of representations – is thus situated at the crossroads of the history of technology, the history of knowledge, and political history.

Finally, the printed political image no longer addressed the 'ignorant' but the 'public' – in fact, various 'publics'. The quite different sorts of difficulty involved in deciphering images, from the directly illustrative vignette to the complex visual metaphor, lead us to new questions about the reception of visual effects. Did all segments of the public see the same objects? Did they perceive the same lessons in them?

These were also crucial questions for those in power and for the party leader, the author, and the printer. Doubtless the specific nature of political printed matter is that it is designed to persuade, to inform, or to prescribe, and it cannot avoid thinking in terms of how it will be received. This focus on efficacy increases the pertinence of our search for signs of the projected reception of a piece in the object itself. But just where do the lines of demarcation fall? Did hierarchies of reception coincide with hierarchies of production and socio-cultural position?

To respond to these questions we have chosen for examination three quite different entries into the dense mass of political printed matter:

1   A form of publication defined by a precise function and use: the political *placard* (broadside), analysed within the circumscribed contexts of the end of the League in 1594 and the assassination of Concini in 1617. Here the image interprets the event by creating an original and directly political configuration of it.

2   A class of symbolic objects, the emblems, studied over long time span and observed as they pass into the book, the point of arrival and of departure towards uses outside the book. The image, in this case, figures a previous metaphorization, and is only political by its application and its derivation.

3   A group of texts, illustrated and not, referring to an event (the end of the siege of La Rochelle in 1628) that for several weeks joined to produce a rich harvest of texts, glosses, and celebrations in a wide variety of publishing formats and writing practices.

7

Readability and Persuasion:
Political Handbills

CHRISTIAN JOUHAUD

Among all mass-distributed print pieces, the *placards* – handbills or broadsheets – show the closest, most vital association of text and pictures. The term *placard* designated a printed sheet that could be posted where it could be read by passers-by. Contemporaries often connected this means of expression with clandestinity. Writings that one would not openly admit to were posted surreptitiously, preferably by night. Thus the word *placard* was often combined with adjectives such as 'injurious', 'defamatory', or 'scandalous'. The most famous were those posted by Protestants over the night of 17–18 October 1534 in several French cities: Paris, Rouen, even Amboise, at the very doors of the private apartments of King Francis I. This 'Affaire des placards' has been thought instrumental in launching the religious troubles that bathed the kingdom in blood for sixty years. Handbills were also used by the government, however, both to inform the citizenry and to prescribe behaviour. That distinction reminds us that there are others: all *placards* were not by any means illustrated, and some were handwritten. Only illustrated, printed *placards* that came from other sources than official agencies will be discussed here.[1]

Should we follow the lead of Naudé and Furetière, who sought the origins of the *placard* in Roman pasquinades?[2] Probably not. Still, it is interesting that the writings of the *beaux esprits* of the seventeenth century mention the cobbler Pasquino, since it shows that for them this mode of expression had something basically 'popular' to it and was not to be considered honourable enough for the members of the 'Republic of Letters'. In any event, if an author composed a *placard*, it was wiser not to let a name appear on it. It would be mistaken to see in this anonymity the mark of a truly popular origin or of some sort of spontaneity. Admittedly, the handwritten posters that sometimes sprang up during the course of popular revolts might pass for spontaneous forms of expression, but recourse to printing implies a preparation and an elaboration incompatible with improvisation.

The *placards* were for the printers one job among others. They required little preliminary investment; they were light in weight and stocks were sold off. When they were illustrated, preparation demanded more time and more work, even when the woodcuts used were inexpensive and had been re-used a number of times, as was often the case.[3] A poster pasted up at night brought no financial return whatsoever, which meant that the *placards* were usually sold by pedlars who hawked them in the streets. Therefore they either had to avoid being too 'scandalous' or their distribution had to be sponsored by a powerful patron.[4]

The first problem is to ascertain whether printed and illustrated *placards* were mass-produced. If we looked only at the criterion of their conservation, the response would be negative, as extant collections are meagre. These were the most fragile of printed pieces, however: a single sheet, printed in a limited number of copies, that often might be dangerous to keep in one's possession. Moreover, the bibliophiles to whom we owe the better part of the original stock of our libraries took no interest in them. Finally, they never figured in the legal registration of publications initiated by Francis I. Thus – and here the scholar confesses to vertigo – one sole collection comes down to us to testify that the League produced a good many handbills. Without this collection, we could conclude that there had been no such thing.

Placards *and Political Action*

That Pierre de L'Estoile's collection of *placards* ever should have survived is nothing short of miraculous.[5] The title that he himself gave to his folio volume of 46 sheets of rough paper merits citing in full: *Les Belles figures et drolleries de la Ligue. Avec les peintures, placcars et affiches injurieuses et diffamatoires contre la mémoire et honneur du feu Roy, que les oisons de la Ligue appeloient Henri de Valois; imprimées, criées, preschées et vendues publiquement à Paris, par tous les endroits et quarrefours de la ville, L'an 1589* ('The fine figures and drolleries of the League. With the insulting and defamatory paintings, placards and posters against the memory and honour of the late King, whom the dullards of the League called Henry of Valois, printed, cried, preached and sold publicly in Paris in all places and streetcorners of the city [in] the year 1589').[6] Printed, cried, sold, preached: L'Estoile speaks of a wide distribution. These were 'all discourses of scoundrels and loafers, the drainage from the dregs of a stupid and rebellious people'.[7] Pierre de L'Estoile was a *politique*, a partisan of Henry IV, and a supporter of a political solution, and certainly not part of the 'dregs of the people', yet he had bought all these pieces, collected them, cut them out, pasted them in his book, and commented on them. Furthermore, his collection, which originally contained more than three hundred printed works of the sort, must have been dear to him indeed if it led him to disobey the orders of his friend the civil lieutenant d'Autry, who had ordered the destruction of all lampoons and other caricatures of the League in 1594. Be that as it may, L'Estoile bears witness, in spite of himself, to the broad distribution of these print pieces in the city. This was reading matter shared over a wide range of social levels.

He also shows us that these texts, far from being consumed passively, were an aspect of political action. Indeed, they were an integral part of it. Proof of this, for example, are the portraits of the duke and the cardinal de Guise, assassinated at Blois in December 1588. Their likenesses were not only put onto *placards* but also carried in the immense processions that took place in Paris early in 1589. They thus became political gestures, taking their place among other like gestures.

Just as significant is the case of a series of *placards* describing in horrible detail the persecution of Catholics in England. They show to what point circumstances helped reading to move people to action. They were posted in the cemetery of Saint-Séverin – church property, where intervention by the police forces would be inappropriate. It was also a place in which people traditionally met, discussions took place, and various transactions were carried out, so it was intensely frequented. The moment was St John's Eve, a time of fine weather and of one of the most important feast days of the year. Shops were closed, workshops were idle, the young had gathered. Everything was calculated and thought of in terms of its effect, and great care was taken to direct opprobrium at both Protestants and moderate Catholics. The King ordered the Parlement to have the *placards* removed, but it was done only 'by night and quietly, for fear of sedition'.[8]

Sixty years later, at the time of the Fronde, methods had not changed. Jean Vallier describes the *placards* posted in November 1650 in several of the most heavily travelled spots in Paris. Portraits of Cardinal Mazarin had been hung on the poles at the end of the streets that served to anchor the chains put up in case of public disorder, actual or threatened. One addition to the likeness showed clearly that its purpose was not to celebrate the cardinal-minister: a cord 'as big as your little finger' had been threaded through two holes on a level with his neck. Under the portrait a text denounced Mazarin's misdeeds and turpitudes and sentenced him to hanging. Vallier adds that this jolly practical joke was accompanied by massive sales of hawked lampoons, which was enough to persuade him 'that it was not people of lesser condition who were the authors of this shameful and cowardly proceeding'.[9]

Both L'Estoile and Vallier accuse powerful people – the duchesse de Montpensier, cardinal de Retz, and others – of having placards made for propaganda purposes. Party chiefs knew very well that print pieces of the sort reached the widest public. It was not even necessary to buy them to have access to them, and perhaps it was not even necessary to be able to read to understand them. In places like the Pont-Neuf that saw heavy traffic, there was always some obliging public reader (who might willingly accept remuneration) or some improvised commentator who would relate the essence of a publication of the sort. Unless of course the *placard* itself had been composed so as to make its decipherment difficult. This is a central question that merits closer scrutiny on the basis of concrete and specific cases. Still, as the essence of the effect of a *placard* doubtlessly lay in the relationship between its text and its image, we need to ask what was an image during the sixteenth and the seventeenth centuries? What was its status; what were the virtues and functions attributed to it? Just like the written word or ideas, the image has a history, the mixed history of its conception, its production, and its varied receptions.

The Virtues of the Image

The image was one of the topics subject to most lively debate in this era, and it gave rise to a good number of theoretical and practical works. The books of emblems or devices stand as proof of this.[10] Three considerations that affected political expression concern us here:

1 From earliest Christian times theologians had been interested in the image. One point seems to have met with universal agreement: image could have a dual function in religious practice. It supplanted letters for the illiterate; thus where sight sufficed for comprehension it was a powerful means for instruction and conversion. But it was also a jumping-off place for contemplation. Like hieroglyphics, it hid from the profane a sacred mystery that could only be pierced through initiation. The virtue of the first sort of image was thus to enlighten; that of the second, to conceal. This distinction served later theoreticians who contrasted 'vulgar' or popular images with highly coded learned images (for the most part symbolic or allegorical). As an implicit consequence, the 'vulgar' images were immediately readable by all – rather, they had no need to be 'read' – whereas the learned ones postulated the existence of two categories of reader, those who would succeed in decoding them and those who would not.

Logic dictated that these clearly distinct categories give rise to products that were just as easily distinguishable, the literature of the pedlars and the catechism on the one hand, learned books on the other. *Placards* lead us to put a different question, however: when the status of the objects was indeterminate or hybrid, as in the case of mass-printed political pieces, what became of the status of the pictures on them? The author and the printer were usually people imbued with this implicit polarity: how did they proceed? To put the question differently, were there not forms of writing and publishing that aimed at producing, in the same time and space and using the same figures, clear images and coded images? If this juxtaposition (or this confusion) existed, we must conclude that between successful and unsuccessful decoding there was room for a great number of partial successes, distorted uses, and contradictory appropriations.

2 From classical antiquity to the age of French classicism, passing through scholasticism and the Renaissance, the image and the imagination held an essential place in the various theories of knowledge. Despite Descartes, the traditional positions reigned until the last decades of the seventeenth century. Furetière still defines the image as a 'painting that one forms [by] oneself in one's mind by the mixture of several ideas and impressions of things that have passed through our senses'. Father Ménestrier, with the clarity characteristic of all his books, writes in his *Art of Emblems* (1662):

We learn only by images what we learn, since nothing enters into the mind in a natural way that has not passed through the senses and through the imagination, the function of which is to receive images of objects and to present them to the mind in order to know them and examine them.[11]

Receiving the images of objects: this formula fully participates in the Aristotelian tradition of the passive eye, revived by Leonardo da Vinci and even by Kepler.[12] Images sent by objects to the brain were contemplated by the mind. All knowledge arose from that contemplation. Imagination preceded intellection and served as an intermediary between perception and thought. The theory of passive vision had an important corollary. If it was objects that project their images towards the eye, then the images that imagination was to present to the mind, according to Ménestrier's formula, necessarily resemble the objects that they represent. This was important for the practitioners of persuasion. Resemblance made of the perceived image a sort of 'painting' of reality. It was that painting that was presented for the examination of the understanding. If it was accepted by the judgement of the mind, it would pass for reality. Its status as representation would be wiped out, for it would take the place of the object. In other words, it could either be rejected as nonsensical or received, which would imply adherence to its order of reality.

In this perspective, the use of an image in material form – drawn, painted, or printed – would have a quite particular function of persuasion. As a representation not of the real but of an intention, it would prompt the same acceptance or immediate rejection as the image of an object. This image would persuade better than a discourse not only because of its pedagogical virtues (as Napoleon reminds us, a good sketch is worth more than a lengthy discourse) but above all because it would permit the transmission of an intention that would be received passively as real. All that was needed was to avoid rejection by providing pleasure: the image had to be 'delectable'. That was the essential thing: in comprehending an image one 'adhered to' what it represents, including the intention instilled it its fashioning.

There was more. Aristotle had insisted that memory belongs to the same part of the soul as imagination. Frances Yates, in her great book, *The Art of Memory*,[13] writes of the thought of Giovanni Battista Porta (1602), 'Imagination . . . draws images as with a pencil in memory.' The memory was simply a set of fabricated and stocked mental images. In order to recall them, things – or ideas or words – must be linked to images. Thus, according to Francis Bacon, 'emblem reduceth conceits intellectual to images sensible, which strike the memory more.'[14] This accounts for the fantastic success of the mnemonic methods studied by Yates, and also for the importance of offering ready-made memory images to people one wanted to persuade and persuade lastingly.

To understand something was to 'adhere' to it, and to adhere to it was to remember. That is why these implicit virtues of the image, which were broadly accepted by theoreticians and practitioners alike, brought politicians to use the image as a means of persuasion.

3 Thirty years after his *Art of Emblems*, Father Ménestrier published a work entitled *La Philosophie des images énigmatiques*. In it he begins by repeating the traditional opposition between learned and popular images. However, since he held to the philosophy of knowledge that we have seen him express so clearly, he could find nothing stable on which to base that opposition. In point of fact, both the process of the production of mental images and their use are the same in all cases. Only modes of circulation, conditions of appropriation, and notions such as genre, which underlay reading conventions, could erect legitimate barriers between trivial images and more noble images. In a sort of codified recapitulation, Ménestrier arrives at the affirmation that trivial images – those appreciated by ignorant people – are illicit or degraded learned images. This degradation threatened all learned images, which must be protected by extremely strict rules of elaboration (hence also of decipherment). Such images could be compared to a citadel under siege. If even one of these rules was broken, the image would sink into triviality.

The long history of the device illustrates this anguished sense of beleaguerment. A perfect figure, the device was 'of all the works of the mind the shortest and the most vivid, the one that says the most and that makes the least noise, which has the most force and takes up the least space'.[15] It joined, body and soul, an image and a motto that, by their association, designated a person metaphorically through an achievement, a goal, or an 'internal concept'. As rules to protect the perfect working out of a device gradually became more precise and more constraining, opportunities for twisting them became more numerous and more frivolous. Devices were everywhere – in festivities, in funeral processions, in public ceremonies, in ballets, and so forth. They also became more and more incorrect. Theoreticians denounced them and compared them to the rebus or other sorts of puzzle found on tavern signs, in disreputable books, and at fairs.

In this manner, an obsessive demand for purity, based on nearly inapplicable norms on which the theoreticians no longer managed to agree, brings us unwitting evidence of the erudite conviction that the image provided an opportunity for shared reading. To that conviction, which we might call epistemological, we need to add the day-to-day reality of an intense presence of the image in the social, political, and religious life of the city. Painted, drawn, embroidered, and sculpted images were on view everywhere. This constant presence obviously had consequences for the readability of printed images and for the idea that their prospective makers may have had of them.

Three commonly held definitions – of objects, of their mode of appropriation, of the functions of the imagination – thus seem to have guaranteed the image a wide reception. We know, however, that its presence in books was not massive. What was its place in the *placards*? L'Estoile's collection permits us to give a satisfactory answer for one specific span of about twenty years. The collection contains 142 items, 76 of which are illustrated by at least one engraving. Fifteen others are manuscript copies or notes in L'Estoile's hand that on several occasions allude to images. Fifty-one have no pictorial material, but two-thirds of these are official proclamations of the king, the ecclesiastical authorities, the Parlement, various entities of the League, and so forth. In all, this means that among the non-official *placards* (those intended to persuade and not merely to prescribe), four out of five are illustrated.[16]

Although the collection is partial and heterogeneous, we can doubtless conclude that images had a dominant place in this type of mass-distributed political printed matter. If we ask the question in reverse order the answer is even more categorical: there were only a very few images that did without text entirely. Furthermore, the few that bore no text were in nearly all cases associated with a text when they were sold. Thus the question of the relation between images and text is necessarily central to the study of the readability of political *placards*. The study of three specific cases should permit a more concrete evaluation of their relative importance.

Three placards *in the Service of the King*

The three pieces to be examined present very different dispositions on the page and degrees of readability, but they share a political logic that makes them comparable as far as their persuasive intent is concerned. In the first piece texts and images are totally interconnected. In the second they are clearly separated in a relation of simple contiguity. The third case could be qualified as mixed, since there is both separation and dovetailing (rather than total interconnection), achieved by a system of references from the images to the text. Three cases obviously cannot be considered to represent all extant associative schemes. An exhaustive study is still lacking.[17]

All three *placards* originated in the corridors of power. One might go as far as to say that they took part, at their level, in the setting up of a representation of the Bourbon monarchy. To this end they used coded images based on humanistic symbolism, which their texts alternately clarify or complicate. Such encoding easily accepts the intrusion of 'trivial' images, however, and a skilfully handled blending of the two levels both opened the way to a hegemonic political discourse and offered an interpretation of it (in the theatrical sense of the term) that aimed at rendering publicization and prescription inseparable.

The Price of Presumptuousness and In Praise of Union

At the top, outside the border, a nearly unintelligible title – *Le Prix d'Outrecuidance, et Los de l'Union* – does little to summarize the thrust of the piece (figure 7.1). To the contrary, only when the general sense of the *placard* is understood can the meaning of the title be grasped. The overall composition is complex: there are a dozen lines or so of printed text at the bottom of the sheet, but text is also scattered here and there about the sheet. Where to begin? There is no clear indication, so one can begin anywhere. Perhaps if we decipher one zone after another the coherence of the whole will become clear. That of course means that the order in the following description makes no claim to reconstruct the reading progression that the author had in mind.

The central motif is a tree, complete with roots, a trunk, and two main branches. Its roots merge into the body of a king, from whom the tree springs. On the other side of the trunk, to the viewer's left, stands a man with his right hand and his left forearm trapped between the two sections of the trunk. Thus immobilized, he is being attacked by dogs or wolves. Indeed, they turn out to be wolves, as we learn from six lines of text directly under his feet:

In this picture allusion is made to Milo Crotoniates, in ancient times a famous wrestler, who too presumptuously attempted to split the trunk of a stout tree with his [bare] hands, where they remained caught and so tightly held that he was eaten by wolves on the spot. A naïve portrait of all criminals of *lèse Majesté* such as J. Clément, J.

FIGURE 7.1 *Placard* from 'Les belles figures et drolleries de la Ligue . . .' (Bibliothèque Nationale).

Chastel and all other like parricides, who have borne and suffered the consequences of their temerity and disloyalty.

The explanation is clear. The helmeted, glaived man represents Milo of Crotona. It is a mythological scene – or rather a semi-mythological scene, since Milo was a historical personage.[18] A correspondence is thus set up with two regicides, the man who assassinated Henry III and the one who attempted to assassinate Henry IV. Jean Châtel's name and the allusion to his sentence permit us to date this placard after 29 December 1594. In all likelihood it dates from the beginning of 1595, during the tense weeks of the expulsion of the Jesuits and the declaration of war against Spain.[19] Beyond Clément and Châtel, the threat of exemplary punishment was valid for all criminals guilty of *lèse-majesté*, including any who might still contest the power of Henry IV in the large fortified city visible in the background.

The king with the flowing beard whose entrails engender the tree is none other than St Louis (as indicated to the right of his crown). His right hand holds the sceptre and his left a book that he is contemplating. It is turned towards him, thus the viewer would have found it difficult to read the text, particularly when the *placard* was posted. The book shows an extract in Latin from Psalm 52, the key phrase in which is, 'I [am] as a fruitful olive-tree in the house of God.'

The tree's two main branches shoot up to the two upper corners of the printed sheet, sending out smaller branches to either side. Inside small circles set along them like fruit are written the names of the descendants of St Louis. At the ends of the two branches are placed two escutcheons, the coat of arms of France to the left and that of France and Navarre to the right. The two last fruits of this prolific tree are Henry III and Henry IV, each holding a sceptre and drawn with enough precision to be easily recognizable. Between them a winged genius, cherub, or cupid (as the inscription *sanctus amor* would indicate) pulls on an anchor cable attached to the two branches to prevent them from pulling apart. In his free hand the symbolic cupid holds a plaque bearing the motto *Ne quid nimis* (nothing in excess).

Within the outside frame there are two other tablets filled with fine print. A framed oval to the right contains a dedication to the king in which A. C. wishes peace and prosperity to the monarch and explains his work. He defines it as a 'picture' that shows 'as in a mirror' the horror of crimes of *lèse-majesté*. He adds that his invention ought not to be 'unfruitful to the people'. To this end, he recalls 'how useful and profitable it is to represent by emblems and hieroglyphic marks some notable doctrine to arrest at a glance and instruct the flighty and inconstant human mind'. To the left, in a rectangular cartouche framed with scroll-work, there are the 'Sapphic verses'[20] of a hymn of joy for a dearly acquired royal peace, the mystic dimension of which is underscored as a Peace of God compared to a sacred ointment assuring triumph over death and hell.

We have needed lengthy description – far from complete, moreover – to describe a layout that the author claimed to communicate to the mind 'at a glance'. In light of the theories discussed above, the time required for a glance – the time of sight – sufficed for the imagination to appropriate the figure. The mind accepted it, 'adhered' to it, and could pursue the decipherment of its various motifs afterwards. Once again, seeing meant adhering to an order of things, to the order communicated and imposed by an image imbued with intent. At the intersection of the vertical plane of history (from St Louis to Henry IV) and the horizontal plane of the space in which the dire act was accomplished and the trap closed, mystical duration triumphed over the criminal moment. Geometry supported intent by constructing the overall image, and it formed a whole with the rhetorical figures of the texts, which delivered a broad assortment of mental images on several levels. There is nothing extraneous in this association (*ne quid nimis*).

In point of fact, there are three trees, three trees in one, that imprison Milo's hands: the genealogical tree, the emblem tree, and the mystical tree. The genealogical tree shows the Capetian family splitting into two principal branches at the height of the sons of St Louis. It clearly shows the right of Henry IV to succeed the Valois, who expired without issue. It also represents the famous Salic law that assigned succession to the crown of France by male primogeniture. The emblem related an action that (as is always the case) was itself a metaphor for a moral teaching, following the principles established in a vast number of books of emblems.[21] Nothing was missing, neither the theme borrowed from classical antiquity, nor the scroll coiled like a serpent around the trunk declaring *Sic Francia divisa coalescit*. Still, the commentary under Milo's feet 'popularized' the emblem by explaining it. As for the mystical tree, the third person in this arboreal trinity, it was the green olive tree of the psalm, the tree of Paradise, and the tree that is the body of St Louis of France. A close connection was established between the saint, the realm and the dynasty (which the canonization of Louis IX re-founded in both spirit and law). The roots were holy, as were the upper branches secured by the *sanctus amor* whose anchor lodged in the heavens – an astonishing figuration of the abstraction of divine right of kings.

But was the only purpose of this handbill to sanctify and celebrate continuity in the French monarchy and the return of peace? The last lines at the very botton of the placard, placed outside the frame so they relate to all the motifs and texts combined, invite us to compare the representation of the two King Henries:

God, wishing to preserve a kingdom from ruin in the highest degree, is apt to call upon a devout and wise prince, and often gives some extraordinary mark of his calling. [This is] what one can observe in our King Henry IV, presently reigning, if one looks at the great persecutions that have tried him from his youth until this age, his constancy, his victories, and [his] singular moderation in the same; also [if one considers] that four sons of King Henry II, being arrived at a man's age and three of them married, all died childless to leave him place; that so many desperate assassins have failed in their wicked designs taken against him. [When one has] verified that his soul was tied to the sheaf of life by the Eternal . . . [one realizes that] this olive suits him well.

This interpretation makes Henry IV not only a legitimate heir but chosen of God. He had triumphed over all his enemies: who could protect him better than God and his armies? Above all, Henry II's four sons had disappeared on his behalf[22] and God had preserved him from assassins. One obviously could not say as much for the unfortunate Henry III, who died without issue under Jacques Clément's knife. Providence had thus commanded this change within the dynasty, making a living branch of the family prosper while an exhausted branch sputtered out. A vital force – the sword that Henry IV, not Henry III, holds in his left hand – points him out to the viewer. The two scrolls simply call greater attention to this force. *Fratrem ne desere frater* adorns the Valois banner, while the Bourbon proclaims, more ambitiously, *Prius ima dehiscat terra mihi*) ('Straight away the earth opened for me from the depths'), an allusion not to the soil that welcomes the dead, but to the soil that nourished this tree, sending its living sap to the figure at the tree's top. When Henry IV ascended the throne he expressed a continuity and a restoration, but also a new alliance between God and France.

Complex forms of learning (history, classical mythology, and iconography) also leave their mark on this *placard*, offering their meaning to the more percipient connoisseurs. The better part of the transmission of the 'message' and the propaganda thrust took place before full decipherment, however. The two intersecting planes, showing the virtues of duration to ward off baleful events, and the triple nature of the tree, had no need of erudite knowledge to be grasped at least on the level of mystery. This was a mystery that university rectors had no special expertise to pierce, for the simple reason that it lay at the very heart of the divine-right monarchy and beyond all profane knowledge.

The Deliverance of France by the French Perseus

If the format of the second *placard* (figure 7.2) remains roughly similar to that of the *Price of Presumptuousness*, one glance is enough to see that the relationship of text to image is completely different. Space inside a heavily decorated frame is divided into two nearly equal parts with a picture in the upper half and fourteen lines of verse in alexandrines in the lower. At the foot of the sheet the bookseller-printer has given his name and address, the privilege obtained, and the date (1594). At the top, outside the border, the unambiguous title, *La délivrance de la France par le Persée François*, describes and summarizes both text and image.

FIGURE 7.2 A further *placard* from 'Les belles figures et drolleries de la Ligue ...'
(Bibliothèque Nationale).

A nude young woman is pictured attached to a wave-beaten rock, offered to a scaly monster with horns and sharp claws, a sort of amphibious he-goat with reptilian feet. A knight, sword raised, rides his flying horse to attack the monster, while a group of oldmen from a city whose towers and ramparts can be seen in the distance look on. The knight quite obviously resembles Henry IV and he wears the famous plumed head-dress of the battle of Ivry. In the sky shines an emblemized sun that served increasingly often to symbolize the king of France. Up to this point, Perseus needed no identification. Recognizing the mythological theme served as a point of departure for a second level of reading. Perseus and Andromeda were often found as a motif in books of emblems or accounts of royal entries. Françoise Bardon notes the importance and frequency of the theme in her *Portrait mythologique à la cour de France*.[23] She also gives three examples of its use in a Lyons edition of Ovid's *Metamorphoses* (1559), another edition of the same work published in Frankfurt in 1579, and a book of emblems that appeared in 1581. The three images are nearly identical and closely resemble our French Perseus.

In each case Andromeda is chained to a rock surrounded by water and Perseus rides up on a flying horse. The study of the differences among them is extremely rewarding, however. The sun appears only in our *placard*, where the spectators also appear much more clearly. Above all, in the three examples presented by Bardon, Perseus' face is not seen, as it is hidden by the lance he brandishes at the monster. In the 1594 engraving the lance is replaced by a sword held high with outstretched arm, thus freeing the face. The French Perseus absolutely had to be recognizable.

The emblem depicted a moral teaching inspired by a specific person or addressed to him in the form of a dedication. This is how Perseus delivering Andromeda was figured in the emblem book of 1581 cited by Bardon. The image, considered as *principis boni imago*, was dedicated to Prince Georges of Liegnitz.[24] The correspondence between the idea and the prince was established through the given name George, Perseus becoming St George slaying the dragon. There was no direct mark in the emblem of the person to whom the dedication was made, as the rules demanded. The correspondence had to be made by decipherment. Super-imposing Henry IV's face on the body of Perseus made an incorrect emblem of impoverished function, trivialized by too great a clarity that failed to respect the reader's acumen. Henry IV's face corrupted the correspondence into a banal resemblance.[25] Even worse, Andromeda was explicitly identified as France in the title of the *placard*. Why did the author accept this loss? Why impoverish the emblem? Was it out of a desire for clarity? Perhaps. It is more likely, however, that this change had a political function and purpose. It introduced something that served propaganda in support of Henry. Françoise Bardon found the key, it seems to me, when she wrote (about another image) that there was an 'ideal resemblance between Henry IV and his representation'.[26]

This felicitous formula gives food for thought. Perseus represented the king, in particular, the king's actions against the League and against Spain. In a normal emblem, the correspondence between Perseus and the king would be established by signs integrated into the image by the motto. When Henry's face was drawn carefully, it was not a sign but the thing represented itself. The thing represented was introduced into the representation and fused with it. To see Henry IV in Perseus was to *recognize* him in Perseus and to adhere already to the logical order of the *placard*. On a handbill put out by the League, the duc d'Epernon is shown whispering diabolical advice to the king.[27] His feet (or rather his paws) are clawed, thus identifying him with the Devil, representing him as the Devil, or presenting him as diabolical. On the contrary, the French Perseus is Perseus in all the attributes of this son of Zeus, but represented as Henry IV. The title is thus perfectly exact: the king is not compared to Perseus; 'Persean' virtues are incarnate in the person whom they represent.

The text is an explicit transposition of the universe of the image into the universe of royal politics. Viewed from afar, three words written in large capital letters catch the eye: FRANCE, HENRY, FRANCE. France had been delivered over to die at the hands of her own and of foreigners, and a Perseus had come to save her. Fourteen lines of alexandrines tell the tale and draw the political moral:

> France, be faithful to him, no longer let [anyone]
> Bind you with doubloons, and believe no more in the abuses
> Of those who have gnawed away the gold of your Diadem.

The text added a more direct appeal, not additional persuasion. France, and by that token the reader, was called upon. Even in its means of expression this was a response to the League, which insisted that religion took precedence over the nation. The text was structured around two comparisons, 'France, like Andromeda . . .', and Henry IV and Perseus ('The monster . . . felt the strength of Perseus' arm as did the Spaniard that of Henry IV'). The straightforward verses make little use of rhetorical figures, as if Perseus and Andromeda should be the only images, shining in splendour in the engraving in the upper portion, a true symbolic figure in action.

The Picture and Emblems of the Detestable Life and Unhappy End of Master Coyon or, Mythology of the emblems of Coyon

On 24 April 1617, Louis XIII got rid of Concini, his mother's favourite, who had been acting as prime minister.[28] This *coup d'état* had been prepared with care by the king and his immediate entourage. Concini was executed. The plan was to arrest him and to kill him at the least sign of resistance, which was what happened. The evening of 24 April he was buried secretly in the church of Saint-Germain l'Auxerrois. The next day, however, 'the people' went to dig up the 'tyrant'.[29] The corpse was exhumed. The mob put a rope around his feet and began to drag him through the streets. As they went, more participants joined this macabre procession until it numbered several hundred people. The body was dragged to a gallows, hanged, mutilated, taken down, and cut into pieces. Several groups then formed and went off in different directions to display these portions of the ex-marshal of France throughout Paris. The festivities continued all day long, accompanied by other jollities. A few pieces of the corpse were given to dogs, people went through the motions of selling others like butcher's meat, and there was dancing, all under the benevolent gaze of the king's guard when they were encountered. Finally, when evening came, what remained of the corpse was burned in several parts of the city.

The *placard* tells of all the events of 24 and of 25 April (figure 7.3). At the foot of the sheet on the right-hand side the two dates follow 'A Paris, chez', replacing what would normally be the printer's name. This was one way among others to emphasize and insist upon the quasi-simultaneity of the actions related and the publication of the piece. It was doubtless a fiction, but it demonstrates a desire to 'stick close' to the event not found in the other broadsheets and lampoons that narrate the affair.

The *placard* was composed of six woodcuts and a text in verse. The presence of two titles, – *Tableaux et emblèmes de la détestable vie et malheureuse fin du maître Coyon* under the pictures as a part of the woodcut, *Mytologie des emblèmes du Coyon* composed in type – hints at the likelihood that the pictures and perhaps the text were distributed autonomously.[30] Be that as it may, at some time close to the event the two were associated, and we see one object on a printed sheet with a system of cross-references between pictures and text. These references point the way to a progressive reading guided by numbers in sequence from left to right and top to bottom, as with writing. The first picture refers to stanzas 1, 2, and 3; the second and the third to stanza 4; the fourth to stanza 5; the fifth to stanzas 6 and 7; the sixth to stanza 8.

Respecting this progression, we can begin with the first picture, a squirrel in a revolving cage. A scroll pointed at the squirrel's head designates him as

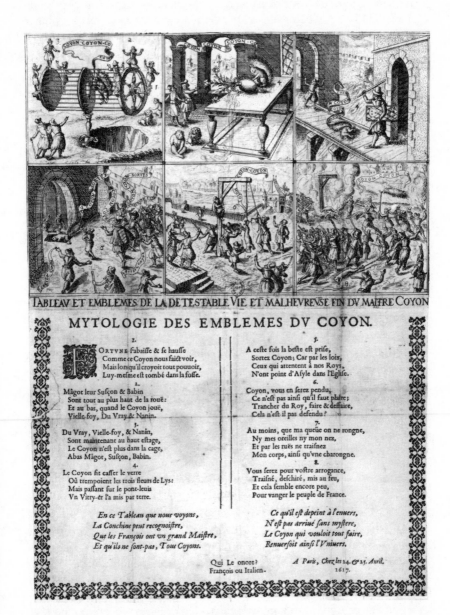

FIGURE 7.3 *Placard*: 'The picture and emblems of the detestable life and unhappy end of Master Coyon' (Bibliothèque Nationale).

coyon. The cage is unsteadily balanced over a pit, called a *fosse* in the last line of stanza 1. The toy-like cage is cylindrical, and each end of the cylinder forms a wheel, which the first word of the text authorizes us to call the wheel of fortune. Thus an iconographical motif that might be styled symbolic is superimposed onto the realistic representation of the cage. Doubtless one of the most commonly used symbols in both pedlars' chapbooks and learned works, the wheel of fortune was an almost inevitable emblem in horoscopes and prognostications.[31] Six tiny figures ride on the two wheels, at the mercy of the squirrel's motion. The text identifies them as Mâgot, Susçon, and Babin on one side; Vielle-foy, du Vray, and Nanin on the other. It is not difficult to pierce these barely transformed names to identify Mangot, Luçon, and Barbin; Villeroy, du Vair, and Jeannin. They allude to changes that took place in November 1616, when the Secretary of State, Villeroy, was replaced by the bishop of Luçon (Richelieu), the Garde des Sceaux, Guillaume du Vair, by Claude Mangot; président Pierre Jeannin at the superintendance of finance by Claude Barbin. These were thus changes authorized by Concini. After his fall, the former ministers were recalled. The distortion of Barbin into Babin and Jeannin into Nanin suggest a use of baby-talk in accord with the representation of the cage as a toy, and with the infantile air of the six small figures. The other transformations of names suggest an implicit judgement of the people involved. We find a *magot* (a monkey or a hoarded treasure) and a *suçon* (a kiss, close to *sangsue*, a leech), which hint of greed or illicit appetites, but we also find old faith (*vieille foy / Vielle-foy*) and the truth (*du vrai/du Vray*).

The two first illustrations are not only contiguous but dovetailed thanks to the presence in both of a dog, a monkey and two men who gesture broadly and function as commentators.[32] The squirrel is perched on a table in the second image and has just overturned a vase

> In which the three *fleurs de lys* were bathing,
> But passing over the drawbridge
> A glazier [*Vitry-er*] set it right again.

The last line of stanza 4 refers to the third image, thus it is the text that assures the dovetailing of image and text this time. Furthermore, the three *fleur de lys* from the vase are repeated in the emblematic form of the arms of France on the glass held under the glazier's arm. He carries a raised stick in his other hand. The glass (*vitre*) obviously refers to Vitry, the captain of the king's guards charged with liquidating Concini, which is repeated somewhat heavily by the *Vitry-er* of the last line of the text. This third woodcut thus represents the assassination of 24 April.

The three other images form a narrative sequence. The squirrel is exhumed to cries of *sortez coyon*, after which he is hanged by his feet. His tail and his ears are cut off. Finally, the mutilated corpse is dragged towards a pyre on which he can also be seen burning, so the last picture shows two consecutive actions. In contrast to the first four stanzas, stanzas 5, 6, 7, and 8 no longer function as explanation for the images. They complement the pictures by allowing the actors in the affair to speak. The second person plural (*sortez coyon*; *vous en serez pendu*) has replaced the third person singular that gave the author of the statement the status of narrator, while the narrative commented on the action that the images symbolized. In the remainder of the text the narrator disappears, and direct address and dialogue take the place of narration. He has entered into the action; he has taken it over, dragging the reader along with him in rhetorical adherence.

Concini as coyon *and as Squirrel*

The motif that continues throughout the six pictures is the representation of the minister as a squirrel and his designation as *coyon*. Neither the juxtaposition of the squirrel and the word *coyon* nor the association of the two terms with Concini is evident, so we need to try to understand them. *Coyon* was an insulting term signifying 'one who professes baseness or cowardice' (Furetière). *Coyon* should not be confounded with *couillon*. They were two quite distinct terms. *Couillon* referred to testicles and more generally to male genital organs, as in Rabelais's famous phrase, *Je vous lui couperai les couillons tout razibus du cul* ('I'll cut the fellow's knockers off so close that never a hair will stand to tell the tail' [LeClercq translation]).[33] François Béroalde de Verville (1556–1623) confused the two meanings in an involuntary jest: 'Like some woman who said that she didn't want anyone to make jokes about her name, for fear of their inventing some *couillonnerie* [by which] she meant some *coyonnerie*.'[34] The latter word was taken in an even broader sense than Furetière gave it, since it is associated here with the idea of imbelicity and foolishness.

Furetière himself explained this broader sense in his examples, and he tells us indirectly why Concini should have been associated with the word *coyon*. His examples say: 'the buffoons of comedy boast of being great *coyons*' and 'the captains who act the brave [man] are always great *coyons*.' If we look up *bouffon* and *capitan* we find: 'Italian comedians are the best *bouffons*'; '*les capitans* are ridiculous characters that are often introduced into comedy, particularly the Italian.' This detour by way of 'buffoon' and 'captain' leads us to Italy, then, and to Concini's country of origin. Two lines of Du Bellay's confirm this interpretation: 'Il fait bon voir (Magny) ces coïons magnifiques, ... Leur Saint-Marc, leur Palais, leur Realte, leur port' ('It is marvellous to see, Magny, these magnificent imbeciles [with] their Saint Mark's, their [Ducal] Palace, their Rialto, their port').[35]

Concini was designated by the term *coyon* because he was Italian. The final puzzle at the very bottom of the sheet added a note of mystery to the denunciation, but it also explained it. It says, in a broad play on words on Concini's wife's given name, Leonora (Galigaï): 'Qui Le onore?' François ou Italien ('Who honours him? [Who is Leonora?] French or Italian').

And the squirrel? If we look to dictionaries of symbols or treatises on heraldry, they tell us that that small animal was usually associated with positive values such as promptitude, enthusiasm, economy, foresight, wisdom, courage, skilfulness, and impartiality. He often symbolized the function of messenger. The only negative note was his tendency to lasciviousness.[36] Was there any specific relationship between lasciviousness, the squirrel, and Concini? If so, at what level did it operate? One answer might be risked by comparing the action depicted in the fifth image, which occupied the central space in the sheet, and the etymology of the word in the ever-faithful Furetière:

Small rust-coloured wild animal who is very light and jumps from branch to branch and who has a long tail. . . . The word comes from the Latin *sciuriolus*, diminutive of *sciurius*, which comes from *schiouros*, composed of *schia* (*umbra*) and *oura* (*cauda*) because this little animal covers himself almost entirely with his tail.

Just the very tail that was cut off with a large knife in the fifth picture. Before then, in images 2, 3, and 4, the scroll bearing the word *coyon* was placed in line with the squirrel's tail. In the last image the tail is carried in procession, triumphantly and at the end of a pole. It is difficult not to see in the central place accorded to the caudal appendage of the squirrel an allusion to the emasculation to which Concini's corpse was subjected in a scene that the memorialists describe as a fit of 'popular' savagery. 'At least, let them not gnaw on my tail / Nor my ears, nor my nose', the *coyon* says, summarizing the mutilations actually carried out on the minister's dead body. The placard concentrates on castration, since the squirrel has neither a long nose nor long ears.[37]

If castration was indeed the central episode in the ritual defilement of the corpse, if it was so perceived by others besides the memorialists, and if the bawdy interpretation of the representation of Concini by a squirrel is the correct one (or was at least offered to the reader as one possible decipherment of the *placard*), then the use of the word *coyon* should be considered an attempt at a double meaning. *Coyon* bore a charge of anti-Italian feeling and lent itself to an obscene play on words such as the one presented by Béroalde de Verville. It also appeared in the title of a lampoon: 'An end brought to the war pacified by the death of Concino Concini, marquis d'Ancre, who was shot, buried, disinterred, hanged, *décoyonné*, dragged about and burned in Paris.'[38] Our *placard* might thus be entitled 'The fable of the *coyon découillonné*'. In this perspective, there was an obvious interest in substituting an animal for the desecrated corpse: it was a way of toning down the savagery of a scene that had actually taken place. It put the event on the level of Carnival festivity. Why, then, bring the question up rather than simply leaving it in silence?

The squirrel was a mask. He both concealed and revealed. He revealed a trait that the author wanted to emphasize. The squirrel was a decoy. He underwent the punishment and attenuated its horror. Can one look at the last image without thinking of the ritual killing of Carnival? From a cadaver cut up into pieces in a spontaneous and savage ritual of judgement, we slip towards a joking festive ritual of the destruction of a fictive scapegoat. The violent image of people acting out (a serious potential threat) was eliminated, while justice was meted out, and vengeance against a minister whom the king himself had qualified as guilty of *lèse-majesté* was emphasized and even exhibited. Under these conditions, the *placard* performed an operation that might be defined as euphemized exhibition.

Emblems had their part in this euphemization, as the two titles remind us (*Tableau et emblèmes*; *Mytologie des emblèmes*). The scroll, the usual accompaniment of emblematic decorations, 'made' an emblem. Furthermore, the word *coyon*, taken with the figure of the squirrel, produced a metaphor founded in an implicit resemblance, which was an emblematic procedure.[39] The first three images could even pass for true emblems. They had a certain autonomy, and they illustrated a moral teaching or action in totally metaphorical fashion. Purists would of course have objected, but even when it was less than perfect, trivialized, and turned into a rebus, the emblem remained an emblem the moment it respected the way emblems functioned. On the contrary, the only emblematic elements in the three final images recounting the events of 25 April were the scroll and, to a certain extent, the squirrel. One can no longer speak of emblems, but only of a tendency towards the emblemization of the narration by the use of emblematic motifs, and by juxtaposition to the three true emblems. This nonetheless indicates an attempt to subject raw narration to emblematic form. The particular property of the emblem was to lead from accident to substance, from the particular to the universal. Thus by that property and by the expectation that its use as a form postulated, it contributed to the process of euphemization. It enlarged; it generalized. It permitted qualifying Concini's actions as an essential disorder by placing them in the category of crimes committed against universal values and against the French monarchy that incarnated those values. Thus it tipped the balance into another universality – that of evil. The moral of the story could be sought and found: all disorder would be punished. The man who had wanted to 'play the king' (*trancher du roi*, stanza 6) had been punished. In fact, he had been punished twice, once by the king in the name of the immanent justice that he personified, and once again by the rioters, by a popular violence that in the name of the law (as the second line of stanza 5 declares) brutally set the world to rights again. 'Le coyon qui voulait tout faire, / Renversoit ainsi l'Univers' ('The *coyon* who wanted to do everything turned the universe upside-down').

Here we touch on the most directly political dimension of the campaign for persuasion in the *placard*. Suggesting that the riot of 25 April was only a sort of carnivalesque explosion manifesting the community's reaction of self-defence against an aggressor accomplished two goals, both related to the activities of the previous day. Concini's assassination was justified. The reaction of Parisians putting the world to rights again showed to what point the fatal act was expected and salutary for the entire community. At the same time, the savagery of their revenge, emphasized by the placard, to some extent wiped out the savagery of the execution itself, which was represented in a purely emblematic manner. The reaction of the citizenry of Paris showed that everyone understood that it was a question of righteous revenge: *vanger le peuple de France*, as stanza 8 says. The accent was on a fundamental solidarity between the king and his people. The act of justice of 25 April repeated the one on 24 April. No court could have rendered justice so well or so unanimously. Exhibition of the riot thus protected the royal act. At the moment of a political change of direction inaugurated by the assassination of a minister, it postulated an infrangible link between the king and his subjects, bound together as if by pact in a common vengeance. In exchange, the riot was legitimized, to some extent authorized, a posteriori. This authorization is depicted in the third image, where Vitry is carrying the *fleurs de lys* under his left arm. In his right hand he holds a stick with which he strikes Concini, the squirrel. In the following pictures the rioters are carrying the same stick. Any riot can be defined as 'an event in search of its own meaning'. The *placard* gave a political meaning to the events of 25 April 1617. In its own way it bore witness 'Que les François ont un grand Maistre / Et qu'ils ne sont pas, Tous coyons' ('that the French have a great Master and that they are not all *coyons*'). The riot lost its subversiveness in this interpretation, since it could no longer be compared to popular upheavals, the very thought of which moved people of wealth and position to terror in their fears of pillage, rape, and the world turned upside down. When it gave such upheavals a meaning by means of the euphemized exhibition of violence, the *Mytologie des emblèmes du Coyon* assured the innocuousness of the grievous public disturbances of 25 April and justified the bloody inauguration of Louis XIII's personal reign the day before.

Readability and Decipherment

All three of these *placards* supported power. They served royal policy; they were a weapon of the state. This confirms how hazardous it is to equate mass-distributed print pieces such as *placards* or *libelles* with oppositional forms of expression. Such an interpretation would be anachronistic. Interpretations that failed to take into account the importance of the context (better, the contexts) would be equally anachronistic. First among those contexts are the conditions of reception of the print form known as the *placard*. In the absence of direct evidence on how such pieces actually were received, we are left with how their authors – writers, engravers, and printers – represented their reception, as seen in the *placards* themselves. Moreover, reading practices, real or postulated, only take on meaning when they are analysed within the context (here, the political context) that gave them an opportunity to function. In the cases before us we have the irreducible specificity of 1594, 1595, and the period shortly after Louis XIII took power into his own hands, and the irreducible specificity of the tactical aims of anyone who produced a print piece in a given political context.

Intellectual environment formed a second contextual circle. Thus opinions on the image widely shared among the lettered go far to explain the mission of persuasion confided to it. Should one go farther, enlarging the circle to evoke 'mental structures' or the 'symbolic universe'? Robert Darnton, in his *Great Cat Massacre*,[40] found symbolic forms in the text that he was studying that seemed to him 'shared, like the air we breathe'. He has been criticized for importing into an analysis of a particular narrative extraneous elements drawn from an uncertain common fund. The squirrel's tail in fact offers an example of a symbol used uniquely for a meaning produced within the text. It becomes accessible only in its manipulation within the printed piece. It was a symbol constructed in reference to a precise meaning linked to the specific conditions of the publication of the *placard*, not to a broad symbolic context.

Nor can readability be analysed outside time. Certain motifs that seem opaque to us doubtless spoke much more directly to people of the late sixteenth century. We have lost contact with the culture of the emblem to which our three placards refer. Why should political tracts, which sought to persuade, have had recourse to the emblem? Emblems were ever-present in city life under the *ancien régime*: they appeared on the walls of public buildings and in churches and *collèges*, not to speak of shop signs. They were everywhere during civic or religious celebrations.

This familiarity is not by itself an answer, however. Even when it was faulty or distorted, the emblem always produced a double meaning. To identify it or recognize it as a form of expression was to admit that there was something to be deciphered in it; a sense to be clarified and conquered; a veil to lift. It was a signal that could be understood in various ways within an extremely wide range of forms that went from tavern riddles to erudite enigmas, from the rebus to hieroglyphics. It was not necessary to decipher them correctly – that is, in conformity with the intent of their creator – to play the game. Furthermore, as our three *placards* show us, a meaning was communicated before decipherment took place, and perception of this meaning was immediate. When the viewer saw the mystery of the three-part tree, the face of Henry IV, or a squirrel in Concini and Concini in the squirrel, it meant that he or she had already been caught in the trammels of an implicit line of argumentation addressed to the eye of the imagination.[41]

Notes

1   For an overall view, see the *Histoire de l'édition française*, gen. eds. Henri-Jean Martin and Roger Chartier (gen. eds), vol. 1, *Le Livre conquérant. Du Moyen Age au milieu du XVIIe siècle* (3 vols, Promodis, Paris, 1982–4). On information put out by government sources, see Michèle Fogel, 'Propagande, communication, publication: points de vue et demande d'enquête pour la France des XVIe–XVIIe siècles', in *Culture et idéologie dans la genèse de l'Etat moderne*, Collection de l'Ecole française de Rome 82 (Ecole française de Rome, Rome, 1985), pp. 325–36.

2   Gabriel Naudé, *Le Marfore, ou Discours contre les libelles...*, A Paris, chez Louys Boulenger, rûe S. Jacques, à l'image S. Louys, M.DC.XX, 22 pp., BN, Lb36 1424; Furetière, *Le Dictionnaire universel* (3 vols, SNL-Le Robert, Paris, 1978, facsimile reproduction of The Hague, Rotterdam, 1690).

3   Copperplate engravings, on the other hand, were extremely costly. See Michel Pastoureau, 'Illustration du livre: comprendre ou rêver', in *Histoire de l'édition Française*, vol. 1, pp. 501–3.

4   Examples can be seen in Christian Jouhaud, *Mazarinades. La Fronde des mots* (Aubier, Paris, 1985), in particular ch. 4.

5   L'Estoile declares that his collection of posters and lampoons made up 'four large volumes'. Only one collection of 46 sheets remains today, housed at the Bibliothèque Nationale in Paris. It remained in the L'Estoile family until Pierre de Poussemothe de L'Estoile deposited it in the library of the abbey of Saint-Acheul, near Amiens, where he was abbot. My thanks to Denis Crouzet, who lent me the microfilm of this collection, which he is studying in detail.

6   G. Brunet, A. Champollion, E. Halphen, Paul Lacroix, Charles Read, Tamizey de Larroque (eds), *Mémoires-journaux de Pierre de L'Estoile* (12 vols, Librairie des bibliophiles, Paris, 1875–96), vol. 4, *Les Belles figures et drolleries de la Ligue* (1888).

7   Ibid., vol. 3, p. 279.

8   Ibid., vol. 4, p. 13, manuscript note by L'Estoile.

9   Jean Vallier, *Journal de Jean Vallier, maître d'hôtel du roi (1648–1657)*, eds Henri Courteault and Pierre de Vaissière (4 vols, H. Laurens, Paris, 1902–18), vol. 2, p. 213. This piece is reproduced in Hubert Carrier, *La Fronde. Contestation démocratique et misère paysanne. 52 Mazarinades* (2 vols, EDHIS, Paris, 1982), vol. 1.

10   See ch. 8.

11   Claude François Ménestrier, S. J., *L'Art des emblèmes ou s'enseigne la morale par les figures de la fable, de l'histoire et de la nature. Ouvrage rempli de près de cins cents figures* (Lyons, 1662; Paris, 1674), p. 12.

12   Michel de Certeau's seminar at the Ecole des Hautes Etudes en Sciences Sociales devoted a session to this question on 23 March 1985.

13   Frances A. Yates, *The Art of Memory* (University of Chicago Press, Chicago, 1966), p. 206.

14   Ibid., p. 371, from Francis Bacon, *The Advancement of Learning*, II, XV, 2.

15   Pierre Le Moyne, *De l'art des devises* (Paris, 1666), p. 10. Furetière gives the following definition of the *devise*: 'An emblem that consists in the representation of some natural body and in some motto that applies it in a figured sense to the advantage of someone; the picture is called the body and the motto the soul of the *devise*.'

16   The term 'illustrated' is taken here in its most neutral sense and in no way is to be taken as prejudging any supremacy of the text over the image in the production of meaning.

17   Such a study would doubtless begin with a detailed inventory of *placards* conserved in the great public libraries of France, since no exact catalogue is available.

18   Milo was a Greek aristocrat, a Pythagorean, and an athlete crowned several times in the Olympic and Pythian Games. He commanded the war launched by the aristocracy of Crotona against Sybaris (*c.*510 BC). Pierre Puget sculpted a *Milon de Crotone* in 1672.

19   The Jesuits were expelled from Paris on 8 January; war was declared against Spain on 17 January. This war lasted until the treaty of Vervins, signed in 1598.

20   Sapphic verse had precise and complex rules of composition. Lines were 'of twelve syllables; the first, fourth and fifth feet are trochees, the second a spondee and the third a dactyl. One puts three lines of this nature in each stanza, which one ends with an Adonic line composed of a dactyl and a spondee' (Furetière).

21   The celebrated work of father Dominique Bouhours, *Les Entretiens d'Ariste et d'Eugène* (Paris, 1671) should be added to the works of fathers Ménestrier and Le Moyne, cited above.

22   Hardouin de Beaufort de Péréfixe repeats this idea in his *Histoire du roy Henri le Grand*: 'for there were ten or eleven degrees of distance between him and Henry III, and when he was born there were nine princes of the blood before him' (cited in Janine Garrisson, *Henry IV* [Le Seuil, Paris, 1984], p. 329).

23   Françoise Bardon, *Le Portrait mythologique à la cour de France sous Henry IV et Louis XIII. Mythologie et politique* (A. and J. Picard, Paris, 1974).

24   ibid., pl. 15.

25   Delicacy consisted in managing to 'represent one figure in another' (Le Moyne). Otherwise, 'this hotchpotch of things is good only for almanac pictures' and was 'more appropriate for tavern signs than for ingenious decorations' (Ménestrier).

26   Bardon, *Le Portrait mythologique*, p. 281.

27   *Le Soufflement et conseil diabolique d'Espernon à Henry de Vallois pour saccager les Catholiques*, a wood engraving 24 cm × 56 cm described in Brunet et al. (eds), *Mémoires-journaux de Pierre de L'Estoile*, vol. 4, pp. 32–3.

28   Many versions of this episode can be found in bibliography old and new. As is
     often the case, one of the most attractive for its precision and the quality of its
     writing is Gabriel Hanotaux, *Histoire du Cardinal de Richelieu* (2 vols, Firmin-
     Didot, Paris, 1895), vol. 2, pt. 1, pp. 185–99.

29   See the *Mémoires* of Pontchartrain (Paul Phélipeaux de), collection Petitot, ed.
     Claude Bernard Petitot, *Collection complète des mémoires relatifs à l'histoire de France*
     (Foucault, Paris, 1824), ser. 2, vol. 17, pp. 232–3; the *Mémoires* of Richelieu,
     collection Petitot, ser. 2, vol. 11, pp. 47–8; Michel de Marillac, *Relation de ce qui s'est
     passé à la mort du maréchal d'Ancre*, collection Michaud-Poujoulat (Joseph-François
     Michaud and Poujoulat, *Nouvelle collection des mémoires pour servir à l'histoire de
     France, depuis le XIIIe siècle jusqu'à la fin du XVIIIe*), 2nd ser., vol. 5 (1835), pp. 451–
     84; and numbers 1052–91 of the *Catalogue de l'Histoire de France* of the
     Bibliothèque Nationale, vol. 1, pp. 485–7 (Lb36).

30   My thanks to Roger Chartier, who brought this particularity to my attention.
     This piece has been the object of two presentations, one by Hélène Duccini at
     the round table held in Rome on 15, 16, and 17 October 1984 (see 'Un aspect de
     la propagande royale sous les Bourbons: image et polémique', in *Culture et
     idéologie dans la genèse de l'Etat moderne*, pp. 221–3), the other by Jeffrey Sawyer
     during a seminar given in March 1985 at the Ecole des Hautes Etudes en
     Sciences Sociales.

31   Examples can be found in André Chastel, *Le Sac de Rome, 1527* (Gallimard, Paris,
     1984) [*The Sack of Rome, 1527*, tr. Beth Archer (Princeton University Press,
     Princeton, 1983)].

32   They provide an iconographic entourage for the wheel of fortune that can also be
     found in certain Tarot cards.

33   François Rabelais, *Le Tiers-Livre des faicts et dits héroiques du bon Pantagruel*, ch. 12
     (Garnier-Flammarion, Paris, 1974), p. 450.

34   Cited in Edmond Huguet, *Dictionnaire de la langue française du XVIe siècle* (Didier-
     Erudition, Paris), s.v. *coyon*.

35   Joachim Du Bellay, *Les Regrets*, CXXXIII.

36   As in Ad De Vries: *Dictionary of Symbols and Imagery* (Amsterdam and London,
     North-Holland Publishing Co., 1974).

37   The mutilations struck the memorialists as well: Richelieu says, 'They cut off his
     nose, ears, and unmentionable parts.' In general, however, they speak of
     mutilations without mentioning emasculation specifically.

38   The full title reads: 'Le définement de la guerre apaisée par la mort de Concino
     Concini, marquis d'Ancre, lequel a été carabiné, enterré, déterré, pendu,
     décoyonné, démembré, traîné et brûlé à Paris, ayant été trouvé atteint et
     convaincu de crime de lèse-majesté, les vingt-quatrième et vingt-cinquième avril
     seize cent dix-sept, selon le recueil fait et augmenté de P.B.S.D.V., historiographe
     du qui fait la réunion pacifique de ce changement délégué à Sa Majesté. De
     l'imprimerie de la voix publique' (Bibliothèque Nationale, Lb36 1030).

39   See ch. 8.

40   Robert Darnton, *The Great Cat Massacre and Other Episodes in French Cultural
     History* (Basic Books, New York, 1984), in particular ch. 2, 'Workers Revolt: The
     Great Cat Massacre of the Rue Saint-Séverin', pp. 75–106.

41   Yates, *Art of Memory*, pl. 15, showing the first page of Robert Fludd's *Ars memoriae*
     (Oppenheim, edition of 1619).

8

〰〰〰

Books of Emblems on the Public Stage: *Côté jardin* and *côté cour*

ALAIN BOUREAU

It was once thought possible to think in images. Towards the end of the seventeenth century, Father Claude-François Ménestrier, in a work of history through pictures, recounted this exploit with delight: 'Louis, Prince of Turenne, after finishing his studies in philosophy at the Collège de Louis-le-Grand . . . presented his thesis in a novel form in which each page was a trophy adorned with Devices, Images, Emblems and Eulogies with magnificent titles.'[1] For his thesis in philosophy the young prince had composed a book of emblems. One might suspect the university authorities of complaisance towards the aristocracy and the court, where the art of the emblem was flourishing, but the prince-doctor was following a fashion that had spread everywhere in Europe, and in two centuries had produced from 2,000 to 3,000 collections applicable to all purposes and uses and printed in all formats and bindings from the humblest to the most luxurious.

Such diversity would seem to exclude any possible overall view of the 'book of emblems' as a genre.[2] Nonetheless, the word 'emblem' (*emblema*, *emblemata*) returns consistently in book titles, so people of the early modern age were conscious of a unity that escapes us. In order to define that continuity, we need to begin, not from the rhetorical or intellectual operation of the emblem or the device, the roots of which are lost in medieval archaeology, but from the reality of the collection of emblems as a publishing phenomenon historically determined by the publication of the *Emblems* of Andrea Alciati in 1531. A good many authors of compilations based their own books on that work, cited it, and claimed to imitate or adapt it; there were more than 150 editions, translations, or adaptations of Alciati, and innumerable prefaces refer to him.[3]

We can thus take as a book of emblems any collection that followed the tradition laid down by Alciati and assembled a series of autonomous brief statements based on a figured or figurable image. Any stricter definition would leave out large segments of the production of emblems. Let me add one important characteristic, even at this level of sweeping generalization: the image is made to *signify*. Also, it was offered for individual appropriation from a book – a novelty, since in the medieval tradition the signifying image was public (statues, stained glass windows, paintings), whereas the private image, the book image, was decorative.

The Emblem and the Device

Our first definition is not totally satisfactory, however, for one specific model of the emblem – the device – gave structure to the genre, though it never absorbed it entirely. The device, which appeared in medieval heraldry at the beginning of the fourteenth century,[4] well before the first books of emblems, exerted a powerful attraction on compilations of emblems. As perfected by the Italian academies in its phase of most complex elaboration in the sixteenth century,[5] the device claimed to associate an image with a person by means of a short proposition, the 'motto' or 'device' proper, that applied perfectly to both elements in the comparison. Father Ménestrier, that great Jesuit master of symbolism in the age of Louis XIV, offers a late but totally regular example. A device that he invented for the king showed a naturalistic lily (not a heraldic flower) with the motto, *Diva se jactat alumna*, which the author translated as *D'une déesse il a tiré sa naissance* (from a goddess he drew his birth). A brief comment explained the image: 'Fables have said that it was from the milk of Juno that the lily was formed.'[6] Anne of Austria stood for Juno and Louis XIV for the lily, with a felicitous assonance between 'Louis' and *lys*, and a convenient return to heraldry. The practice of the device thus consisted in finding flattering predicates that worked in both the literal and the figurative sense and in joining them to two themes, one immediate and figurable, the other absent.

Practice of this attractive exercise tended to transform the emblem into a device. Alciati's original emblems bore no motto and the title of the image played no part in constructing a metaphor. Beginning with the 1540s, however, devices and emblems began to intermesh. On the one hand, devices were borrowed from occasions for magnificent display (on decorations, on clothing, in works of glorification) and were organized into series (the so-called heroic devices). On the other hand, emblems of general significance not aimed at any one individual adopted the triangular metaphorical operation of the device to become what were called moral devices. There was room beside the heroic device, with its brilliant instantaneity, for the moral emblem, where the rhetorical form was extended by a sometimes lengthy exegetic gloss. From then on, most books of emblems were in fact collections of moral devices.

What interests us is to comprehend the cultural behaviour that this immense output induced, in particular on the public scene and in relations of demonstration and persuasion. Even a rapid description has shown that books of emblems could play quite different roles according to how they were envisaged. The conception and perception of a device could demand rhetorical and intellectual competence; or the reader could find himself or herself immersed in the powerful flow of images and figures that invaded the age of French classicism; or again those who held political power might use emblems as an ingenious, agreeable, and prestigious technique for glorification. Without neglecting the continuity of the genre, we need to examine the emblem, first backstage, where its intellectual elaboration took place, then as it placed the world on both sides of the stage, both *côté jardin*, in the smiling garden of images, and *côté cour*, in the court, where all converged.

An Intellectual Technique

It seems that the western world has created a unique art of decipherment in the emblem. In the last twenty years, highly elaborate publicity images have preferred metaphor to description. The consumer has collaborated willingly in flights of fancy that use a formula pertinent to both the object and its figuration to extract qualities from representations of landscapes or objects and attribute them to the product advertised.[7] Whether the setting was the subtle exercises of sixteenth-century Italian academies, the complex celebrations of absolutism of the seventeenth century, or the rapid fire of ingenious proposals that surrounded Napoleon in the Council of State as it sought a motto for the new imperial seal on 12 June 1804,[8] people became practised at conversing and suggesting by means of the emblem and the device. We need to look beyond play and toadyism, to take the emblematic relation seriously; we need to consider it as a way of perceiving the world and as a cultural point of convergence. We need to explore that relation (taking 'relation' in the sense anthropologists give it, as, for example, a 'joking relation') and to grasp a domain of expression that reveals, classifies, distributes hierarchical positions, and facilitates thought for the group and about the group.

Particularly when the book of emblems absorbed the device, it practised, and created a public for, the intellectual technique of metaphorical evocation of the world. Mastery of how emblems work should not be confused with the art of 'reading' pictures. Unlike allegorical or descriptive illustrative images, in which a portion of all possible meaning is grasped out of a continuum of forms, the emblem produced a meaning (or a determined set of meanings) with nothing left over. The image was there merely to give material form to figures latent in the metaphor.

The history of publishing offers confirmation that a material image was not essential to the genre. Claudie Balavoine offers scrupulous proof of Alciati's indifference to illustrations in his collection.[9] He expressed regret only over the errors in the text of the first edition of his work, published without his authorization in Augsburg in 1531, and he corrected them carefully for the edition of Chrétien Wechel in Paris without rectifying the manifest errors of iconography. When his complete works were published in 1547 and 1548, Alciati removed all the illustrations from his *Emblems*. Alciati's most illustrious successors showed a similar lack of interest in images: the privilege given in 1544 for the first edition of Maurice Scève's *Délie*[10] states that 'the present book treating of Loves, entitled Délie' could be printed 'either with Emblems or without Emblems'. The decision to illustrate the book with emblems seems to have been the printer's and to have depended upon commerical tactics. In 1540, the same printer, Denys Janot, put out the *Hecatomgraphie* of Gilles Corrozet,[11] one of the first French compilations of emblems, in two versions, one with illustrations and the other without.

The book of emblems was no different in that respect from the other illustrated books published in the sixteenth century, in which the publisher reserved the right and the option to illustrate the book if it proved successful. Another important collection, *Le Théâtre des bons engins auquel sont contenus cent Emblemes*, printed by Denys Janot in 1536, first appeared without illustration, before Janot endowed it with luxurious images, as did later publishers in Angers (1545), Lyons (1545), and Paris (1554). Even when illustrations appeared in the work from the start, the author often paid little attention to them. Thus an author as concerned about the quality of his texts as Guillaume Guéroult allowed Balthazar Arnoullet to publish his *Premier Livre des Emblèmes* (Lyons, 1550) using the same woodcut to illustrate two quite different adages: *Trop enquerre n'est pas bon* (Too much curiousness is not good) and *En putain n'ha point de foi* (Don't put your trust in whores).[12] The image, in these cases, only acts as an illustration and a possible but optional application of a metaphoric procedure. The subsequent publishing career of the book of emblems confirms the indifference to illustration: one-third of known works have no pictorial matter. In 1679 our Jesuit, Father Ménestrier, who should have been immune to technical difficulties and problems of publishing costs, thought it unnecessary to provide illustrations for his *La Devise du Roy justifiée. . . avec un recueil de cinq cens Devises faites pour Sa Majesté et toute la maison royale*.[13] It is true, as we shall see, that the image influenced the emblems produced, but in its founding principle the emblem served more for thought than for depiction.

The emblem, taken as a cognitive exercise, invented a relation between the concrete and the abstract, between the visible and the intelligible. To borrow an example from the collection of *Emblemata* published by Jean Mercier in Bourges in 1592: a vignette represented an ox and a donkey harnessed together under the motto, *Nusquam convenient* (They will not go anywhere together). A brief gloss at the foot of the page gave the meaning of the metaphor: *Vel nulla vel una in populo religio est semper* (In a people one finds but one religion or none). At the moment of the last spasms of the League it was obvious that the two animals figured Catholicism and Protestantism, the parallel existence of which seemed impossible to Mercier. The emblem established a relation between two themes (a heterogeneous team and the presence of two religions), one visual (or figurable) and the other verbal, by the intermediary of an ambivalent predicate (*convenire* could mean go along together, step by step, or, in the figurative sense, get along together). Emblematic competence thus resided in the formulation of an ambivalent predicate, in the discovery of a relation that expressed the meaning by structuring the themes already present in new fashions. The themes figured pre-existed the operation and formed the elements of current ideological sets or subsets in the fields of religion, politics, or morality. The themes that did the figuring came indifferently from the real world or from fiction: they could be natural or fabricated objects, animals, plants, zoological scenes, or historical, Biblical, legendary, or mythological references. In other words, the world of the figures did not reveal anything; it offered an opportunity for an encoding or a formalization. I cannot see any neo-Platonism at work in emblematics, in spite of certain professions of faith that arose from simple cultural habits, nor any belief in the plenitude of meaning in being. Elements from the real world, chosen and combined, offered themselves to the art of the author, the great predicate-maker of being. Jean Mercier did not labour on the land in his native Berry, but worked in the jubilation of words and the joyous explosion of finding the right one.

This instrumental grasp of the world as expressive matter for thought can be clearly perceived in one of the principal modes of classification of the books of emblems, classification by types of figure. Camerarius was doubtless the first to make use of this criterion, passing from plants (*Symbolorum et emblematum ex re Herbaria desumtorum centuria*, Nuremberg, 1590) to quadrupeds (1595), to birds and insects (1597), to fish and snakes (1604).[14] The famous Jesuit rhetorician Nicolas Caussin, in his *Polyhistor symbolicus electorum symbolorum et parabolarum historicarum stromata XII libris complectens* of 1628,[15] lists all figurable beings under twelve headings including world, gods of the Gentiles, birds, quadrupeds, fish, serpents and insects, plants, stones, and manufactured objects. The presence of manufactured objects on the list, corroborated by their frequent use in books of emblems, shows that it was not a question of finding traces of divine design and meanings in the physical world. The author of emblems organized being by his stage direction, as indicated by the theatrical metaphor encountered so often from Guillaume de La Perrière (*Le Théâtre des bon engins*) to Giovanni Battista Cacace (*Theatrum Omnium Scientiarum*, Naples, 1650). The emblem materialized thoughts (the *bons engins*, or clever inventions of the mind).[16] The plasticity of the figure, and its nature as a sign, appear clearly in the reutilization of the same figures with different mottos and meanings. Among many examples of this, Jakob Bornitz, in his *Emblematum ethico-politicorum sylloge prior* (Heidelberg, 1664), borrowed images from the *Emblemata moralia et bellica* of Jacobus à Bruck (Strasbourg, 1615).

Thus the emblem served to classify, to manipulate, and to order something in function of thought. Another classificatory effect of emblematic literature can be seen in the glosses, which occasionally take up considerable space. An immense treasury of citations, Christian and profane, of observations, and, narrations found thematic and methodological distribution through them.

The ingeniousness of the emblem could demote the genre to the somewhat frivolous minor arts, among them, fanatical pastimes like crossword puzzles, but in reality the emblem played a capital role as a transitional object of reason. As is known, Donald Woods Winnicott has defined the transitional object as the maternal object that, while it keeps the real or symbolic imprint of the mother, structures the participation of the child in the real world. In the same fashion, the emblem, in its glosses and its figures, amassed and safeguarded the immense treasure of knowledge and beliefs of the Middle Ages, and distributed it according to rational categories formulated by its predicates and turned to new thoughts. The emblem did not reject belief in the virtue of one object or another in the physical world, or in the reality of one legend or another; it made that belief nonessential or secondary. It instrumentalized belief. Meaning passed from themes to relation. While it wrote the 'prose of the world', as Michel Foucault termed it, the emblem treated that prose as a set of arbitrary signs or signs of arbitrary motivation. Doubtless it was instrumental in an essential way in the transition from 'resemblance' to 'representation',[17] from the symbolic to the formal.

From Resemblance to Representation

The intellectual process of metaphorical interpretation of the given has roots in medieval Christian exegesis. From the earliest ages of Christianity, the Church had to deal symbolically with the sacred text of the Old Testament. Both tradition and truth had to be safeguarded, for where rite and dogma were concerned, christological and apostolic teaching contradicted the letter of the Book; thus that letter had to stand for something else besides itself, it had to *represent*. Until the birth of antisemitism in the twelfth century, the principal complaint addressed to the Jews by the Christian Church was blindness, and the Synagogue was figured as a blind woman.[18]

The Jews failed to see the meaning hidden behind appearances; the condemnation of sin behind the literal prohibition of eating pork.

This symbolic prefiguration of representation in the early modern age, in which Christian cosmogony still guaranteed the validity and the meaning of the metaphoric operation, found its most complete expression in the virtuosic practice of quadruple exegesis constantly applied in sermons from the twelfth century on.[19] In this sense, the treatises on mariology of the twelfth and thirteenth centuries constitute prototypes of the emblematic collections of the Renaissance.[20] Thus the *Liber marialis* of Jacobus de Voragine, written around 1292, treats in alphabetical order all the symbols (plants, animals, objects) that represented the Virgin. This slow preparation of the representative system gave birth to the emblem when the figured themes multiplied and became secularized. A more flexible and complex predicative relation was needed between the two series of variables, and the 'word' – the motto – replaced the simple, uniform predicate of *stat pro . . .* or *est* (as in, the rose, balsam, the elephant, 'stand for' or 'are' the Virgin).

Transitional forms existed between those two methods for creating metaphors. The case of the English friar Nicole Bozon is particularly interesting. His *Contes moralisés*,[21] written in the fourteenth century, relate 'properties of the world', fables, and anecdotes to a morality by turn strictly religious and secular. Moreover, the connection between what was figured and what did the figuring was crystallized in a Biblical citation that prefigured the 'motto' in a religious key. Thus Nicole Bozon speaks of the stag hiding when it sheds its antlers and then applies this characteristic to bailiffs, who were *bauds et fiers* when in power, but lost their haughtiness when they no longer held office. A verse of Psalm 87 applied to both the stag and the bailiff: *Exaltatus autem humiliatus sum et conturbatus* (And being exalted have been humbled and troubled).[22]

It is conceivable, then, that the vast output of heroic devices and books of moralizing emblems, after Alciati, captured the intellectual heritage of Christian representation and brought it out of its closed world. Formal thought on relation offered infinite possibilities: through metaphors, laws and adages could be discovered in all domains.

It would seem that the great intellectual revolutions in the West were produced by successive moves away from appearances, treated as means of interpretation, as forms of meaning. I am thinking here of the scholastic exegesis of the twelfth century, of the nominalism of Ockham of the fourteenth century, of the Cartesian formalization of the seventeenth century, and of the universal diffusion of symbolism at the end of the nineteenth century (Freud, Saussure, Walras, and Pareto).

The role played by the minor art of emblematics in the formalization of thought appears more clearly if we note that emblematic literature developed during the period in which François Viète (1540-1603) was discovering the formalization of relation in mathematics.[23] As is known, he invented the notation by letters of undetermined constants. Before him, unknown variables in equations were of course denoted by letters (our x, y, z), but no one had thought to represent constants in that fashion (our a, b, c, in second degree equations such as: $ax^2 + bx + c = 0$). Solving an equation before Viète had to be done numerically every time. After Viète, one could use algorithms (which Viète called *logistiques*) and pass on to calculation of the algorithm and to the generative rule. Notation by letters introduced a form that could be used in thought and manipulated in calculation. In this it was similar to the figure of the emblem: 'The numerical *logistique* [algorithm] is the one that is exhibited by numbers, the specific [one] by species or forms of things, such as the letters of the alphabet.'[24] The 'specific logistic' *represented*, which is why Joël Grisard proposed 'representation' as a translation of *species*.[25] The same processes were at work in Viète and in emblematics: thinking about the general, in itself inaccessible and mute; using the means of the particular without losing one's way in occurrences; speculating on forms and representations by neutralizing their relation to concrete reality. As in emblematics the 'motto' established a relation between separate universes (society, religion, politics, morality, and so forth), so Viète's *logistique spécifique* accounted for homologous relations in domains that direct description (in numbers or figures projected in space) could not comprehend together (algebra and geometry). Viète's second translator, Vaulézard, glossed his *logistiques* as others had commented on Alciati, by explaining, in what he called an 'exegesis', the concrete applications in numbers and 'in lines' of Viète's work.[26] The science of the concrete discovery of abstract relations, which Viète calls *zéthèse*, compared the given and the required (the figure and the figured of the emblem) by finding the algorithm (the 'motto') that connected them.

A century later, after a similar penchant for formalization and calculation based on form had inspired the mathematical physics of Descartes, Leibniz extended it to the empirical world. In his *Dissertatio de arte combinatoria* (1672-6), he proposed to invent the *characteristicae*, notional signs that would permit calculation in all domains, theology included. As Frances Yates put it,

Leibniz envisaged the application of the calculus to all departments of thought and activity. Even religious difficulties would be removed by it. Those in disagreement, for example, about the Council of Trent would no longer go to war but would sit down together saying, 'Let us calculate.'[27]

The later fortunes – all the way down to Lévi-Strauss – of this dream of formalization and homologous operation on empirical givens is well known.

Before we leave Viète, I should note that his career outside the mathematical domain was itself an emblem of transcription as efficacious grasp of the world. A jurist by profession and an impassioned mathematician by avocation, his great specialty was cryptography. After his studies he served as preceptor to Catherine de Parthenay, the daughter of his protector, the sire de Soubise. When Catherine was obliged to correspond secretly with her mother to arrange a divorce – a delicate affair necessitated by her husband's impotence – she and her tutor devised a code system using secret inks and Latin and Greek citations.[28] Later, when Viète had become *maître des requêtes* of Henry IV, he gained fame for deciphering the coded messages of the Spanish. Not only did he discover the coding cipher of the correspondence but also the code's principle of permutation. He proudly signed himself, 'Interpreter and Cipherer of the King'.

The book of emblems thus learned the concrete art of abstract formalization. This role did not last, however, due to the very triumph of reason in classification. Formalization created its own domain, modestly announced by the *Logique ou art de discourir et raisonner* (1603) of the polygraph Scipion Dupleix[29] and crowned by the establishment of the notion of method. Frances Yates has noted that in 1632, five years before Descartes's *Discourse on Method*, the Académie du Bureau d'adresses assigned method as the topic for discussion at its opening session.[30]

Emblematics and Politics

The novelty of emblematic culture was not limited to a 'zethetic' grasp of the physical world and of knowledge. The themes figured also took a new turn. It appeared difficult to form a coherent corpus of all the precepts that emblems had turned into images, for we find previous knowledge either scattered thinly or fragmented. The Christian religion rapidly took hold of the genre and turned it to the purposes of an illustrated and worldly catechism. Books of religious emblems represented nearly one-half of the total corpus. The themes figured had no autonomy in these works, and the formalism was reduced to pedagogical ingenuity. Between 1580 and 1640, however, before the genre was captured by the court, a specifically emblematic knowledge is recognizable outside the production of religious works. It was based on a multitude of fragments and aphorisms, and its principal domain was public life, signalled by the many collections of *Emblemata politica* or *civilia*.

At first sight the lessons taught by such emblems seem disappointing: within the same collection they pass from an extremely conformist political morality to unbridled praise of cynicism. On this level as well, however, formalization was more important than content. These encyclopaedias of power categorized by image or by theme, rather than by persons and doctrines, the traditional entries of the *Miroirs des Princes* or the *Traités*. The emblem commented on figured *action*. By that means, political thought was introduced into the world of representation (understood as simulation or reconstitution of a process). The demonstration of an adage operated before the reader by a *scene*, potential or actual. It mattered little whether one depicted (or imagined) a nettle or someone about to be stung by nettles:[31] in both cases action founded the device, *Leviter si tangis adurit* (Touch it lightly and you will prick yourself). That was what made emblematic representation different from the allegorical process in the arts of memory by which the reader contemplated an image already in place and already invested with meaning. The distinction was an important one for the domain of political thought by means of images. In one instance, the emblem instilled a system of permutations among demonstrative signs that raised the reader to the status of a potential participant in public affairs; in the other, utopia built a preconceived model of a doctrine. In 1602, in fact, Tommaso Campanella intended his *City of the Sun* (published 1623) to be a *modello* for a city and for a mnemonic 'place'. Figuration lightened the mind's burdens, but did not have the heuristic value of the emblem.

An intermediate form appeared with the *Civitas Veri* (City of Truth) that Alfonso Del Bene published in Paris in 1609 from the manuscript of his father, the Florentine patrician Bartolommeo Del Bene.[32] We are still in the realm of utopia, since the text and images in this work depicted a circular city that embodied a political morality derived from Aristotle. Thirty-two figures show details of the overall architecture of the city, which is presented on one outsize plate, each figure, indicated by a letter of the alphabet, referring to an allegorical meaning. In other words, meaning was produced by exhibiting an illustrated doctrinal corpus, not by display of one occurrence of practical reason. An emblematic turn is hinted at, however, in the arrangement of the book and in the moral anthropology it outlines. The work is divided into thirty days (*dies*), thus evoking an itinerary, a latent action akin to an adage uttered by someone reading or receiving an emblem. The neo-Aristotelian formulation of the political principles of the city constructs a theory of the political man as a man of action. Del Bene distinguishes three sorts of human life: the sensual life, the active or political life, and the contemplative or philosophical life. The citizen of the City of Truth followed the second life. He entered into the city by one of its five gates, which represented the five external senses. He made his way towards the central citadel of 'political beatitude', passing by the palaces of the 'virtues of action' (or of morality): courage, daring, magnificence, magnan-imity, mildness, affability, taste for truth, urbanity, public justice, distributive justice, equity, heroism, and friendship with equals and unequals. When he arrived at the centre of the city, after having avoided the palaces and the labyrinthine back alleys of the vices, he could reach the citadel of beatitude by mounting slopes that figured the three 'internal senses', common sense, force of imagination, and memory. At the summit of the citadel, he could linger in the five temples of the virtues of the mind (science, art, prudence, intelligence, and wisdom) before he attained the third form of human life, the contemplative and philosophical life. Method is more important here than the model. Using allegory, Del Bene theorizes on the formal approach of emblematics: beginning with the natural (the external senses, the virtues of action) one passes on to what is intelligible with effort (climbing the slopes) through the use of common sense, imagination, and memory (empirical abstraction, schematic figuration, and commentary presentation). Action was schematized into a method for political reflection in Del Bene's text, even though it was still enmeshed in mnemonic and allegorical illustration.

When it founded precepts on the dynamics of observation of and reflection on the world, the emblem legitimized secular political thought in the sixteenth century and rescued it from the dilemma that had trapped it between Machiavelli and St Thomas Aquinas – between efficacious thought with no intelligible basis and well-founded thought with no practical application. Political acumen was not a science in that age but an art, in the sense of the terms so admirably stated by Ernst Kantorowicz.[33] It was a human technique capable of taking inspiration from nature to form new principles adapted to changing times. The emblem functioned like the law, by demonstration and homology. A good number of the great emblem-makers between 1531 and 1640, beginning with Alciati, were in fact men of the law. Unlike the treatise on jurisprudence or politics, however, the book of emblems, thanks to its atomized composition, reflected different points of view and successive *visions* (in the root sense of the word) of the world. They all came down to the same concern: how were people to live together? The structure of the genre implicated both the subject (the citizen) and the Republic (the common good).[34] The subject or the citizen was, in fact, the moral person who profferred or might proffer the adage figured by the emblem. This statement, explicit in the case of the device or in certain dedicated emblems, attributed particularity to that person in the public micro-space constituted by the gallery of devices or the collection of emblems. A communitarian function of the kind is clearly perceptible in the *album amicorum*,[35] a collection of civic emblems of which an early example is Nicolaus Taurellus' *Emblemata physico-ethica, hoc est naturae morum moderatricis picta praecepta* (Nuremberg, 1595). Each emblem was dedicated to a burgher of Nuremberg, whose arms were figured in the frame of the image and whose name followed the text of the device placed at the head of the page. These devices, taken together, constituted the political morality of the city.

A Heroic and Republican Culture

It is hardly surprising that a culture both heroic (through individual choice or merit) and republican (through participation in the commonweal) should have developed particularly well in the Low Countries and in the free cities of the Rhineland, far from the great monarchies. The political book of emblems between 1580 and 1640 had something Erasmian about its communitarian spirit ('Ego mundi civis esse cupio, communis omnium', Erasmus said), as well as a touch of cultural universality.

The compilation of emblems of Julius Wilhelm Zinkgref, *Emblematum ethico-politicorum centuria*, published in 1619 in Heidelberg, can aid in comprehension of how practical reflection on political action joined with interest in the common welfare. Zinkgref, a jurist by formation, was an administrator in the service of the Count Palatine. His work was well received and went through five printings from 1619 to 1665. His collection differed from the *alba amicorum* in that emblems were not applied to specific persons. I have no intention of establishing any strict order in its composition, as that would be inappropriate considering the nature of the emblem, and would account poorly for contradictions and contrasts. I might observe, though, that the thirty-five first figures portray means to power (1–3: vigilance; 4–6: terror; 7–8: undefinability; 14–15: ruse; 19–21: incessant activity; 30–35: union of forces). In this section the individual signified is rarely (in three instances) a sovereign; more often a 'we', a 'you' or a 'one'. Here power, undergone or imposed on others, concerned all subjects. The following group (36–57) seems to depict the manifestations of authority from the viewpoint of those who govern (36–41: the majesty of monarchy; 42–3: the role of public expenditures) and from the viewpoint of the governed (48–50: punishments and rewards; 51–7: love of the leader and obedience). A third group (58–97) exalts the public merit of the wise man who knows how to stand up to fortune. The final antithesis between merit and fortune – so important in heroic culture – seems to lead to a shift from power to authority. Force with ruse gives power; power in the service of excellence gives authority. Just comprehension of authority assures the stability of the state, a lesson that appears clearly in the last four vignettes, which exalt the common welfare. The comment on emblem 99, a trumpet, is, 'The voice, the advice [and] the regard of a citizen is as much in the state as I am in battle.' The last figure (100) depicts the emergence of the *public*: a handsome image shows a beehive placed in a farm courtyard near a wide-open door, beyond which we see open fields. The legend, *Privati nil habet illa domus*, is translated into French as *Communauté entière*, and glossed by the quatrain:

Who has for himself only birth
And profits nothing but himself
Does only harm to the public
And his good causes only confusion.

What we see here is that the emblem provided a corpus of predicative statements (which could change into single utterances), the grammatical subject of which is the public individual (sovereign, hero, courtier, magistrate, or simple citizen) and the predicate some given attribute (force, strength, authority, power). The governors and the governed, aside from any individual or external ends they may have had, shared in a general orientation to the common welfare – to the state.

Diversity in access to the common good could be compared to the conception of liberty that was developing in the Protestant milieux. Most of the authors of political collections from 1580 to 1640 in fact come from such circles. Since Wycliffe Reformation (or pre-Reformation) thought in the fifteenth century saw predestination as coinciding with positive liberty realized in the just act. Individual predestination gave a particular content to every freedom. The world necessarily gave an imperfect image of that predestination, for nature, without grace, was composed of objects, human and other (the 'pre-known'; *praesciti*), not of free agents (the predestined; *praedestinati*). The emblem suggested predestination, but it figured it by what was already given: the pre-known, an image in the thrall of the chosen material. Grace, predestination (or, in lay terms, civic reason), operated by the *choice* of one adage or another, precisely in what it professed or denounced. The individual could go towards the figure, the image as part of the pre-known, just as the tyrant could. The true subject was not figured, however: it did the figuring.[36]

The individuation of the emblemized subjects – the variables in human society – was forged by action, by figured action and enunciated choice. The emblem then became part of a heroic and republican culture profoundly new in what it refused (rank, social status, doctrine) and what it called for (publicly manifested individual merit). The common welfare, or the Republic, was constructed as a juxtaposition of such heroic desires. They were all different, but were all turned towards the same object. The windowless monads in this secular theodicy had the same aim: the state of the citizens.

The Return of the Image

Up to this point, the book of emblems has been discussed in its cognitive function as a symbolic object, and as a propaedeutic to theoretical reasoning on relation and to practical reasoning on the notion of the common good. As this cognitive essence became manifest, however, it was often in the form of an illustrated book, an *album*. The image, forced out of the picture by discovery of the practical powers of reason, made its return.

The image existed as a decorative element from the very beginnings of emblematics in 1531, when Alciati's first publisher, Heinrich Steyner, decided to illustrate Alciati's epigrams. Seeing the success of the pictorial emblem, other early printers followed suit. Publishers, often counter to their author's intentions, reaffirmed a continuity between the device (which remained popular in courts and academies) and the emblem, treated as a decorative element. When the device was used on furniture or on pieces of clothing, it enclosed the figured element, a small pictorial motif, in a heavy, prominent framework. When transferred to the printed page, it gave way to pleasing effects, as can be seen in the collection of Battista Pittoni, the *Imprese di diversi prencipi, duchi, signori e d'altri personaggi et huomini letterati et illustri* (Venice, 1562). The emblematic scene occupied no more than about one-fifth of the image, and even the poetic gloss was lodged in a second heavily decorated framework.

After 1540 some printers arranged emblems by decorative and rhythmic criteria that placed images in series parallel to the text, thus putting the text to the service of the image (see plate XVII). Heinrich Steyner placed woodcuts along a continuous series of emblems, between the title and Alciati's epigram, with no interruption between two emblems. Chrétien Wechel (1536) devoted a page to each emblem, with the image occupying the upper third of each page. In his edition of Alciati (Lyons, 1548), Guillaume Rouillé placed each emblem on a page with a border imitating complex marquetry work in relief, while the richly worked engravings of Macé Bonhomme occupied more than one-half of the page. Little by little, the image won primacy.

The first edition of Maurice Scève's *Délie*, published in Lyons by Sulpice Sabon, gave the image a dual role in the internal rhythm of the volume. An image opened each group of nine ten-line stanzas, which corresponded to each two-leaf signature. The illustrated page carried one full stanza and one half stanza, and was followed by three pages of two ten-line stanzas and one half stanza. The heavy borders of the image followed a regular succession of geometric forms: rectangle, circle, rhombus, horizontal oval, triangle, vertical oval.

Serialization could even be carried over from one compilation of emblems to another to produce an effect of a collection of volumes. This was how Denys Janot printed the *Théâtre des bons engins* of Guillaume de La Perrière and the *Hécatomgraphie* of Gilles Corrozet, in Paris, in 1539 and 1540. In both cases we find full-page illustrations on the left-hand pages and verse commentary on the right-hand pages. The richly worked borders were the same for all illustrations and nearly identical in the two works, which also have the same format and binding.

From Representation to Allegory

The *Théâtre* and the *Hécatomgraphie*, which are presented explicitly as books of emblems, are also alike in their shift away from the device. The 'motto', missing or reduced to a title, plays a minimal role. The richness of the illustrations reinforces an allegorical tendency that was only latent in Alciati. The image, massive and enigmatic, compels attention; the commentary simply unfolds its meaning, in a direct antithesis to 'zethical' (heuristic) emblems. The relation between text and image is reversed: the illustration absorbs both the reader's gaze and the meaning, and the commentary is reduced to a simple explanation. Far from breaking free of its closed circuit between the literal and the figured meaning, the notion adheres to its figuration: the emblem becomes an *effigy*.

To return to Guillaume de La Perrière (see plate XVIII), emblem 63 on 'Opportunity' presents, on the left-hand page, an image with neither title nor device: a nude woman holds a razor in her left hand. Her hair is falling forward; her feet are winged. The commentary explains these mysterious details through a series of questions: 'What is the name of the present image? ... Who was its author? ... And what is she holding? ... Why?'. The presentation of the *Ymage d'occasion* in Gilles Corrozet's collection is less purely illustrative, since the image bears a title and a brief caption in verse. The accompanying text, however, has the same allegorical function as in La Perrière:

> Opportunity, if one happens to inform oneself
> Of your manner, of your portrait and form,
> And one asks you what really signifies
> What is seen in your effigy,
> You will answer saying in this sort ...

The text is a hermeneutic guide following the sinuous detours of an overburdened effigy. The allegorical image, quite unlike the emblematic schema, crystallizes and absorbs meanings. Inspired by descriptions of ancient paintings and by the figures of the arts of memory, it follows the vogue for hieroglyphics prompted by the success of the *Horapollon*, an ancient compilation of the Egyptian language published in 1505 in Venice by Aldo Manuzio,[37] a success that had been prepared in 1499 by Francesco Colonna's 'language by images' in his *Hypnerotomachia poliphili*.[38] The image was valued in the pseudo-hieroglyphic genre because it promoted retention, and the text simply satisfied a curiosity that had been piqued by the illustration.

To return to our books of emblems: when a printer like Denys Janot emphasized the printed image, he seems to have been taking advantage of a fascination for the rhebus seasoned with vague reminiscences of neo-Platonism. He inaugurated a publishing practice, which eventually led to the contemporary art book, of pictures luxuriously reproduced and offered to the curiosity of a reader who scarcely pauses over the commentary. In a similar vein, Abel L'Angelier published a very handsome folio in Paris in 1615, containing 64 full-page engravings by excellent engravers (Léonard Gaultier, Caspar Isac, and Thomas de Leu), *Les Images ou tableaux de platte peinture des deux Philostrates sophistes grecs et les statues de Callistrate*. The commentary, overwhelmed in this edition by the spendour of the pictures, was a reprinting of an allegorical text that Blaise de Vigenère had published without illustration at the end of the sixteenth century. It seems to have been the printers who promoted the illustrative effigy, profiting from the immense vogue of the emblem. Nowhere in emblematic literature of the early modern period do we find a real conjunction of image and text. Emblematic 'zethics' did without (or could do without) figuration; the illustrative effigy stifled the text.

Didactic Aims

Nonetheless, some in the early modern age believed that knowledge could be imparted through the interaction of text and image. Images of this sort, which I call *vignettes*, followed the ancient tradition of Gregory the Great, by presenting tedious or inaccessible truths by gradual introduction to the text. When Georgette de Montenay composed her *Emblèmes ou Devises chrestiennes* (Lyons, 1571), she had that sort of pedagogy of the image in mind:

> The hundred portraits will serve as prods
> To awaken the resistant cowardice
> Of slumberers in their lascivity.
> Alciati made exquisite emblems,
> Seeing which sought by many
> Desire possessed me to begin my own,
> Which I believe to be the first Christian [ones].
> One must search everywhere
> For appetite for these jaded people:
> One will be attracted by painting,
> The other will add to it poetry and writing.[39]

Christian pastoral literature, both Catholic and Reformed, made extensive use of the emblem in the sixteenth and seventeenth centuries as a vehicle for attracting worshippers to the faith by the combined prestige of the image and the genre. Among a thousand examples, let me note Johann Mannich's *Sacra Emblemata LXXVI in quibus summa unius cujiusque evangelii rotunde adumbratur* (Nuremberg, 1624), a collection that closely resembled the missal in that it presented emblems for the Gospel readings for Sundays and feast days. The first phrase of the reading for the day served as the motto for the device.

Besides its prestige, what did the emblem as a form bring to the banal mechanism of didactic illustration? It brought an amusing pastime and a metaphorical function, both of which had a lasting influence on pedagogical publishing. The disconnected composition of the book of emblems allowed readers to open it at any page, letting themselves be guided by the attractiveness of the image. Pictorial paths to knowledge through entertainment developed, as if by compensation, at a moment when pedagogy, following Petrus Ramus, declared itself entirely rational and renounced the fantasies of the arts of memory in favour of ordered reasoning.[40] As early as 1507, Thomas Murner had attempted to teach logic by means of a book taking the form of a deck of cards, the *Chartiludium logicae, seu logica vel memorativa* (published in Paris; reprinted in 1629). A similar text, *Charta lusoria, tetrastichis illustrata*, was published by Jost Amman in 1588. Inspired by cards, it resembled a book of emblems in its treatment of the image and its use of short accompanying texts. In the seventeenth century, universal history was taught in the elementary schools of Port-Royal by means of decks of cards, and in our own day 'educational games' pursue an honourable career.

On the other hand, by the metaphorical connection that it suggested, the emblematic principle created a new relation between the text and the image, and one might wonder whether the use of nondescriptive legends under an illustration did not proceed from the emblematic mould. Legends of the sort had a didactic purpose, and they established a triangular relation between the figure, the text, and the motto. Fragmentary arrangement prompted thought, since it benefited from both a separation of the images and continuity in the text. I am thinking in particular of Jean-Jacques Rousseau's captions for the engravings illustrating *La Nouvelle Héloïse*, which raised the image to a level of greater generalization than was given in the text, as in, 'Inscription for the 6th plate: Paternal Force'.[41] Even in pedagogical manuals in France today, photography takes on a symbolic cast when it generalizes the particular. The underlying designation of meaning as the point of convergence in a two-way movement between the idea and the fact, the universal and the particular, the context and the event, is based on emblematic form. But if we accept the idea that a distracted or curious perusal of a schoolbook, jumping rapidly from one image to another, reflects a long-standing divorce between the text and the image, we subscribe to the disillusioned point of view that Georgette de Montenay expressed in 1571.

In the 'Avis au lecteur' of her *Emblesmes*, she recognized that contemplation of the image did not necessarily lead to reading the text, and she complained of the frivolous consumer of images:

> As I have already seen, in my presence,
> With no regard for the maxim,
> One [person] notes a face or some hat
> That might be better made Huguenot style.

The Emblematic Icon

Illustration was favoured in the emblem by something else besides the exigencies of publishing and didactic purpose: a persistent belief in the meaning of *resemblance*. By a curious switch, the portrait found a place in emblematic literature, as shown in the mixed composition of Théodore de Bèze's collection of emblems, *Icones id est verae imagines virorum doctrina simul et pietate illustrium* (Geneva, 1580). The work was composed of thirty-eight portraits of famous men and forty-four emblems placed after the portraits and treated in the same manner. The *Icones* of Jean-Jacques Boissard (Paris, 1591) contained portraits of Greek, Turkish, and Persian heros, all presented emblematically. The left-hand page gave the person's biography; on the right-hand page, his portrait in a medallion was framed by a decorative border containing a distich in all respects like the 'motto' of heroic devices. At the bottom of the page a short poem connected the image to the couplet and the biographical text. In his introduction Boissard stated that he was publishing the 'medallions of various men with Emblems and heroic devices enriched with French sonnets for the elucidation of the Latin meaning and explanation of the painting and the figures'.

The development of the emblemized icon (in the sense in which Bèze used the term) presupposed a belief in physiognomy – in the significance of specific human physical characteristics. One can of course see the icon as a new version of neo-Platonism, but also as an exaggerated form of the heroic vision of the world, present, as we have seen, in particularized adages on the commonweal. This sign of the emergence of modern individualism was suggested as early as 1562 by our mathematician, François Viète, who concluded his *Mémoires de la Vie de Jean de Parthenay-Larchevêque, sieur de Soubise* with a description of the 'humour' of his hero. He explains:

I wanted to tell you what [I have written] above in passing concerning the humour of the Sieur de Soubize because I see that it is one of the things that historiographers who write the lives of some seek to do diligently, even to remarking their form, their stature and the traits and lineaments of their face, to that effect putting their [model's] effigies and medals at the beginning of their books.[42]

Confidence that physiognomy had meaning found powerful support in the theory of the expression of passions in Descartes. Passion could be explained by 'agitation of the blood' and 'animal spirits' and was expressed in 'corporeal actions'. Taking inspiration from Descartes and from Cureau de la Chambre,[43] Charles Le Brun wrote a *Conférence sur l'expression des passions*, based on lectures that he gave in 1668 and 1671 on 'the expression of the passions' and on 'physiognomy',[44] in which Sébastien Le Clerc's engravings clearly depict the grammar of facial expressions and their relation to forms found in nature, in particular in animals.

In the emblemized icon, the motto and the gloss were expected to transport the individual out of particularity to the heaven of universality. This is what Henry IV expected when, in his rustic naïveté, he asked a learned courtier to suggest a device for the portrait of himself in armour beside a nude Gabrielle d'Estrées. The courtier, according to Tallemant des Réaux, disrespectfully proposed: *Baisez-moi, gendarme* (All right, give it to me, officer).[45] The courtier obviously did not accept the principle of the icon! Iconic derivation, which benefited from the techniques and the prestige of the emblem, played an important part on the public stage under examination here. Existence in a political sense was shown and perpetuated by the emblematic icon. Between 1623 and 1631, Daniel Meissner published two volumes in Frankfurt that contained 832 icons of European cities. His was a resolutely descriptive album, but nonetheless it derived from the emblem, as its title indicated: *Thesaurus philo-politicus hoc est emblemata sive moralia politica figuris aeneis incisa et ad instar albi amicorum exhibita*. To pick one of its icons, his description of the city of Hoorn in Holland shows a totally 'realistic' general view accompanied by the arms of the city (a hunting horn, *horn* in Netherlandish), surmounted by a device (*Natura optima dux et magistra*) and explained by a bilingual inscription in Latin and German under the image.

Sovereigns merited an emblematic icon. In 1609, Jacques Le Vasseur composed a gallery of royalty in the form of a book of emblems, *Les Devises des Roys de France, Latines et Françoises tirées de divers autheurs, anciennes et modernes avec une brève exposition d'icelles en vers françois et la paraphrase en vers latins*. When the album absorbed the icon it lent itself easily to courtly eulogy. When this occurred, all that remained of emblematic form was an overall composition by autonomous units and the use of annexed devices. That is what appeared in an immense and magnificent folio of Francesco Terzo, *Austriacae gentis imaginum pars prima*, published in Innsbruck in 1558. The work contained no text. It was a series of plates, each devoted to an illustrious member of the house of Austria, pictured as a monumental statue. Above them were heroic devices and scenes from the subject's life; below there were armorial trappings and allegories. The device, a minor element in the icon's procedures for indicating character, returned to the decorative and signaletic status of its heraldic origins. It is significant that the German word *Stammbuch* designated both a collection of emblems and a book of heraldic crests.

We see here a fourth element furthering the primacy of the image over the text: the heroic device used in the book of emblems to turn illustration to courtly flattery. Thus when Jacob Typot published his *Symbola divina et humana pontificum imperatorum regum* (Prague, 1601-4), he arranged 930 devices on plates of six images each, classified by rank and by nation, on the exact model of books of arms.

During the first hundred or so years of its existence (1530-1650), emblematic literature combined two uses that were distinct but in constant interaction. Cognitive use of the emblem exercised the formal capacities of authors and readers by the practice of metaphoric schematization; on the public scene, emblematic centuries contributed to the definition of a heroic and state culture, in which the common welfare was constructed from a series of complementary individual actions that could be figured because they were public. Decorative and celebrative use of the emblem attracted the reader-spectator by an increasing proliferation of the image in a variety of forms (the effigy, the vignette, the icon, the court device), all of which were imbued with a doctrinal or courtly political spirit.

Absolutist Arrogation

After the 1660s the absolute monarchy took over and monopolized the production of both sorts of emblem and turned them to its own profit. The monarch became indistinguishable from the commonweal (*L'Etat, c'est moi*), and at that point heroic manifestations of individual merit for a common purpose could only compete to celebrate the king, as witnessed by the abundant production of court devices in the last decades of the seventeenth century. Emblematic practice thus came to have the dual function of educating the subject of the emblem, who had become a subject of the king, and celebrating the king. Quite obviously, absolutist use of the emblem found its highest expression at the court of Louis XIV, but Spain under Philip IV and Charles II, or the empire of Ferdinand III and Leopold I, provided occasions for similar glorification.

The court device taught, expressed, and proved fidelity to the monarch. The famous dialogue of Father Dominique Bouhours, *Les Entretiens d'Ariste et d'Eugène* (1671),[46] shows this new role clearly. Bouhours's text is a veritable guide to aristocratic communication at court. Ariste (etymologically, the Noble) and Eugène (the Well-born) speak first about the sea, since they are at the seaside, then about the French language, secrecy, fine wit, the undefinable *je ne sçay quoy*, and finally devices (sixth conversation, pp. 371-581). This final position placed the art of the device at the summit of the hierarchy of cultural practices of the court aristocracy: 'As for myself, I call it the science of the Court' (p. 575). Skilled practice of the device in fact *distinguished*, in all senses of the word. It assured a privileged communication ('it must not be understood by the people; and it is only intelligent persons who should penetrate its secrets', p. 446) that was transmitted by osmosis. Collections of devices were 'the books of subtle scholars whom the schools have not ruined and the world has polished' (p. 378). It permitted initiation into aristocratic and royal spheres: 'If I had to instruct a young Prince, I would like to do it by the Device' (p. 580). As we have seen, the Prince of Turenne practised philosophy by the device; according to Father Ménestrier, the same was true of Louis XIV: 'And M. de Gomberville of the French Academy presented to him the Doctrine of Manners in Emblems, taken for the most part from the Poems of Horace and accompanied by handsome figures.'[47]

The entire realm of figuration and thought converged in the court and in the device ('the objects of all the sciences and of all the arts are in a certain sense its domain'). This domestication of emblematic individualism was carried out to the exclusive profit of the royal glory. As what was figured shrank from the common welfare to the monarch, the pool of figurable motifs shrank as well. Thus one figure was reserved for Louis XIV – the sun with the multivalent device *Nec pluribus impar* that Louis Douvrier had forged for him around 1663. That emblem, repeated *ad infinitum* on objects and buildings, was now a sign of recognition and no longer a sign of knowledge. Emblematic agility could only be displayed to vary the predicate in the thousand ways to relate the sun to Louis XIV. Except in the religious domain, non-royal devices drew their validity from dependence on the great device and in a combination of the appearance of heroic glory, formal regularity, and expression of submission to monarchic order. Thus '*Uno sole minor* (Only the sun surpasses me in greatness) accompanying a moon was composed for Monsieur, the king's only brother ... *Soli paret et imperat undis* (I obey the Sun and the sea obeys me) under that same Body for the duc de Beaufort, the admiral of France' (pp. 436–7).

A Courtly Art

The courtier's ingenious art of the device reached its height in a brief work written by Father Ménestrier in 1659 to honour Chancellor Séguier, the *Devises, emblèmes et anagrammes à Monseigneur le Chancelier*. Twenty or so devices celebrated Séguier's judiciary function, his role in the organization of the French Academy, the protection that he accorded the Church, his entire submission to the king, his name, the sign under which he was born, and his arms (the ram). Rhetorical refinement went so far as to create an emblem from the anagram of his name in Latin (Petrus de Seguierus): a sceptre with the motto *Pure regis est usui*, for 'as the sceptre serves only for kings, he employs his life only for the service of his King.' It was a clever mix of maximized personal aggrandisement ('appearance') and monarchical devotion.

Since the monarch had absorbed heroic mobility and instantaneity, all his movements had to be recorded, endowed with meaning, and perpetuated. Indeed, devices engraved on medals permit a reconstruction of the royal scenario that was just as important as historiography and substituted for it. The great debate between history and poetry as a medium for glorification that Charles Le Brun had posed in the Academy of Painting in 1667 was resolved,[48] and the founding of the Academy of Inscriptions in 1662 was intended to connect true history with the making of inscriptions – that is, with legends and devices.[49]

The best example of a poetico-historical use of emblems can be found in a work of Father Ménestrier published in 1689 and reprinted in 1693 and 1700, his *Histoire du roy Louis le Grand par les médailles, emblèmes, devises, jettons, inscriptions, armoiries et autres monumens publics*. This handsome folio volume, decorated with 61 plates, set down the history of Louis XIV by reproducing and commenting on the medals struck on the occasion of the major events of his reign: his birth, his majority, his marriage, the birth of the dauphin, his victories, his decisions, and so forth. The two faces of the medal permitted the presentation of both the icon of the ruler and the device that raised the event from the anecdotal level to that of an essential and prodigious action. *Prodige*, which kept its medieval sense of the miraculous, marked and proved the sacred nature of the royal person: 'Providence . . . had promised to make the course of his life nothing but a chain of prodigies' (p. 9). But Providence only permitted and sanctioned the prodigy, which was primarily a dramatization of the royal act and an emblematic form of the event. What was important was to draw essence from accident, which would in return show that the accidental was not fortuitous. Glorification did not escape from a tautological affirmation of power; it gave it 'appearance'.

There was a direct and perceptible causal connection between the prodigy and the emblem (whether it was recited, written, painted, or sung) as a singular and universal action, a transmutation of heroic individuality, and a manifestation of the commonweal. Indeed, the monarch was celebrated more than the monarchy, the action more than the institution. Thus Ménestrier explains the relative neglect of the crowning of the king in representation of his deeds through medals: 'The crowning adding nothing to the right of the crown; it is but a ceremony' (p. 33). The ceremony was to the prodigy what the allegory was to the emblem: it pre-existed the person of the king, whereas the prodigy was born under his feet as he moved (on the condition that empirical existence was ritualized and held at a distance by figuration). As we have seen, Terzo's sixteenth-century icons presented fictive statues of the dukes of Austria, and not their portraits 'from nature'. In Ménestrier's work, the prodigy was twice removed from the subject, by being reproduced on the medal and by the commentary in the form of a device. The prodigy thus existed only when represented, just as the action of the emblem had no existence outside the cognitive intervention of the predicate. The prodigy reconstituted particularity and delivered it over to a memorable apotheosis.

An anecdote of Saint-Simon, in which, admittedly, the word 'prodigy' is platitudinous, shows well just how far such emblematic ritualization of the insignificant could go. Represented, it became a trace and mark of devotion and thus of majesty revered. In 1707, d'Antin received the king in his château of Petit-Bourg. Saint-Simon reports,

It is a prodigy [to see] how far d'Antin went in paying his court down to the last detail during this visit. He did so down to the humblest valets, winning over those of Madame de Maintenon, while she was at Saint-Cyr, to enter her rooms. He made a map of the disposition of her bedroom, of her furnishing, even of her books [and] of the disorder in which they were found shelved or thrown upon her table, even of the places in the books that were found marked. Everything was discovered in her rooms at Petit-Bourg precisely as at Versailles, and this refinement was much remarked.[50]

The true history that Ménestrier preached was thus made by collecting representations of prodigies, 'military parades, feasts, solemn entries and amusements' (p. 2), which were themselves represented in devices and medals. A dramatization of the sort implied immediate and constantly renewed figuration. The life of the monarch was lived and viewed in the Cartesian and heroic mode of continual creation. The device, forged in the illumination of the moment, doubled by the icon that depicted the ruler from year to year, was admirably appropriate to the representation of prodigy: 'It is also the only sort of Monument that represents the King to us in all different ages' (p. 18). Unfortunately, material execution could not keep up with the lightning rapidity of the royal acts, and Ménestrier had to break off the fine chronological order of his *Histoire*. In the final two-thirds of his volume he was obliged to reproduce the last devices of the royal life out of order and with lacunae. He states,

I follow the order of these actions and of these events to give [an order] to these figures, which had none other in their creation but the diligence of the workers and the determination of those who make the Types and the Legends and who are now still working on the Medals for the first years of the Reign. Thus there are many famous actions that have not yet appeared as historic Images and that await being rendered the honour that is due them. (p. 39)

The Monumental Image

Ménestrier countered the tide of time that made emblemizing impossible by erecting a *monument*. He himself designated the medal and the device as monuments, but beyond that generic designation, the device was central in a veritable culture of the monument that guided people's minds as well as their gaze. In Ménestrier's work, representations of monuments of his fabrication punctuated an indefinite chronological succession of prodigies: the series of plates crowded with the greatest possible number of medals with devices was broken here and there by a large monumental image. The device became a motif for detail, a raw material for building a complex whole that fascinated by its grandiose construction or the refinement of its particulars. The reader-spectator could indulge in beatified contemplation of the whole or in wonderment at minutiae.

One of these monumental plates, the 'Birth of the King until his majority' (plate XIX) is presented as a solar effigy. At the centre there is a stylized icon of the royal face wreathed with a legend that is a device in Latin. Twenty-four round medallions, each encircled by a Latin inscription, surround the central face in a double circle of devices; sun's rays go out from them in all directions. Two circular devices at the head of the image and four at the foot frame this triple-circled sun. The thirty devices in medallion all figure the sun in different situations and with different landscapes, explained in the Latin legends. Each one bears a number that refers to a glossary and translation by Ménestrier placed at the end of the volume. The order of the numbers is more bewildering than helpful, since they switch back and forth around the central icon.

Later in the volume, the devices for the royal marriage form a nuptial bouquet. Even later (plate XX), Ménestrier solemnizes the Revocation of the Edict of Nantes in a monumental page resembling a temple to Religion and Piety, represented allegorically around a double figuration of Louis XIV in allegorical effigy (as Hercules killing the Hydra of Protestantism) and in icon (the obverse of a medal). Below, there is a large rectangular panel containing twenty-four circular devices, six of which represent the sun. They are explained at the foot of the page, under the picture: 'These emblems and these devices represent what the King has done for Religion.' The book also immortalizes temporary monuments built for an entry or a celebration. One plate shows the triumphal arch erected in Lyons on the occasion of the birth of the dauphin. It shows four devices, among a great proliferation of allegories and effigies. Another plate celebrating the same occasion shows the fireworks display in Regensburg. A boat in the form of a dolphin (*dauphin*) ploughs through the waters of the Danube, its cockpit covered with indecipherable devices and emblems. The emblematic monument, with its profusion of detail, shows that there is something to see rather than actually showing it, something like a tourist taking in a curiosity at one glance because he or she knows that its grandeur is attested by a detailed description that can be read in the guidebook.

Ménestrier's book of devices fitted neatly into the symbolic strategy of Louis XIV. It appeared in 1689, the very year in which the budget for royal buildings collapsed, following reversals in the monarch's fortunes. It was as if it were important to find a substitute on a reduced scale for the gigantic effort of monumentalization of public life undertaken between 1660 and 1689. 'Your Majesty knows that when the brilliant actions of war are lacking, nothing marks the grandeur and the spirit of Princes better than buildings', Colbert told the king.[51] In point of fact, on 31 October 1660 the king solidified his monopoly of construction by forbidding building in Paris and its immediate surroundings without his express permission. It would be superfluous to recall the great royal public construction projects: let me simply note that the gates of Porte Saint-Denis and Saint-Martin were erected in the 1670s (when they no longer had any practical function) on the exact model of temporary triumphal arches for royal entries. The monument neither showed nor demonstrated: it occupied, filled, and saturated the world with the royal presence, and when the world was unattainable through lack of time or money, the book-monument – the book of devices – took its place. In 1688, a year before the publication of Ménestrier's work, Donneau de Vizé offered the king an allegorical *Histoire de Louis le Grand* in the form of a miniature monument in brown tortoise-shell inlay tinted in red, generally attributed to Boulle. During the same epoch, Thomas Le Roy offered the monarch a vermeil-bound copy of his *Devises pour les Tapisseries du Roy*.[52]

In the absolutist symbolic system, the emblem and the device gave intention to construction; they made a vast text of the world. Paradoxally, it was in emblematics in the service of absolutism that the relation between text and image seems the strictest. The text of public inscriptions took on pictorial value to speak of magnificence, whereas the image became saturated with meaning and constantly returned the viewer to a protocolary text. The book of emblems then became the libretto for a monarchical opera, in which recognizing the arias and following the general direction of the plot-line were more important than the details of the story.

What remained of this monumental culture disseminated by the book, the only place in which synthesis and temporal concentration of the outsize universe of the prodigy could take place? Doubtless not much. After the eighteenth century, the monumental was reduced to decoration. The rooms of Austrian castles and monasteries (Gaarz, Vorau, Ludwigsburg, etc.),[53] which are entirely covered with emblems, belong to the Rococo aesthetic, for they lack tension between the event and eternity, between the person and the universe.

286 ALAIN BOUREAU

Thanks to the diversity of its uses and published forms, the book of emblems probably did not teach the early modern age how to see the image or to master its operations. We have encountered the imbalance between text and image constantly. Nonetheless, the heritage of this noteworthy genre seems important, above and beyond its historical role, for its elaboration of a dual involvement of the subject and the state, and for the definition of the absolutist symbolic system. The emblem exerted metaphorical capacities and, on a variety of levels, contributed to the installation of the *epistémè* of representation. It renewed, diversified, and refined the principle of illustration in publishing. Finally, the book of emblems facilitated, on various social levels, individual appropriation of the *signifying* image. Like books and pamphlets on royal entries, it instituted a many-layered and diversified private grasp of the public collective imagination.

Notes

1 Claude-François Ménestrier, *Histoire du roy Louis le Grand par les médailles, emblèmes, devises, jettons, inscriptions, armoiries et autres monumens publics* (Paris, 1689). I cite from the 2nd edition, 1693.

2 The bibliography is immense. There is a bibliography of collections in the classic work of Mario Praz, *Studies in Seventeenth-Century Imagery* (Warburg Institute, London, 1939; 1947). A revised bibliography can be found at the end of the 2nd edition (2 vols, Edizioni di storia e letteratura, Rome, 1964), pp. 231–576; addenda and corrigenda, Rome, 1947. An excellent bibliography of secondary works on emblematics is in Arthur Henkel and Albrecht Schöne, *Emblemata; Handbuch zur Sinnbildskunst des XVI. und XVII. Jahrhunderts, Supplement* (J. B. Metzlersche Verlagsbuchhandlung, Stuttgart, 1976). For political aspects of the emblem, see Alain Boureau, 'Etat moderne et attribution symbolique. Emblèmes et devises dans l'Europe des XVIe et XVIIe siècles', in *Culture et idéologie dans la Genèse de l'Etat moderne* (Ecole Française de Rome, Rome, 1985), pp. 155–78.

3 See Henry Green, *Andrea Alciati and his Books of Emblems: A Biographical and Bibliographical Study* (Trübner, London, 1872), an older work but still useful.

4 See Michel Pastoureau, 'Aux origines de l'emblème: la crise de l'héraldique européenne aux XIVe et XVe siècles', in Marie-Thérèse Jones-Davies (ed.), *Emblèmes et devises au temps de la Renaissance* (J. Touzot, Paris, 1981), pp. 129–36.

5 See Robert Klein, 'La théorie de l'expression figurée dans les traités italiens sur les Imprese, 1555–1612', in his *La Forme et l'Intelligible, écrits sur la Renaissance et l'art moderne* (Gallimard, Paris, 1970), pp. 125–50.

6 Ménestrier, *Histoire*, p. 15.

7 See S. B. Wurfel, 'Emblematik und Werbung. Zum Fortleben einer Kunstform im 20. Jahrhundert', *Sprache im technischen Zeitalter*, April–June 1981. See also Claude Françoise Brunon, 'De l'emblème à la publicité. Le monde construit et déconstruit', in *Le visible et l'intelligible*, forthcoming.

8 For my analysis of this session of the Conseil d'Etat, see Alain Boureau, *L'Aigle. Chronique politique d'un emblème* (Editions du Cerf, Paris, 1985), pp. 167–70.

9   Claudie Balavoine, 'Les emblèmes d'Alciat, Sens et contre-sens', in Yves Giraud (ed.), *L'Emblème à la Renaissance* (Société d'édition d'enseignement supérieur, Paris, 1982), pp. 49–59.

10 See Fernand Hallyn, 'Les Emblèmes de *Délie*: propositions interprétives et méthodologiques', *Revue des sciences humaines*, 51(1980), pp. 61–75.

11 See Barbara Tiemann, *Fabel und Emblem: Gilles Corrozet und die französische Renaissance-fabel* (N. Fink, Munich, 1974).

12 Eneas Balma, 'Le cas de Guillaume Guérould', in Giraud (ed.), *L'Emblème à la Renaissance*, pp. 127–35.

13 Paris, 1679.

14 Admittedly, this mode of classification had less success than the great lexicon of figured themes, the *Iconologia* of Cesare Ripa (1593) [new edition, Pietro Buscaroli (ed.) (2 vols, Fogola, Turin, 1987)].

15 The work, which classifies figurative themes, completed a previous text of Father Caussin, *Electorum symbolorum et parabolarum historicarum syntagmata* (Paris, 1618), which was a compilation of ancient and medieval bestiaries from an emblematic point of view.

16 See Jean Nagle, *La Métaphore de la théâtralité au XVIe siècle*, forthcoming.

17 Michel Foucault, *Les Mots et les Choses. Une archéologie des sciences humaines* (Gallimard, Paris, 1966) [*The Order of Things: An Archaeology of the Human Sciences* (Pantheon, NY, and Tavistock, London, 1970)].

18 See Bernhard Blumenkranz, *Le Juif médiéval au miroir de l'art chrétien* (Etudes Augustiniennes, Paris, 1966).

19 See Henri de Lubac, *Exégèse médiévale. Les quatre sens de l'Ecriture* (2 vols, Aubier Montaigne, Paris, 1959).

20 After Bernard of Clairvaux, mariology became a veritable mystical and rhetorical discipline (in Conrad of Saxony, Anthony of Padua, Matthew of Aquasparta).

21 Lucy Toulmin Smith and Paul Meyer (eds) *Les Contes moralisés de Nicole Bozon, frère mineur, publiés pour la première fois d'après les manuscrits de Londres et Cheltenham* (Firmin-Didot, Paris, 1889). My thanks to Yvonne Régis-Cazal for calling these texts to my attention.

22 ibid., pp. 123–4.

23 On François Viète, see Joël Grisard, *François Viète, mathématicien de la fin du XVIe siècle. Essai bibliographique*, unpublished thesis, Paris, 1986, p. 30. (This thesis is available for consultation at the Alexandre Koyré Centre in Paris.)

24 François Viète, *Introduction en l'art analytic*, tr. Vaulézard (Paris, 1630); see also *La Nouvelle Algèbre de M. Viète: précédée de Introduction en l'art analytique*, tr. and commentary Vaulézard, ed. Jean-Robert Armogathe (Fayard, Paris, 1986), p. 30 [*The Analytic Art: Nine Studies in Algebra, Geometry, and Trigonometry from the Opus restitutae mathematicae analyseos, seu, Algebra nova*, tr. Richart Witmer (Kent State University Press, Kent, Ohio, 1983)].

25 Introduction to the translation of Viète's *Zeteticorum libri quinque* (*Zétiques*), in vol. 2 of Grisard, *François Viète*.

26 See Vaulézard, 'Avertissement au lecteur', in Viète, *Introduction en l'art analytique*, p. 71.

27 Frances A. Yates, *The Art of Memory* (University of Chicago Press, Chicago, 1966), p. 383.

288 ALAIN BOUREAU

28 The anecdote is given in Girard, *François Viète*, p. 10. Girard refers to André Jean Baptiste Boucher d'Argis, *Principes de la nullité du mariage pour cause d'impuissance* (Paris, 1750).

29 Republished (Paris, 1984) on the basis of the 2nd edition.

30 Yates, *Art of Memory*, p. 369.

31 The second emblem in Nicolaus Taurellus's collection, cited above, p. 271.

32 See Hermann Bauer, *Kunst und Utopie. Studien über das Kunst- und Staatsdenken in der Renaissance* (W. de Gruyter, Berlin, 1965).

33 Ernst Kantorowicz, 'La souveraineté de l'artiste. Note sur quelques maximes juridiques et les théories de l'art à la Renaissance' (1961), in his *Mourir pour la patrie et autres textes* (Presses Universitaires Françaises, Paris, 1984), p. 357.

34 See Boureau, 'Etat et attribution symbolique'.

35 See Max Rosenheim, 'The Alba Amicorum', *Archaelogia*, 62(1910), pp. 250–308. The *album amicorum* often left blank spaces for the notes of the recipient.

36 See on this topic the course notes of Alexandre Koyré, *De la mystique à la science. Cours, conférences et documents. 1922–1962*, ed. Pietro Redondi (Ecole des Hautes Etudes en Sciences Sociales, Paris, 1986), pp. 143–5.

37 See Claude Françoise Brunon, 'Signe, figure, language: les *hieroglyphica* d'Horapollon', in Giraud (ed.), *L'Emblème à la Renaissance*, pp. 29–47.

38 See, on this famous 'Dream of Poliphilia', Giovanni Pozzi, 'Les Hiéroglyphes de l'Hypnerotomachia Poliphili' in Giraud (ed.), *Les Emblèmes à la Renaissance*, pp. 15–27.

39 Dedication of the Lyons edition of 1571.

40 See Yates, *Art of Memory*, ch. 10, 'Ramism as an Art of Memory', pp. 231–42.

41 Jean-Jacques Rousseau, *Oeuvres complètes*, eds Bernard Gagnebin and Marcel Raymond (4 vols, Gallimard, Paris, 1955-), vol. 2, p. 766. All of Rousseau's 'Sujet d'estampes', ibid., pp. 761–71, merits analysis.

42 François Viète (supposed author), *Mémoire de la Vie de Jean de Parthenay-Larchevêque, sieur de Soubise*, ed. Jules Bonnet (L. Willem, Paris, 1879). Bonnet's attribution of the anonymous text to François Viète is convincing.

43 René Descartes, *Les Passions de l'âme* (Paris, 1649). Marin Cureau de la Chambre, *Les Charactères des Passions* (Paris, 1640).

44 The lectures were published in 1698. See also *Expressions des passions de l'âme représentées en plusieurs testes gravées d'après les desseins de feu M. Le Brun* (Paris, 1727).

45 Tallemant des Réaux, *Historiettes*, ed. A. Adam (2 vols, Gallimard, Bibliothèque de la Pléiade, Paris, 1960–1), vol. 1, p. 77.

46 Page numbers in the text refer to the edition consulted, the 'new edition' of 1691.

47 Ménestrier, *Histoire de Louis le Grand*, p. 17.

48 See Bernard Teyssèdre, *L'Art au siècle de Louis XIV* (Livre de Poche, Paris, 1967), pp. 203–5.

49 See Louis Marin, 'L'Hostie royale: le médaille historique', in his *Le Portrait du roi* (Minuit, Paris, 1981), pp. 147–69. All of Louis Marin's works are pertinent to this study.

50 Saint-Simon, *Mémoires*, ed. Yves Coirault (3 vols, Gallimard, Bibliothèque de la Pléiade, Paris, 1983–4), vol. 3, pp. 34–5.

51 Cited in Teyssèdre, *L'Art au siècle de Louis XIV*, p. 8. On the relation between

building and royal celebration, see the works of Gérard Sabatier, in particular his
'Versailles, un imaginaire politique', in *Culture et idéologie*, pp. 295–324.

52 Teyssèdre, *L'Art au siècle de Louis XIV*, p. 250.

53 See Wilhelm Harms and Helmut Freytags (eds), *Auserliterarische Wirkungen
barocker Emblematik in Ludwigsburg, Gaarz und Pommersfelden* (Fink, Munich,
1975).

9

Printing the Event:
From La Rochelle to Paris

CHRISTIAN JOUHAUD

In Paris at the end of 1628 and the beginning of 1629, texts and publications in a variety of forms focused on one event – the end of the siege of La Rochelle. This study hopes to show how print pieces dealt with this event, how the different sorts of printed matter that revolved around this event related to one another, and how the relationship of text and images in these publications came to dominate the production of meaning.

On 29 October 1628, after fourteen months of siege and blockade, La Rochelle surrendered. The old Protestant city opened its gates to the King to avoid a catastrophic final assault. It escaped being sacked, but the winners imposed harsh conditions: the destruction of the fortifications and the suppression of the city's ancient privileges and municipal liberties.[1] The citizens of La Rochelle saved only their property and their liberty of conscience.[2] This event – the taking of a fortified city generally considered impregnable – prompted an enormous reaction. Throughout France, *Te Deum* Masses were celebrated, bonfires flickered, and public rejoicings were organized. Only the Protestant strongholds of the south of France held back in consternation.

Paris celebrated the victory with a *Te Deum* Mass and offered a solemn entry to the King on 23 December. During the eight weeks preceding his entry a quantity of pamphlets, brochures, broadsheets, and books appeared – several hundred of them – celebrating the glory of Louis XIII. Rather than calming down with the festivities of the end of December, the excitement seemed to gather momentum. News, exaltation of the victory, guidelines for proper celebration, descriptions of public festivities: echoes of the event and efforts to interpret it were indissolubly mingled. Printed matter played all roles at once. It is impossible to construct an exact chronology of all the publications that appeared. The king's entry on 23 December served as a central point from which they radiated, some texts preparing the moment, others becoming an integral part of it, and still others prolonging it. Before we pass on to them, one question needs to be asked: what did people in Paris know of the end of the siege of La Rochelle on 23 December? What had they been told during the preceding weeks?

Information and Celebration

We know when and how the news of the victory arrived in the capital. The very day of the surrender, Louis XIII dispatched the chevalier de Saint-Simon to the two queens, Marie de Médicis and Anne of Austria. The King charged them to announce the news they had received to 'his servitors'. He limited his further efforts to addressing a letter to the *prévôt des marchands* (the chief magistrate of the city) and the city's *échevins* to order them to organize the first celebrations, and another letter to the archbishop of Paris to thank him for his good prayers. In the days that followed, other letters left La Rochelle, in particular one addressed to the Parlement with the articles of peace granted to the rebellious city.

As soon as the *échevins* had received the news, during the first days of November, their letter was printed by Pierre Rocolet, 'imprimeur de Messieurs de la Ville de Paris', whose name formed the anagram, 'Porte ce lorier' (bear this laurel).[3] The letter proclaimed, 'The inhabitants of our city of La Rochelle have had recourse to our clemency as their sole refuge.' The archbishop had his letter published as well,[4] and the articles of peace went through at least four editions in Paris.[5] The latter were prefaced by the royal declaration recalling the grievous error of the citizens of La Rochelle and their total submission. They 'have most humbly begged us to pardon the crime that they committed' and the city was returned 'to our hands to dispose of it according to our good pleasure and will and prescribe what manner of life that we shall esteem [proper] for the future, without other condition than it please us to accord them out of our goodness'. Let there be no doubt, the victory had been total.

Another set of texts took it as its explicit aim to inform Parisians of just what had occurred at La Rochelle. These texts did not originate – at least not apparently – from governmental sources but belonged to the vast category of narrations and descriptions (see plate XXIII). Three samples will serve, all published in mid-November at the latest.[6] They came from the presses of three different printers, on each occasion 'with permission'. These three pamphlets have the same format (octavo) and approximately the same number of pages (16). The typographic characters used are of different sizes, however. One publication changes print size in the middle of the press run, from twenty-two lines of thirty-five strikes to about twenty-seven lines of forty strikes, as if the printer had been unwilling to increase the number of pages when he discovered that he was running over. This is eloquent witness to how hastily made these pamphlets were. Both for economic reasons and for commercial and political efficacy, they had to appear as rapidly as possible.

The texts of the three narrations show a certain number of similarities. They insist on the same facts and emphasize the same key moments. Beyond their differences in writing style, they thus contributed to persuading the people of Paris of a particular version of the surrender of La Rochelle.

All three of these publications describe at length the rituals of submission and pardon. Two of them insist on the interview between the king and the deputies from La Rochelle; the third accentuates the initial meeting between the deputies and Richelieu. All of them note carefully the attitude of the representatives of the vanquished city: 'They, tears in their eyes and confusion of their error in their mouths ...',[7] 'they knelt and bowed low, requesting and crying for pardon and pity, [and] one of them ... speaking for all [and] trembling, made his speech'; 'having finished his speech in these terms, with a voice trembling with respect and fear ...'.[8] The contents of this famous 'harangue' are given in the three texts, but in three different versions. Another account, purportedly published in La Rochelle, offered yet a fourth version.[9] If the words changed from one version to the other, the substance remained the same: the citizens of La Rochelle recognized their error, begged for pardon, and appealed to the clemency of the king, swearing eternal fidelity to him. There is of course no mention of the fact that the contents of this discourse had been the object of negotiations and were finally dictated by Richelieu. The king's answer also inspired three slightly different versions. He reprimanded his wayward subjects: 'I know well that you have always been malicious [and] full of artifice and that you have done your utmost to shake off the yoke of obedience to me';[10] he then gave his pardon.

The three accounts give different versions of another scene in the ritual of submission as well: the attitudes and the cries of the citizens at the king's entry into the city. Only when we refer to the *Histoire du dernier siège de La Rochelle*, published twenty years later by Pierre Mervault, a Protestant from La Rochelle, can we question the spontaneity of these demonstrations. Mervault says, 'Following the commands that had been given them by My Lord the Maréchal de Schomberg, they fell to their knees as the king passed, crying "Long live the King and have mercy on us."'[11] Louis XIII was not making a 'joyous entry'. In full armour, escorted by his troops, he entered a vanquished city. Still, it was important that the defeated citizens accepted participating as actors in the spectacle of their own defeat. It was a spectacle that the victors bestowed upon themselves and that the defeated citizens bestowed upon them. Printed matter prolonged and publicized this spectacle in Paris and doubtless throughout the realm. The dramatization of punishment and pardon restoring the pact between the king and his subjects required narration of their submission. All these texts accorded only minimal space to liberty of conscience – an essential clause on the religious or the political level. It was mentioned, but that was not what was going to impress Parisians.

The horrors of the siege, the contrary, were described at length. There is not the least hint of denunciation of the cruelty of the assailants, nor horror before the fifteen to twenty thousand people killed during the siege. Such descriptions had a two-fold aim. First, they were to illustrate the hell into which heresy and rebellion plunged the city. Second, they were to underscore the royal leniency. The king's victory was unquestioned and the city, bled white, would have been easy to take by assault; instead, the king had pardoned, he had ordered bread distributed to the survivors. More than other authors, the recital of horrors fascinated the author of the *Remarques particulières de tout ce qui s'est passé en la réduction de La Rochelle*. He saw 'in bringing low a proud and haughty people hard pressed by death and reduced to sighs and moans', a mark of the 'justice of God and of His saintly judgements'.[12] It was a miserable Babel that had fallen. Within its walls cannibalism had been rampant: dead bodies had been devoured, a mother had eaten her daughter alive, and a little girl the fingers of her infant brother. Famine had been a sign of God's judgement, and when one is abandoned by God, there are no longer limits to how low one will stoop. The king, by his clemency and by the re-establishment of the Catholic cult, had reconciled the city with God.

The description of the horrors of the siege had more than a purely political function, however. It belonged in the tradition of the *occasionnels* that reported monstrous occurrences.[13] An urban public (larger than has been thought) habitually consumed narratives of the sort and bought the booklets that told of them. Authors and printers adapted the story of the siege of La Rochelle to this market. There are two sensational texts in this vein, entitled *Mémoire véritable* and *Mémoire très-particulier*, that list the prices of foodstuffs sold in La Rochelle during the month of October 1628 (see plate XXIII). Politics was never far off, however. As these lists reported prices, they also communicated an interpretation of the misfortunes of La Rochelle in total conformity with the official line. An examplary moral lesson, the exorbitant prices of foodstuffs 'illustrate the desolation of a city that rebellion had turned against obedience to its ruler'.[14] Their obstinacy 'deprived a good number of them of life, and made their souls sink to the bottom of hell, where they were reduced to a horrible and perpetual servitude'.[15]

The same foodstuffs appear on the two lists, which leads us to conclude that although they came from two different and apparently competing printers they had a common origin. The *Mémoire particulier* states that the text was sent to the queen mother, who was thus supposedly responsible for its publication. That is perfectly possible, but we need to take claims of this sort with a good many precautions. They were frequent and in all cases unverifiable. Whatever the case may be, the two texts were copied from the same manuscript. When the prices they give vary, the divergences always seem to be tied to reading errors or problems in decipherment. Thus on one list a cow cost 200 *livres* and in the other 2,000. Did a pint of wine cost 3 *livres* and a 'pound of cured beef skin' 7, or was it the other way round? Did an ounce of bread cost 32 *sols* or 22? Did two cabbage leaves cost 5 *sols* or 10? Was a pound of grapes 18 *sols* or 18 *livres*? The printers evidently did not question the accuracy of the figures, and we know from other contexts that negligence was current practice with pedlars' chapbooks. In the stocks of the *Bibliothèque bleue*, we can find manuals of elementary arithmetic that provide sample problems that do not add up.[16]

Two things counted. First, the enormity of the figures cited: 8 *livres* for an egg, 24 *livres* for a pound of sugar, 800 *livres* for a *boisseau* of wheat![17] The wildest and most murderous records for speculation in times of scarcity were pulverized, and the city was plunged into the realm of the incredible and the unreasonable. Second, unbelievable prices were set for horrible foodstuffs: a pound of dog meat for 20 *sols*, 100 *livres* for a *boisseau* of vetch, 20 *sols* for pieces of dry ox-skin, 32 *sols* for a pound of donkey meat, 10 *sols* for a dog head. Undoubtedly the people of La Rochelle ate donkey, dog, and worse, but to list prices for them as strictly and precisely as for the city-controlled tariffs – as if shops to sell dogs stood beside shops selling apples, eggs, or cabbages – gives the impression of setting sail on a ship of fools.

Official letters, the articles of surrender, speeches, narrations, and sensational memoirs were all offered to Parisians at the conclusion of the siege of La Rochelle. That is what information was at the time. Early in 1629 the annual issue of the *Mercure françois* appeared.[18] It brought nothing new, outlining once again the triple political lesson of the event in the completeness of the king's victory, the greatness of his clemency, and the incredible misery of the people of La Rochelle, which God had visited upon them in punishment for their revolt. It insisted a bit more, however, on the evil intent of the foreigners – the English – who had abandoned those who had put confidence in them.

Information was soon submerged in the rising tide of hymns of joy, cries of victory, and praise dedicated to the king. These pieces soon accounted for the better part of the presses' output. For the first time since the League, increased production could be explained by a combination of religious fervour and political fervour. Victory over La Rochelle exalted both monarchical order and Catholic order. The *Apostrophe à La Rochelle rendue à l'obéissance du Roy* makes use of a long series of imprecations against the Protestant city to celebrate the alliance between power and religion for the greater glory of the 'most Christian King . . . today marvellously triumphant amidst the heretic people'.[19] Political divisions within the Catholic camp, which had left such profound marks on the history of the realm since the League, seem to have become suddenly obsolete. Ultramontane Catholics and 'politicians' were reconciled in a hymn of glory to the Most Christian King. This was an illusion to anyone who knows what happened after, but it is basic to an understanding of the atmosphere that reigned in Paris at the 'triumph' of Louis XIII. Professors from the Royal Collège, the Collège de Clermont, and the Collège d'Harcourt, barristers in the Parlement, parish priests – all the authors who signed the lengthy poems published came from various levels of the academic and institutional world, but by and large they came from the same milieu.

The Jesuits threw themselves heart and soul into this profusion of writings. Their *collèges* put on theatrical performances and ballets to the glory of the king; they organized poetry competitions, the results of which were then published. One of these was an anthology of 188 pages of Latin, Greek, and French verse composed by the pupils of the Collège de Clermont that appeared at the end of December.[20] Among the French poems in this collection was *La Rochelle aux pieds du Roy*, which was reproduced in an appendix to the great book of the royal entry of 23 December by the Jesuit Jean-Baptiste Machaud. The involvement of a personality as eminent as Father Nicolas Caussin, who published a work entitled *Le triomphe de la piété à la gloire des armes du Roy, et l'amiable réduction des asmes errantes*,[21] shows how intensely the Society of Jesus was mobilized. This work benefited from a remarkably speedy publication: the approval of the Sorbonne was given on 4 December, that of the provincial of the Society on 9 December, the royal privilege on 11 December. A preface entitled 'The author's designs' makes his aims explicit: 'I intend to raise up here a trophy to divine Providence that made piety finally triumph in the arms and the conquests of our great monarch.' The book is 334 pages long and most of it was written between the beginning of November and early December. Although the fall of La Rochelle was expected, the circumstances could not have been known ahead of time.

The Jesuits had no monopoly on poetic fecundity. Thousands of lines of verse were written for the entry of 23 December. One of them, entitled *Le génie de la France au Roy. Sur l'heureux retour de Sa Majesté en Sa Ville de Paris*,[22] 185 stanzas long, was dedicated to the *prévôt des marchands* and the *échevins*. In its dedicatory epistle the author attempted to tie his work in with the monumental festivities organized by the municipality in the king's honour. He says:

Seeing the laudable care that you bring to preparing (as far as possible) for His Majesty an entry to some extent presentable[23] and corresponding to the glory of his triumphs, I would have thought to attack the honour of the Muses if by a criminal silence I had remained silent before such a handsome subject for speech and amid the public acclamations of all France.

This immodest profession of faith is dated 19 December. Three days sufficed to print and sell this brochure of 43 pages. It was ready in time to participate in the triumphal choir, just like a 'talking painting' atop a metaphorical arch of triumph. I should note in passing that this quite traditional vision of poetry was in harmony with the philosophy of images. Figures of rhetoric produced images just as well as pencils and brushes did. In the festive context rhetorical images combined with material images, 'mute poems', while the images traced by words were 'speaking paintings' that came to the aid of the artist or painter whose art was doomed to fall short of the perfection of the object he was transposing.[24]

How, indeed, could one paint on canvas or carve in marble the promise of a triumph that had become universal? The problem was expressed by Louis Gaborot, *avocat* to the Parlement and *maître particulier des eaux et forêts* of Dourdan, in a work entitled *Triomphes du Roy Louis le Juste sur son heureux retour de La Rochelle en sa ville de Paris*:[25]

> The world, not being able to sing
> Your exploits nor praise you enough,
> Great King, sole equal to yourself,
> Publishes in its extreme delight
> That Mars does not equal your virtue.
> To you who never have been brought low
> It will suffice me to say
> That the world is under your empire.

Louis XIII master of the world? This was a courtier's flattery, but also shows a desire to play the game of the classical triumph, in which the triumphant hero was identified with Capitoline Jove for one day. In this case, print takes on meaning only through its association with gestures, acts, and political rites.

The Royal Entry of 23 December

With the *sacre* (coronation), the *lit de justice* (King's Justice), and funerals, the royal entry was one of the major ceremonies of the French monarchy. In principle, it took place when the sovereign entered one of his 'good cities' for the first time.[26] Under Henry IV and Louis XIII, however, a good many other occasions for solemn entries were seized.

The entry was presented as 'a rhetorical combination of gestures, words and objects designed to manifest the alacrity and the joy of subjects subjugated to the sovereign, whom they acclaim'.[27] This excellent definition tells the essence. Gestures, words, and objects were combined that usually remained separate, even in the most intense moments of public life in the cities, cut off from one another by divergent and autonomous practices and different modes of circulation and expression. Print pieces held an increasingly invasive place among such objects.

The ceremonial entry was above all a procession all through the city. It remained nearly unchanged during the sixteenth and seventeenth centuries. The city's principal governing bodies and crowds of burghers went to meet the king outside the walls. The city fathers saluted him and made speeches. Then the king's entourage and the citizens joined to form a procession, entering the city gates after the mayor or the *échevins* had presented the keys of the city to the king. The procession then moved towards the cathedral or the principal church of the town through richly decorated streets. In the early sixteenth century a new model, the triumph of classical antiquity, gradually supplanted the medieval version of the *Fête-Dieu royale*.[28] The new model was particularly well adapted to the celebration of a royal victory, as in Paris in 1628. Jean Boutier writes, 'The final *Te Deum* recalled the sacrifice in the temple of the Capitoline Jupiter. The streets themselves, where all sign of practical activity had been removed, became as many atemporal "sacred ways" decorated with arches of triumph.'[29] The subject merits more detailed commentary, but I shall only remark briefly on the political significance and repercussions of the entry. The ritual solemnized and dramatized the pact that linked the city and the king: submission was exchanged for explicit or implicit recognition of urban privileges and ancient municipal liberties. The ceremony embodied quite different interests, however. Some concerned life within the city. The entry recognized and fixed social and political order under the king's gaze. Position in the procession served the basic function of social classification and established a precedent for the future. The order it showed before the onlookers became the order of society itself.

From the point of view of monarchical power, the entry gave an opportunity to put new life into the mystical link between person and function. A certain man on horseback, a body, a face was indeed the king; the function inhabited the person. In return, a successful exhibition of the person threw light on his function. Finally, the entry provided a theatre for power. On this stage everyone was by turns actor and spectator. The burghers looked at the king, and they looked at the more powerful of their own number parading by at the king's side. The king contemplated the city through which he rode, which had put on its best face for him. What he saw were emblems of his own royalty carpeting the streets and decorating the triumphal arches like 'as many images of his own reflection'.[30] He saw his city looking at him and he saw its gaze materialized in the decorations, while the city saw him seeing himself in it. This was the source of the demonstrative force of the ceremony and it explains the care taken to prepare it, in most cases according to the indications and under the supervision of agents of the monarch.

Pictures, written matter, and print pieces were everywhere in an entry, posted up, carried and waved, pasted on available surfaces, sold, distributed. After the festivities were over, the publication of a pamphlet or a book narrating the entry made it live again. This could be a slim publication of only a few dozen pages, but there was a tendency to consider that the length of the narrative should be proportional to the brilliance of the festivities themselves. In many cases genuine books appeared, often richly illustrated. This was true of the publication for the entry of 23 December, the *Eloges et discours sur la triomphante réception du Roy en sa Ville de Paris après la réduction de La Rochelle; accompagnés des figures tant des arcs de triomphe que les autres préparatifs*,[31] a large quarto volume of 182 pages, with appendices and many engravings. A book of the sort was necessarily costly, if only to cover expenses, and could not easily pass for a mass-distributed book. Still, it deserves a closer look. It summarized the pieces posted and exhibited on the occasion of the entry, and hence it offers an opportunity to measure the differences from other print pieces that circulated on the day of the festivities and sought to interpret them.

One might easily object that the book on the entry could never reconstruct perceptions – or even any one perception – of the real spectacle. In presenting a coherent narration, its very mobility decomposed what one pair of eyes saw from one fixed position. It 'stands back from reality perceived as a whole'.[32] Basically, the book confounds the eye of the reader with that of the principal actor, the king, who was the only one who could ride through the entire adorned city and profit from the totality of the spectacle of its tableaux and monuments in logical sequence. And even then, the king could only pass from one motif to another and one high point to another, missing the preliminaries and the sequels. Worse, he could not see himself as he went.

In reality, the book of an entry was usually far from being a complete narration of the event. It was the official version of a celebration of power. One might even hazard a paradox by suggesting that the entry was presented to the spectators as the acting out of a book that had not yet been published. How was it planned? The king began by informing the municipal government of his wish to receive a solemn entry. The municipality then chose one or more scholars who were charged with drawing up a proposal. If it was accepted the scholars became project directors, gathered a team of assistants, and, generally under the supervision of an appointee of the king, moved on to carrying out their proposed programme. At that point paintings, sketches, devices, and decorations of all sorts had to be created, dramatic interludes written, and so forth. Later, the 'inventors' of the entry would take charge of writing the book that would clarify the subtleties of the plan and the profundity of its intentions, thus giving (or restoring) the true sense and coherence of the festivities. This key to comprehension of the initial intentions would open wide the door pushed ajar by the spectators' partial decipherings and bar the way to false interpretations. One might wonder whether any totally successful decipherment of an entry existed other than in the book. We shall return to this fundamental question.

The Book of the Entry

Eloges et discours sur la triomphante réception du Roy en sa ville de Paris was thus the title of the one and only exhaustive and official work on the entry of 1628. It was printed by Pierre Rocolet, the printer by appointment for the city government, and it bears a privilege dated 11 February 1629. The dedication to the king is dated 25 April, which means four months elapsed between the celebration described and the publication date. This letter is signed by the *prévôt des marchands* and the *échevins*. The author's name (which we know from other sources)[33] appears nowhere, since this was the official version, bearing the city's seal. In the dedication the municipal government makes its intentions clear: 'Your Majesty has vouchsafed to have taken such pleasure in the efforts of your subjects to receive you, glorious and triumphant, in your good City of Paris, that we have thought you would approve our design to leave a record of it to posterity.[34] The writers added, 'It is not reasonable that the triumphs due to you should be imprisoned within the walls of one city and terminated by the brevity of a single day.' Thus their intention was to assure publicity over both space and time for this royal reception organized and paid for by the city. We might note in passing the word they used to define the contents of the book: *discours*. The book was to be a discourse – *the* discourse – on the entry. Their aim was not only to use description and narrative to provide an interpretation of the entry, but indeed to offer *the right* interpretation, the one that conformed with the original intention and was thus the only legitimate interpretation. The *prévôt des marchands* and the *échevins* conceded that one might reproach them 'with having waited much' before seeing to the publication of the book. Their answer was that the king's new victories (in Italy) only reinforced the pertinence and the political lesson of the entry and gave it more lasting echoes.

The anonymous preface, *Au Lecteur*, completes the explanations given in the dedication and sketches a portrait of an implicit reader, presented as needing explanations and awaiting them impatiently:

You have greatly desired that the explanation of the paintings made for the reception of His Majesty appeared immediately after the day that [the reception] took place, [so that] each piece, being presented in all lights simultaneously and being instantly explained, would bring continuing pleasure.

It is important to note here that someone reading the book about the entry could not be considered as having completely understood it on 23 December. On the contrary, the 'paintings' were conceived as remaining partly mysterious, hence new light needed to be shed by the book. Its potential public, then, was the population of Paris, the men and women who had watched the entry, and people from the provinces or abroad who would discover it within the pages of the book. The organizers had contracted a debt with Parisians, repaid by the publication of the *Eloges et discours*. They offered a long-delayed pleasure.[35]

The preface was also intended to justify the principal choices made in the organization of the entry, beginning with the one most freighted with consequences: why was classical antiquity taken for a reference and a guide? The answer might seem surprising:

As we make more use of the ways to arrange a Triumph properly, they are also more widely known. The people who sees them sees the thought behind it more easily and the joy that it takes in the public good. Modern inventions still need several centuries to make themselves known before they are received in a use that must render joy just as public as the common good for which it is sought. This is what has obliged [us] to prefer ancient pieces to those of our own times.[36]

We have not strayed from the question of decipherment: to know was to some extent to recognize, to discover anew figures that, if they were familiar, were at least (in the minds of those who contemplated them) well adapted to the circumstances in which they were found. In this perspective, the brilliance of a success, far from being measured by criteria of novelty or surprise, would lie in the virtuosity with which traditional motifs were manipulated. This allowed the organizers to concentrate on expectation, and on the nearly universally shared representation of the perpetuation of the common good entrusted into the hands of the kings of France. The word 'people' did not refer to a clearly defined social entity. It was used more as *populus* than as *plebs* and its use was not intended to set up any barrier between those called to understand and those who would be excluded. On the contrary, it insisted (indirectly) on shared participation, or at least on the intention of such a sharing. The pleasure felt on contemplating the figures (to which only, and only later, the book of the entry would give the key) was in itself an indication of their efficacy. They evoked a particular form of 'adherence' to the representation of the *bien public* that they illustrated. Emotion was put at the level of comprehension. Need I point out that once again this implies a specific philosophy of the image and of the functions of the imagination in intellection?

All in all, the book was to be useful in two ways. It would shed light on obscurities in the motifs used in the decorations; it would 'make painting speak'. In the entry itself, however, pictures were almost always associated with fragments of text, which must be included in the global term 'paintings'. It was the association of image and text that produced meaning by fabricating a metaphor that would strike the spectator. Far from commenting freely on the decorations or paintings, the book offered solutions and defined norms. It provided readers with an opportunity to judge the pertinence of the organizers' decisions. It called upon them to appreciate how well the motifs chosen were adapted to the situation and how their high tone (and complexity was part of elegance) reflected the victory being celebrated. By enabling its readers to evaluate the organizers' sagacity, the book proved that the construction of the entry was a political success.

Finally, the book contributed to 'making this triumph immortal in the memory of future ages'. Those who had not seen would see, and it was the role of the plates to make them to see. It was for them that 'the firm touch of a rich burin' had been 'implored'. Those who had been struck by the 'true' figures (paraded through the city) would have no need of these copies. They had already been persuaded and the book could do little more than explain the mechanisms of the various figures.

After the dedicatory letter to the king and the preface to the reader came a preliminary description of the 'general plan and summary of the city's preparations'. This text combines two different genres. It includes a true summary inventorying the contents of the book, chapter by chapter, but there is also something resembling a plot summary in a 'brief overture to what will be explained after in greater length'. This blending of two quite different genres is made possible by establishing a totally parallel movement as the king moved through the city from triumphal arch to triumphal arch and the reader progressed through the book from chapter to chapter. An invention governs this parallel advance, embodying the political aims of both the entry and the book. The theme was celebration of the twelve royal 'qualities' that triumphed in the person of Louis XIII. Why twelve? Because there were twelve signs of the zodiac, which embraced the entire universe, and twelve labours of Hercules. Each 'quality' corresponded to a monument erected for the entry, to a halt of the royal cortège, and to a chapter of the book. Thus the book reproduced the progress of the entry, but it also traced a path towards knowledge – one might almost say, the path of initiation.

The argument was organized around Basilée, 'the tutelary goddess of States'. She was surrounded by her ladies in waiting, twelve great ladies who bore the names of the royal virtues. Each one ruled over part of the palace of Basilée: 'halls, galleries, gardens'. By visiting them one after the other, one learned the secrets of the goddess, the secrets of politics. The palace was presented as the ideal domain of politics. A space contained and imposed order on a system of knowledge. The entry was but a transposition to the scale of the city of this specialized knowledge. If we think of the artificial memory systems studied by Frances Yates,[37] we see that they shared the association of an idea or a system of knowledge with structured spaces, each containing one or more images. This was more than a simple pedagogy of memorization. By respecting a logical progression from place to place one ended up discovering unknowns. These new truths and unknown terrains were conquered from the starting point of acquired knowledge totally mastered. Such systems of thought show the strength of neo-Platonic currents of thought during the Renaissance. Decades earlier they had been the object of violent controversy, and they had been picked up in watered-down form by the *collèges* as simple mnemonic techniques. In the seventeenth century they were no longer seriously viewed as interpretation grids. As is often the case, what was abandoned by metaphysical speculation or scientific reflection continued to circulate, but by other means and in other circuits; here, in politico-festive practices.

All these considerations contribute to the primacy of discourse, of the written word, and of knowledge over the traditional ritual of the entry. More exactly, this ritual was imbued and surrounded with political intentions that transformed it and made the strictly urban dimension – the civic and contractual dimension – of the ceremonies pass to a secondary level. Furthermore, the entire *ordre de la réception du Roy et de son entrée dans la ville* that is the direct narrative of the entry occupies only six pages of the book. Its 'order' is reduced to a strict minimum: the organization and composition of the municipal cortège, its route as it went to greet the king, the 'harangue' of the *prévôt des marchands*, the marching order of the city militia. 'This done,' it concludes, 'His Majesty mounted on horseback and everyone took his place to enter, in the order that follows, by the faubourg Saint-Jacques.' Each person's place is carefully and precisely noted, but as briefly as possible. What was indispensable was to recall position and costume. The king's route and the actual ceremonial of the entry, which took several hours, is summarized in one sentence: 'In this order, His Majesty came to Notre-Dame, stopping at all the gates, triumphal arches and musics distributed in various parts of the city, in which he took singular pleasure.' One page sufficed to describe the *Te Deum* and the evening celebrations.

The Entry of the Book

The primacy of discourse over action or even the narration of action can also be found in a paradoxical place. The first page of the book is taken up by an engraving (see plate XXII).[38] It shows the king seated with several persons in long robes kneeling at his feet. Others are standing next to the royal throne. Two windows in the background open onto a scene of warfare; next to them there is a wall covered with a tapestry. Troops, seen through the window, are marching outside and cannons are firing. No commentary accompanies the engraving, but at the very bottom of the page a few lines give the names of the engravers and the printers and tell where all the 'figures' of the book could be purchased.[39] This announcement makes one think that the aim of the engraving was to attract customers by showing the high quality of the work right on the first page. But why should there be no explanation of the scene represented? Perhaps it was considered explicit enough to do without.

When the same engraving was published alone it did bear a title, however: 'Le Roy harangué au nom de la Ville par le prévost des marchands.' The historian François de Vaux de Foletier thought this title erroneous.[40] For him, these were the deputies of La Rochelle at the king's feet on 29 October 1628. The proof, he suggested, lies in the landscape seen through the two windows: it must show the end of the siege of La Rochelle and the royal troops filing by to enter into the vanquished city. De Vaux de Foletier was right on that point, but not in his interpretation of the scene. For one thing, what is portrayed corresponds perfectly to the brief description of the arrival of the *prévôt des marchands* and the *échevins* before the king, given later in the text without reference to the engraving:

The said lord governor, *prévôt des marchands*, *échevins*, *procureur*, court clerk and treasurer dismounted and went up to the hall, which was well decorated and tapestried: in which, having found the King accompanied by my lord the duc d'Orléans his brother, my lord the comte de Soissons and other Princes and officials of the crown and lords most richly dressed and especially His Majesty magnificently and royally dressed being in his chair, the said lords of the City put themselves on their knees and the *prévôt des marchands* made his Speech.[41]

Those named in this passage are recognizable in the engraving; Richelieu, who had remained in La Rochelle, is notably absent. There is another indisputable proof. The features of the persons kneeling are drawn with precision. Parisians would have recognized them immediately as the *prévôt des marchands* and the *échevins* of the city.[42] As for the war scene, De Vaux de Foletier did not admit any doubt that an engraving as precise as this could be other than 'realistic', hence a trustworthy reflection of the actual scene or backdrop. For him, as for many others, there are only two sorts of image, documentary depictions of a perfect transparency and coded images that require deciphering. This too categorical opposition is exactly what our engraving makes us question. In the seventeenth century, precise drawing in no way precluded abolishing space and representing La Rochelle in Paris, if that was what was intended. Abraham Bosse gave pictorial form to a causal connection. The *messieurs de la ville* can be found kneeling before the king because the victory had taken place at La Rochelle. The causative act is inscribed in the representation of the scene that it produced. It also

establishes a correspondence with the tapestry that shares wall space with
the windows. It shows ancient heroic combats taking place among armoured
knights with plumed helmets. There is continuity in glory through combat
from the tapestry to the landscape seen through the windows. Continuity of
another sort – dynastic and monarchical – is established by the duc
d'Orléans, easily recognizable on his brother's left.

An interplay between opposites is established between the scenes inside
and outside that expresses the dual dimension of royal authority: power and
paternal affection; the punishment of a rebellious city and good words for a
faithful and zealous capital. The faces of the kneeling men are serene with
slight smiles. The king is speaking to them. One element is surprising,
however: only four of the eight *messieurs de la ville* are looking at the king,
while the other four turn their eyes, not towards La Rochelle but towards the
readers, towards us. It is impossible to avoid their eyes. The procedure was of
course often used by engravers and painters of the period.[43] It rivets the
viewer's attention and it projects out from the engraving, towards interpretation, something invisible that could not be represented by any other means.
Looking at the readers (who are looking at them), the city fathers reflect the
gaze of the king fixed on them. It is a remarkable exaltation of their role as
intermediaries between the royal power and its subjects. The importance of
their function finds here a representation that tempers their submissive
attitude of kneeling men. This circular exchange of looks also contributed to
the perfect meshing of the entry and the book of the entry. These gazes were
directed at people holding the book open and contemplating it, not to those
present at the municipal 'harangue'. The readers were present at one of the
central scenes of the entry. The next page bears the dedicatory letter to the
king, which, in its own way, was also a 'harangue'. The king received the
dedication of the book just as he received the dedication of the entry from
the mouth of the *prévôt des marchands* and the *échevins*. But when we pass
from the actual speech made on 23 December 1628 to the book of April
1629, we have shifted from an event to a purpose, from a ritual to an ideal
representation.

Triumphal Arches

Concerning triumphal arches, I can do no better than subscribe to the
generous humanism of Frances Yates when she said, 'My attitude towards
the reader . . . has always been the humane one of trying to spare him the
more awful ordeals of memory and I shall therefore . . . present only a few
selections.'[44] I shall not inflict on the reader a detailed description of the
twelve triumphal arches that punctuated the advance of the royal cortège
through the streets of Paris. My purpose is not to write a history of the entry,

but to study the functions of print in the entry, the interpretations of the entry that it offered, and the relationships that it established between its own version of the event and those of the actual participants. The citizens of Paris saw and deciphered the inscriptions and the images posted up, in particular on the triumphal arches. Print offered its own vision and its own decipherment. It determined an interpretation and defined ways of reading. A study of the first and the fourth arches should be enough to give a general idea of them, on the condition that we keep in mind that the entry was perceived as a whole. Every segment, every halt was part of a system that made use of the twelve qualities of the King, Louis XIII, to promote a particular discourse on the French monarchy. The twelve qualities, personified by twelve emblematic virtues, were clemency, piety, renown, love of the people, justice, felicity, prudence, authority, strength, honour, magnificence, and eternal glory. The actions of the king – and in particular his recent victory – made these qualities triumph, but their triumph needed the king as a symbol of their unity.

The first arch was dedicated to clemency. We have an engraving of it (see plate XXI) and twelve pages of text to guide our understanding. The monument was erected at the head of the faubourg Saint-Jacques. Very probably constructed out of wood, it was covered with paintings, inscriptions, and a variety of decorations (branches, ribbons, and so forth). In the book the engraving depicting it is placed before the text. The chapter itself, however, ends with the words, 'The figure of the first arch is the following.'[45] The decision to place the engraving at the head of the chapter was thus taken after the text had been printed. Someone decided that it was better to see before reading and understanding; that, in the last analysis, the publishing venture should comment on an engraving rather than illustrate a text. Perhaps this person decided that it was better to keep as close as possible to the spectators' impressions of the entry. The decision must have seemed important if this prestigious book, which claimed to bring back to life the perfection of a great political celebration, was allowed to risk being perceived as imperfect and being accused of error or confusion.

General reflection on clemency as a royal virtue brought in Tacitus, Xenophon, Pliny, and Cassiodorus to expound on the first triumphal arch. Next came a few lines on architectural orders, which were supposed to set an overall tone, grave Doric for the martial, as opposed to the gaity of the Corinthian. The arch to clemency was of the Doric order, 'as can be seen by the bases, the capitals, the mouldings'.

At the centre (and my description follows the book's), an isolated 'frame' represents 'the triumph of the king's clemency'. A large, nearly square picture indeed shows in the engraving under a statue on a large pedestal and above a decorative frieze placed over the archway. The usual procedure in describing a picture is to name the objects or the themes it contains first and interpret them afterwards. This is, in any event, the manner of catalogues and historians of art. This is not what happens in our seventeenth-century book. The interpretation is given immediately, before the description, in a mode of decipherment that eliminates the stage of recognition and searching for clues.

The elimination of that intellectual operation makes the status of the representation itself less clear. We do not see Louis XIII *as* a triumphant warrior, but a triumphant warrior who *is* Louis XIII. This echoes the remarks in chapter 7 concerning the *placard* entitled *Le Persée françois*.[46] Françoise Bardon reflects on the meaning of the notation *où étoit dépeint le Roy* in a picture showing Neptune hunting sea monsters.[47] Just how far can commentary go in this game of substitution? To what point can it obscure the operation of representation with impunity, using intent to identify the thing or person represented with the representation of it? The immediate reading of intent imposed an order of importance: in order to produce meaning, what was represented dictated its law to what was representing it. The book, which restored intent, overwhelmed the actual entry; the celebration itself was reduced to a simple illustration – also in the sense of 'what renders illustrious' – of a pre-established discourse, thus of the book that transmitted it. That the engraving was ultimately placed at the head of the chapter did not weaken this hierarchy. Less explicit, it was even more effective and could be imposed as a presupposition.

To return to the illustration, 'His Majesty is in the chariot itself' drawn by four white horses. Here there was no confusion with Neptune; the king's face was easily recognizable. The royal lineaments were manifest proof of the insertion of Bourbon virtues into the apparatus of the ancient triumph.[48] The king's eyes *were* clemency. Soldiers brought before him standards that showed proof of his warlike exploits in the reconquest of the Ile de Ré and La Rochelle. A winged victory, crowning him, hovered over his head. The vanquished – unruled passions and a long troop of vices – followed after him in chains. Impiety tamed, perfidy vanquished, audacity brought low, impudence captured, fury garrotted, bloody cruelty in chains, pride and arrogance broken, lubricity in irons – three furies sufficed to represent all of these, since the number three was gifted with the virtue of symbolizing infinite numbers and unlimited sums. Threes returned constantly: three paintings adorned the arch, three others were placed between the columns on either side, a trio of emblems surrounded the large picture. Two corniced entablatures to the right and the left of the central picture established the base of a triangle, while a statue holding up a rainbow formed its summit.

The entablature under the cornices was decorated with two pictures showing 'two effects of the iris, which the Ancients and the Holy Fathers have always estimated the hieroglyph of clemency'. In the seventeenth century, the word *iris* served to designate the rainbow, the flower, the female divinity of classical antiquity who was messenger to the gods, and a precious stone. The two emblematic pictures played on this plurality, since each one evoked several levels of meaning and several types of knowledge (mythological, religious, symbolic, botanical, and so forth). The association between the image and the short text accompanying it at the base of the entablature underscored correspondences among the various properties and produced the metaphorical meaning of the emblem. Everything revolved around 'iris' as rainbow, superimposing its natural properties on its function as a divine sign in Genesis ('the first sign that God deigned to use to assure men of their pardon').

To the left, the goddess Iris was stopping the rain. To the right, iris, the rainbow, perfumed nature. According to the commentary, two properties of the rainbow were shown. In the first case, however, it was the goddess who embodied the symbolism. Iris was seated on a storm cloud, binding Jupiter with three bands in the three principal colours of the rainbow. The motto was *Pluvium ligat aera*. The picture on the right showed the result of this action: the rain had stopped, the rainbow was shining, and where it touched the grass irises (the flowers) had sprung up to perfume the air (*perfundit odore*). These two emblems of royal clemency were linked by another rainbow of much more imposing size, 'which reigned over the whole work', held up by a statue personifying clemency. The four lines of Latin verse at the foot of the statue (translated in the book) drew an explicit correspondence between the statue of clemency and the end of the siege of La Rochelle. The arch itself also bore a motto, *Coelesti principis Clementiae*, which 'contained the consecration of the triumphal arch to the clemency of His Majesty'.

The ensemble made up of the large painting and the triangle of emblems that surrounded it had a double thrust. It marked the appropriation by King Louis XIII of general values of a triple resonance, cosmic, mystical, and mythological. It also transferred the concrete action that had just taken place at La Rochelle to a level of generalization that integrated it into universal and transcendent values. This was a transposition of the principal political function of the entry discussed above: the reciprocal investment, by means of ritual and spectacle, of the function and the person of the king. The king's actions were *authorized* by the values of the French monarchy of which he was the depository, and these values were in turn revitalized by the brilliance of his actions. Three devices on the frieze elaborated this correspondence.[49] Altars were depicted at either end, one with Jupiter's lightning bolts and the motto *Innoxium* to show contained force, the other heaped with gifts, sacrifices to appease the royal ire, with the motto *Placabile* (easy to placate). At the centre a globe was circled by olive branches and bore the motto *Magni custos clementia mundi* (clemency safeguards the universe). The olive crowns the world and crowns the king. 'It is enough for me to say / That the world is under thine empire', adds one of the many poems that circulated at the time of the entry.

The upper section of the triumphal arch was held up by four columns on either side of the central archway. Between the columns six cartouches bore other devices. The book presents them three by three, beginning with the upper left-hand one.[50] The first showed an elephant surrounded by sheep with the motto *Mitis majestas* (translated as *débonnaire majesté*). The comparison between the elephant and the king was based on the naturalists' characterization of that animal's nature. It was a colossus, fierce in combat and omnipotent, but gentle with the weak (here, the sheep). Beneath this came the figure of the king on horseback holding a laurel branch 'as was done in Rome on the return of victorious rulers' (*adventus optimi principis*). The cartouche at the lower left showed love replacing the sword of justice with an olive branch, symbol of peace (*vindicta ex armata*). The cartouche on the upper right bore a club from which green and living branches were sprouting. It was Hercules' club, an invincible arm, and was made of olive wood: 'In fact, naturalists remark that among all trees only the olive, being dried, indeed, even worked, and returned to the soil grows green once more and takes root.'[51] Thus the king's clemency reflowered: *Prona clementia* (inclined to clemency). Next came a motif that we are told comes from an antique medal 'that the Romans engraved . . . for the happy return of their Rulers', showing fortune as a Roman matron with a horn of plenty (*Fortuna redux principis*). Below that, a goddess held wheat stalks with one hand and the helm of a ship with the other (*laetitia fundata*: joy well installed). If we consider the devices two by two horizontally (upper, middle, lower) instead of vertically, left and right, we can see that each level corresponded to a different *motivation* of the symbol.[52] The top was the province of the naturalists and the natural properties of the symbolizing motif (the elephant, the olive tree) used as a basis for comparison. At the middle level the motivation was historical, involving events and ancient coins. On the lower level – but hierarchically superior – were two emblems of indirect, totally implicit motivations. These were the ones that purists thought the best. Only combined decipherment of the image and the motto lent them meaning. Recognizing the motif was indispensable, of course, but it was no longer enough. One had to be practised in the art of the device.

A final element completed the decoration of the arch: a large marble plaque at the top of the arched opening bore an inscription in Latin dedicating the triumphal arch 'to the most Christian King for having delivered France from rebels and foreigners'.

Piety followed clemency on the route of the royal cortège, but we will move on to the fourth arch, devoted to love of the people (see plate XXII), and will consider only one aspect of it. The formula is misleading. What was meant was the love that the people showed for the king. How could that be considered a royal virtue? It was proof of the king's ability to arouse and merit their love. The decoration, which was simpler than the first arch, was composed of three principal elements. At the very top were three hearts with flames coming out of them. Under them, six shields bore the arms of the king, the queen, the queen mother, the governor of Paris, and the city of Paris. To each side the people 'was represented in various poses' behind a balustrade.[53] Roses were scattered everywhere. How, though, did all these elements function as a coherent whole?

The shields presented a strict hierarchy embracing the king, his mother, his wife, the governor, and the city as a civic body. They were placed inside rosettes, turning them into inhabited roses. Why this insistent presence of roses? The rose was the flower of love. More concretely, its omnipresence on a triumphal arch recalls a laudable custom of the city of Paris:

Every year at the St John's [Day] bonfire, six long garlands of roses are prepared, the first of white roses and the others of scarlet roses: the white one is for the King, if he cares to touch off the fire, or for the one who takes his place in this ceremony; the five red ones are for the *prévôt des marchands* and the four *échevins*, for love is the great knot of States and the mystical chain that keeps all parts of the political world in their duty.[54]

This commentary enables us to understand the engraving and to grasp that the harmony of the political system was represented in these escutcheons, roses, and flaming hearts. Roses surrounded the escutcheons just as the garlands of St John's Day circled round the king or his representative and the *corps de ville*. But a second mystical chain surrounded the roses: the love of the people. That love was symbolized both by the three flaming hearts at the very top of the construction and the 'balustrades of columns between which the people was represented in various poses'. Something important was being presented here: this effigy-people occupied two positions *at once*. It was part of the mystical chain. It had its role in the perennity of the political system. But it was also part of a vertical hierarchy, where it occupied the lowest position. Thus the arch became a figure to show the mystery of organic order.

Discrepancies

The *Eloges et discours sur la triomphante réception du Roy en sa ville de Paris* offered a reading of the royal triumph that was both perfect and legitimate. Other texts point to the existence of other readings. One of these was the *Traduction française des inscriptions et devises faites pour l'entrée du Roy*.[55] The work is a pamphlet of fourteen pages published without date, place of publication, or printer's mark. It was a sort of 'instructions for use' of the inscriptions, pictures, and, more generally, the different motifs of the entry. It is written in the past tense ('Lower, the king was depicted . . .'), which leads one to suppose that it appeared after 23 December. Prudence is called for, since it may have been written in the past tense to ensure continued commercial success and larger distribution to the work. In any event, it must have appeared at or soon after the event if the text were to be of any use, as the reader needed to have the organization of the festivities in mind to understand what he or she read. No illustration nudges the memory, and the descriptions of the pictures and statues are extremely elliptical.

The text is interesting for its differences from the official book. In the description and analysis of the first arch to the king's clemency, for instance, the order of description in the *Eloges et discours* was very clear: it began with the large picture, passed on to the entablatures on either side, and then described the rainbow and the statue that held it up at the top. It continued with the frieze and the cartouches between the pillars to the two sides, and finally described the dedicatory inscription. The *Traduction* began at the top (rainbow and statue), moved on to the principal picture, then to the inscription, which was presented as something like a caption to the picture instead of relating to the whole. This means that the frieze was described as placed underneath the marble inscription, in contradiction to the book and the engraving. The greatest discrepancy appears in the interpretation of the lateral motifs, however. The entablatures and the cartouches were placed on the same plane. The second cartouche to the left was described first, followed by the upper left tablet. Then the author changed to the right-hand side, describing the entablature first, then the middle cartouche, then the upper cartouche. The author then returned to the left-hand side to describe the entablature. The two lower cartouches are completely ignored. It is of course possible that the text described the arch as it really appeared and the book of the entry completely transfigured it. It seems more plausible, however, that the *Traduction* twisted reality and sinned by omission. In the book, the arrangement of the motifs worked to construct a unified and coherent representation of political power. Everything formed a system. In the *Traduction*, that unity disappeared and the author seemed content to sum up the elements. What is more, as we have seen, the author focused on the centre, and the sides were not clearly described, which skewed the global significance of the arch as a monument celebrating a royal virtue.

Still, the author of the *Traduction* identifies the motifs with no hesitation: 'At the top of a portal was clemency holding up the sky', 'lower there was the King painted as Mars, triumphant in his chariot'. He never falters, even when he proposes sizeable differences from the official interpretation. The description of the large painting mentioned furies in chains following the royal chariot. But it added that before the chariot were the captains, the standards, the mayor, and the principal citizens of La Rochelle who 'were kissing the feet of the horses of the said chariot'. The book mentioned only

pictures brought before the king representing his exploits. It was unthinkable for the *Traduction* that the vanquished not be portrayed,
particularly when the marble inscription was made to serve as a caption for
the large painting. The inscription read, 'He avenged France against both
the rebels and the enemies foreign to the kingdom.' In shifting meaning, the
Traduction showed that it cared little for the rules (to which the book
attached great importance). In an allegorical painting everything had to be
transformed; the triviality of the real must never appear. It was represented
by symbols, signs or allusions.

Lower down, the globe that the *Eloges* called *camayeu d'un monde couronné
de branches d'olivier*,[56] in which the word *camaïeu* – cameo – was borrowed
from the technical vocabulary of art, in the *Traduction* became *un globe bleu
couvert d'une branche d'olivier*, focusing on a more pedestrian visual
perception. Incidentally, the author of the *Traduction* was well enough
informed to tell – in Paris – an olive branch from a laurel. The discrepancies
in how the cartouches were 'read' are even clearer. The knight in the second
panel on the left was immediately identified as 'the King on horseback' and
the motto was cited accurately (*adventus optimi principis*) in the *Traduction*.
The same is true of the elephant: 'an elephant in the middle of several sheep'
(top left). But why an elephant? What did it represent? No interpretation was
suggested. The same was true for Fortune with the horn of plenty (centre
right). The discrepancy becomes greater still with the *riche devise sur la massue
d'Hercule*, as the *Eloges* called it, which becomes *un arbre penchant chargé de
fruits* (a bending tree loaded with fruit). As for the pictures on the two
entablatures, the connection between them (described above) is totally lost
in the *Traduction*. For the right-hand painting it gave, 'there was another half
rainbow on a meadow dotted with flowers'; for the left-hand one, 'there was
also the air in the figure of a man seated on an eagle, tied by a woman.' The
polysemy of the word *iris* was not noted; the man sitting on the eagle was
not identified as Jupiter. We must take care not to jump to the conclusion
that the author simply failed to figure out the arch. The central operation in
the process of signification – the representation of the elements by human
figures – was correctly given ('the air in the figure of a man'). Only the
intervening steps had disappeared.

The same sort of thing can be found in the translations. The Latin mottos
were transcribed just as they had appeared in the *Eloges*. The translations
seem correct and are free of actual mistranslations. Still, they are so literal
that they weaken the correspondences set up by the association of text and
images. Thus for the goddess Iris and Jupiter, *pluvium ligat aera* was
translated as *Il lie l'air pluvieux*, substituting 'he' for 'she', which contradicts
the description of the picture, in which the woman was binding the man.
Was this negligence? Perhaps not. The *Traduction* skipped intermediate
steps. An implicit relationship was established between the image and the

king; the motto was taken as a direct reference to the hero of the festivities. The interpretation was all the more plausible since Louis XIII was often represented as Jupiter riding an eagle.[57] This 'real presence' of the king in his representation under the allegorical 'species' of Jupiter and Iris permitted total and reciprocal identity between the sovereign and the virtue being celebrated.

Many other examples of discrepancies and distorted translations could be cited. The *Traduction* rendered *adventus* as *avènement* (taking the throne) rather than the *arrivée* found in the *Eloges*; *redux* (in *fortuna redux principis*) as *retournée* rather than *de retour*; *mitis* as *douce* rather than *débonnaire*; *prona clementia* as *facile clémence* rather than *enclin à la clémence*. All these small adjustments affected the overall interpretation of the motif. They altered the relationship between the word and the image, but they did not falsify anything essential. The device was not perceived as a cultural construction and much of the information that it contained and its rich allusions were lost, but neither the king's majesty nor his clemency suffered. The notion of shared culture became concrete. An inaccurate reading of a motif ended up producing a correct decipherment of the intention that gave rise to it.

The *Eloges et discours* also took liberties with the Latin maxims, but they were of a different sort. How could four lines of Latin verse produce ten lines of alexandrines in French?[58] The book championed an unequalled complete-ness. It developed all possible meanings, even marginal ones, and it backed them up with a host of cultural references. For the arch of clemency alone, the paths of decipherment wound through a forest of citations to more than thirty-five ancient authors.

Some of them – Aristotle, Plutarch, the two Plinys, Tacitus, Virgil – would surprise no one and may have stirred up memories of university days for some present at the royal celebrations. The same was true of the familiar passages from Scripture. But who would recognize inscriptions from Claudian, St Paulinus of Nola, Publius Papinius Statius, or Cassiodorus? Or from the Greeks Alcinoüs, Apollonios of Rhodes, Libanius, or Themistius? Still, those who liked that sort of thing needed to know that these trails existed. They must also have known that they would never grasp them all, which explains their reported impatience to consult the book for which they had had to wait four months. The book made the entry complete. The appetite for completeness had already been teased by the monuments of the entry. The decoration of the arches of the Châtelet was an excellent example of careful organization, to prompt decipherment and arouse expectation.

The arches of the façade of the building had been transformed into a temple to strength. The viewer first saw (to begin at the bottom) fourteen statues of generals of classical antiquity, all conquerors of maritime cities, grouped seven Greeks to one side, seven Romans to the other. Each statue of an illustrious general from the past offered a few words to King Louis XIII in French or Latin verse. For Alexander it said, 'Must I confess that since your victory is without loss for your own, it was what was lacking to the height of my glory?'. Between each of the statues and above them was fire, real and symbolic: 'Some represented by emblems to make the force and the virtue of the King more brilliant by hieroglyphics for fire, the others true and natural, which, shining at [the tips of] white wax tapers, chased away the darkness of that place and the shadows of the night.' The emblematic fires, obviously, combined a painting and a motto. The arches themselves were richly decorated. A sky 'burnished with gold' showed an 'infinity of golden stars', giving an impression of peaceful depth. Fourteen figures stood before this backdrop, arranged in two decorative groups (once more using the virtues of the number seven). Eight allegorical figures were depicted, the four cardinal virtues to one side, the four parts of the state (politics, army, commerce, agriculture) to the other. Three angels were figured on either side, sweeping down from heaven and holding the emblems of the French monarchy (the crown, the sceptre, the hand of justice, the insignia of the two orders of Saint-Michel and le Saint-Esprit) in their outstretched hands. They seemed to float, suspended mysteriously above the spectators. The 'fourth part' of the temple of strength was music: 'Music must not be lacking; that would too indiscreetly take away the voice of this great body.' Music was not a supplement but an integral part of a whole and of a representation. Nothing was more important than making people grasp the coherence of this whole. The Greek and Roman generals were symbols for strength and royal virtue. They addressed the king as their peer. But as he surpassed them in perfection, their words were filled with enthusiasm and flame, symbolized by the emblematic fires (and fire was recognized as the 'symbol of strength, royal by nature'). It was the task of the fine white wax tapers to give material form to the fires of history shown in the emblems and to transmit their light to the infinite quantity of stars that decorated the arch. The stars thus seemed to light the angels and the royal attributes descending from airy realms.

The *Traduction* gave its interpretation of the temple to strength as well, moving once again from top to bottom. Instead of rising slowly, deciphering one element after another and moving towards the cosmos, it began with the angels and descended to the statues of the ancient generals. As it was with the arch of triumph, the correspondences that effect the transitions within the system were not grasped. Still – once again – the essence was there. Steps along the way to decipherment were skipped over, but the link between history, the cosmos, and the king was tightened.

This connection was an Ariadne's thread for wending one's way through the entry of 23 December, but it also aided in finding the way from the festivities to print. One last monument shows that this thread could lead to unexpected figures in mobile monuments. Three chariots took their places in the cortège and followed the king through the city before the eyes of the citizens of Paris. They were richly decorated 'machines', the first, *à la rustique*, representing the Age of Gold; the second, the Roman circus; the third, the city of Paris. An engraving of the Roman circus hints at a rather strange function and 'most high meaning'. For the first time (and on page 169), the book speaks of an engraving as such, without immediately connecting it with what it represented. The author carefully indicates that it had not actually been possible, for lack of space, to harness more than one horse to each of the chariots depicted as rolling around the Roman circus, and that the engraver had taken the liberty of showing three horses harnessed to each chariot. For the rest, however, he claimed that the engraving described the rolling Roman circus of 23 December faithfully.

The race was run according to the rules: four chariots on a track turned around *metae*.[59] At the centre stood an equestrian statue of the king. On the side, in the foreground of the engraving, were the royal arms: the escutcheons of France and Navarre, the crown, and the neck chains of the two orders. A large 'L' recalled that the person and the function of the king could not be separated.

The 'significance of the circus is lofty, celestial and divine, taken from the very establishment of the world.'[60] The circus represented the heavens, its twelve gates the twelve mansions of the zodiac. The four colours worn by the champions and their supporters represented the four seasons. It is certain that the vision was directly cosmic. Still, while the cosmic explanation was advanced for each one of the motifs, the text appears to treat the statue of the king differently, using a purely historical explanation: in Rome the statues of conquering generals were erected on the central separation of the circus in this manner. The race that took place around the statue did indeed reproduce cosmic movement: 'At the two ends of the circus were the *metae*, or the guideposts, around which the racers made their turns to express the regular motion of the sun on its elliptical course.' This was the geocentric, Ptolemaic cosmos. However, at its centre was the king, so often designated as the sun. The book of the entry even stated four pages later, 'Paris, after having been deprived of its sun for an entire year ... now sees that beautiful star shining with glory appear on its horizon.'

To read the rolling monument and the engraving in an emblematic sense would thus lead to a heliocentric Copernican interpretation of the cosmos. It is tempting to point out the contradiction between the text, which expressed the traditional representation of the universe, and the image, which purportedly showed the initiate a Copernican (thus heterodox) cosmos – even better, exhibited it two steps from the king in an official cortège. I would certainly be reproached for projecting onto a political celebration in 1628 a scientific debate that contemporaries, even if they had been aware of it, would never have thought to compare with what they saw in the streets of

Paris or the book of an entry. The innocence of the 'engineer' who constructed the chariot could be argued, or that of the engraver or the Jesuit father who wrote the book. I might nonetheless note that it is surprising that the author used cosmological interpretations for all motifs but the king, and recall that it was a burning question at the end of the 1620s. Be that as it may, the three wagon-drawn 'machines' apparently met with enormous success 'with the people'.

Les Eloges ends with two pages bearing the same text in Latin and French, written in large capital letters. The page layouts and the type characters used show an intention to imitate inscriptions in stone, making a dedication to the king, as we have seen on the arch of clemency. The book seems to be offered as the last monument of the entry, one that would contain all the others and would alone be capable of putting a true and fitting end to the celebration.[61]

Lifting the Veils

Beyond their divergences, the *Eloges et discours sur la triomphante réception du Roy en sa ville de Paris* and the *Traduction française des inscriptions et devises faites pour l'entrée du Roy* both contribute to an enterprise that might be called publicization. Many similar texts were published. As early as 31 December, François Pomeray published a 166-page *Histoire de la rébellion des Rochelois et de leur réduction à l'obéyssance du Roy*,[62] which soon went through three printings. The work, the original version of which was destined for 'the foreign nations', was a translation from the Latin by Jean Baudoin, 'translator of foreign languages by appointment to H[is] M[ajesty]'. Within a few weeks there appeared works entitled *Traduction française du Panégyrique du Roy Louis le Juste*[63] and *Eloge du Roy victorieux et triomphant de La Rochelle*.[64] The *Panégyrique* had been given as two lectures at the Sorbonne by 'the professor and orator of the Greek language' in the Collège d'Harcourt, on the occasion of a solemn session of the university advertised by posters 'put up in the public places of the University of Paris'. These translations brought the reading public the purest products of erudite sociability from university and literary circles. They joined with other forms of dissemination of information, some distributed on the very day of the entry. Spectacles, for example, were immediately publicized in printed descriptions.

The most sumptuous of those spectacles was doubtless the fireworks display on the evening of 23 December. As was always the case at the time, this was not simply a display of rockets tracing more or less conventional figures. It was a veritable narrative: the fireworks told a story. On this occasion the *maître d'oeuvre*, Horace Morel, chose the familiar story of Perseus and Andromeda and adapted it to the theme of the day's festivities.

Morel's description of the fireworks had already been published, both to whet the appetites of his spectators and to make his pyrotechnical choreography 'readable'.[65] The same text was published later in Nantes, with only minor modification of the title, as a narration of the fireworks that had taken place some days earlier.[66] Instead of the fireworks 'that Morel is to make for the arrival of the King', the Nantes printer changed the title to read, 'subject of the fireworks made at the entry of the King into his city of Paris'. He changed nothing else.

It is scarcely surprising to discover that Andromeda was attached to her rock as an offering to the monster, who obviously spurted a great quantity of flames from his nose, eyes, and throat. A flaming Perseus swept out of the night to deliver Andromeda, whose rock then burst into flames. After this summary description, Morel explains 'the mystical meaning of this fable'. Andromeda is La Rochelle; the monster, the English; Perseus, the King. Note that the monster does not represent heresy but foreigners, which was hardly appropriate for the entire length of the siege. Morel gives a somewhat coarse justification of his choice: Andromeda was a virgin; La Rochelle had never been taken; Perseus was the son of the greatest of the gods, and Louis XIII of the greatest of kings. The true intent was apparently to affirm a continuity, to attach the celebration of 1628 to other royal celebrations in which the theme of Perseus and Andromeda had already appeared.[67] Memory was needed for full comprehension of the fireworks display, and printing came to its aid.

The central concern of the author of the *Chariot triomphant du Roy à son retour de La Rochelle dans sa ville de Paris*[68] was doubtless remembrance of the celebration and of the themes that served to exalt the royal person. The book went as far as to offer its readers a canonical memory system. Using the central theme of the entry, the author constructed a chariot to honour the triumphant king. Each of its components referred to a royal virtue, thus combining all the triumphal arches of 23 December. The four horses were named prudence, clemency, fortune (or felicity), and magnanimity. The charioteer was glory, the axle, Fame, the wheel, pomp (or magnificence), its spokes, trophies, its hub, strength. The platform and the seat of the chariot were victories, honour, and praise.

Anyone who managed to imprint this chariot of virtues on his or her imagination would then see it when he or she thought of the entry and, when one virtue was mentioned, immediately recall the disposition of the decorations on the triumphal chariot. Recall of both the celebration and the discourse on power that it offered was programmed into one effort of memory and imagination.

The ambition of a good entry programme was to make the mysterious universe of substances and first causes visible, using artifice to make it appear within the universe of accidents and appearances. It had to embody ideas by transposing them into the trivial language of sense perception. Tools existed to achieve that end. They were signs more charged with substance than others, signs that began by addressing imagination in order to speak to the intellect. Mass-distributed printed matter did not hesitate to use them.

The *Prédictions tirées de l'octonaire ou nombre de huit* was part of a rich publishing tradition.[69] The first of its two parts was supposedly written before the surrender of La Rochelle, which it predicts; the other just after. The publication appeared early in 1629. It obviously gained credibility from the brilliant success of its first predictions. The text pursued two goals: it demonstrated the benificent influence of the number eight everywhere and in everything, and it looked for it wherever possible in the world of Louis XIII. Its most striking demonstration came from the name of Jesus: written in Greek it constructed a perfect octonary. If the letters of the Greek alphabet were used to make numbers, the letters in the name Jesus totaled 888. Like Jesus, Louis XIII was under the sign of eight. Eight names made of eight letters could be attributed to him. Furthermore, he was born in October, the eighth month of the old calendar; he was 28 years of age in 1628, and he was the sixty-fourth king of France, which was 8×8. As a result, the surrender of La Rochelle would take place 28 October 1628, 'the sun in the sign of Scorpius, which is made of eight letters'. In this sort of literature the conclusion is always turned towards the future: the surrender of La Rochelle inaugurated and manifested the coming of the age of the octonary and announced long years of happiness for France and its king.

The science of anagrams is close – at least in its perception of reality and its procedures for forming meanings – to that of prediction. Both belonged to the ancient family of *griphes* or enigmas.[70] People played at and with anagrams everywhere from disreputable taverns to the most fashionable salons. Not everyone made the same use of them, but, like the triumphal arches, everyone shared in their operations for the production of meaning through decipherment. In *Les Triomphes de Louys le Juste, et le Victorieux découverts dans l'escriture saincte en un Psaume, que l'eglise chantoit au jour mesme de la reduction de La Rochelle à son obéyssance* (see plate XXIV),[71] Father F. Bon, 'priest attached to the church of Saint-Jacques of the Butcher's market in Paris', set himself the task of seeking out mysterious affinities with Holy Scripture. From one single verse of Psalm 60 he drew no fewer than seven anagrams in Latin, using all the letters, concerning the taking of La Rochelle – no mean feat. He defines them as 'so many paintings in which different images and representations are seen depicted on the same background', thus underscoring the obligatory passage to the imagination through the eye. He likened the art of the anagrammatist to that of the painter or sculptor, but here the artist was working on inspired materials bearing a truth that was intangible, even if it was seen in different lights. These anamorphoses of the truth, produced by creative art, unveiled verities available but concealed in the Biblical text. They were as many statements (otherwise illicit) to portray the mystery of an immutable truth, representable, however, only in the multiplicity of its possible and infinitely numerous metamorphoses. Each of the anagrams thus became a hieroglyph of Holy Scripture.

Les Triomphes de Louys le Juste was not published until the beginning of 1629, but when the entry took place Father Bon had already published one anagram taken from Psalm 60, 'the one that the Church – O marvel! – chanted on the day of your victorious entry into La Rochelle'. The form he chose was an illustrated *placard* (see plate XXIV).[72] The famous dike appears in both the picture and the text: 'Au point que mon coeur estoit le plus angoissé, vous m'avez relevé en la pierre' (When my heart was in anguish, thou hast exalted me on a rock; Psalm 60.3) became 'Voicy le Roy Louys Treiziesme qui borne, et arreste la Mer par une digue de pierre' (Here is King Louis XIII who bars and stops the sea by a stone dike). The dike had figured on two of the triumphal arches, but fairly discreetly. The anagram gave it a new status as part of the working of Providence. It was Providence that had enabled the King to put 'a net in the mouth' of the ocean and to control the power of storms. On the engraving we see the king on a celestial throne. He wears full coronation regalia: robe, crown, sceptre. Two angels float on clouds, bearing scrolls on which the texts of the psalm verse and the anagram are written. A third angel crowns the victor with a laurel wreath with one hand while the other holds a pen to write the psalm verse and its anagrammatic translation. The royal throne is perched on a rainbow. The arch of clemency had appeared again, now that the warrior's might had imposed his will. The two lower ends of the rainbow are resting on the dike. The space between contains the towers, houses, ships, and ramparts of La Rochelle. It is a strange dike, formed like two keys: it imprisons the city, the better to open it to obedience to its sovereign.

From La Rochelle to La Rochelle

On 15 January 1629, Louis XIII left Paris for Italy. The party was over. Parisians continued to hear of the King's victory at La Rochelle, however. They received printed reports of the celebrations that had taken place abroad and, in France, in Troyes, Dijon, Chalon-sur-Saône, Mâcon, and Grenoble as he made his way to Italy.

From Rome there arrived a *Récit véritable des actions de grâce et réjouissances publiques faites à Rome pour la réduction de La Rochelle*.[73] Next, from Venice, the *Harangue prononcée devant la Sérénissime Seigneurie de Venise et l'Ambassadeur du Roy par Rémond Vidal gentilhomme françois sur l'heureux succès des armes de Sa Majesté*.[74] Both works were printed in Paris.[75] The first, dated 11 January, takes the form of a letter. It related the fireworks display over the Tiber, the illumination of the embassy, the street-corner fountains running with wine, all offered by the ambassador. It spoke of the solemn Mass in the church of San Luigi dei Francesi, attended by Pope Urban VIII, of the mitigated joy of the Spanish, and of the poetry contest held ('all the best wits immediately set to work to publish verse on the subject', samples of which the author offers). The second text, the formal 'harangue' in Venice, gives an idea of the triumphal tone that seemed appropriate for envoys of the French monarchy (or at least for their image in Paris):

The entire world had become a theatre; Pontiffs, Emperors, Kings and Republics, all the rulers and all the peoples of the universe were spectators: France was the stage on which the bitter tragedy of the rebels of La Rochelle was finally performed and concluded to the immortal glory of our King.

Parisians heard about the entries offered to Louis XIII by his 'good cities' through two sorts of publication, of nearly identical contexts but of quite different publics and patterns of distribution. Two texts written in Troyes and Dijon were soon reprinted in Paris: one was a narration, *La triomphante entrée du Roy dans sa ville de Troyes*[76] and the other a description of the triumphal arches in Dijon. Information on festivities in other cities appeared in the 1629 *Mercure françois*, which published brief accounts of the entries in Mâcon and Grenoble and longer ones on those in Dijon, Troyes, and Chalon, using pamphlets published in those cities and even reproducing all fifty pages of the Troyes publication.[77]

Anyone who had attended the 23 December entry in Paris would have found the description of the five triumphal arches in Dijon perfectly comprehensible. The first arch expressed the 'good wishes and the submission of the city of Dijon'. The second evoked the defeat of the English, the third, the taking of La Rochelle, the fourth, the King's clemency, the fifth, his triumph (in classical terms). Although the arches were all related to the event being celebrated, they sought to 'cover' the royal high deeds of recent months rather than forming a coherent whole.

Since its Parisian counterpart has been discussed, it is tempting to pause over the Dijon arch to the King's clemency. It stood next to the Jesuit *collège* and the Jesuits perhaps had a hand in its construction and decoration. It reserved few surprises. There were four columns, capitals, and a cornice. On top stood Hercules; beside him was 'a cupid taking away his club'. Although the description is not clear on the point, it seems this was a statue on a pedestal rather than a painting. There was also a painted emblem of clemency (a lightning bolt separating an eagle and an olive branch), a dedicatory inscription, *Clementissimo Victori*, and, under the pedestal, a few lines of Latin verse. That was all: a statue, a painting, an inscription. The description might of course have been incomplete. Still, Parisians had nothing else to judge by, and they could take pleasure in the disproportion between celebrations in the Burgundian city and in the capital of the realm. The paucity of motifs was partly compensated by a poem in French, twenty-four octosyllabic lines long, that hung from the vault of the arch. If the poem was to be readable the paper would have to have been rather long. Similar poems were posted on the other arches.

Paris had had two months to prepare its entry; Dijon and Troyes had barely two weeks. In an atmosphere of improvisation, the citizens of Troyes could make only three arches. For the rest, pedestals were installed on which statues were erected, accompanied by emblems and paintings. Time was so short that only a part of the plans could be carried out.

The mayor, the *échevins* and those who were charged and commissioned [to do] the porticos, sculptures and paintings . . . were greatly saddened not to be able to finish some of the said works ordered, begun, and settled on, because of the brevity of the time [and the] rigours, cold spells, frosts and snows of the season, no matter how much diligence the workers brought to it, during the ten days they worked even at night.[78]

Why should this tract have made such a show of the city fathers' difficulties? Unlike the Paris book, the Troyes text (or at least its first part) was evidently interested in showing what went on behind the scenes and in evoking constraints and efforts. The entire process of decision making and preparation was described: the arrival of the news that the king would come, the meetings of the city council, the choice of 'several fine minds versed in History and Poetry' who drew up an outline and then supervised the carpenters, painters, drapers, ribbon merchants, and others who carried out their plans. We learn that they made use of old decorations as well as making new ones. Above the ancient drawbridge at the entrance to the city, there was a large escutcheon honouring Henry IV's entry in 1595. The decision was made to repair it, repaint it, and decorate it, adding a new inscription, *fleurs de lys*, and crowned intertwined 'L's. Similarly, an equestrian statue of the King with perfectly recognizable features was placed in a niche above the main entrance to the Hôtel de Ville. Details were furnished on removal of windbreaks from the streets, on problems of street maintenance and rubbish removal, on the sand that had to be brought in, on *grosse guillemette*, one of the city's cannons that needed repair, on the poor (to be kept away), on torches, lanterns, and foodstuffs. It was as if the principal objective of the text, beyond the political celebration, was to show the efforts of the municipal government, its zeal, and its talent for organization – and to show the initiative of certain wealthy private citizens.

The description of the decorations and the account of the celebration came in the second part of the work. The greatest 'hit' was incontestably the the mechanized float in the shape of a galley, moved by 'artifice and springs'.

It stood, richly decorated, in front of the City Hall. A nine-year-old girl chosen 'for her good grace, beauty and assurance' (but also because she was the daughter of a *conseiller de l'échevinage* and member of the city council) was perched on this 'machine' to await the King. When the King's carriage arrived, the galley moved forward, leaving the little girl on a level with the door. She curtsied three times and offered the King a heart of gold, a present from the city. The heart opened with a spring mechanism, and the King could see inside a gold enamelled *fleur de lys* bearing a crown, placed on a double 'L' and two laurel branches, also of gold. The little girl then recited:

> Sire, the flower of Kings and the heart of France,
> This heart that encloses a Lily which with all our heart we offer you,
> Is the heart of our hearts and nothing do we breathe
> But the Lilies and the honour of your obedience.

The King then replied, 'My pretty little one, I thank you, you have done well.' The book quotes his words, and it reports several times on the King's expression: 'he seemed satisfied'; 'he laughed'. This was not the point of view of an ideal spectator, an abstract eye, as in Paris, but rather of an authorized witness, perhaps some Troyes notable.

The golden heart gleamed throughout the entry. Platforms – *échafauds*, as was said at the time – had been set up at strategic points in the city. On them children, adolescents of both sexes, young ladies, and women sang or recited verse as the King passed by. They prepared the gift of the city's heart, the word *coeur* returning unflaggingly, reiterated continually. Furthermore, the history of the gift had not ended when the King had the heart in his hand. The text tells the epilogue in great detail:

And the said day, after His Majesty's supper and [after] he had several times handled, looked at and considered the heart that had been presented to him before the City Hall and had shown it to the lords and gentlemen who were then near him, he put it into the hands of My Lord the chevalier de Saint-Simon and ordered him to transport it that very hour to the lodging of the sieur de La Ferté, the father of the young lady who had given him this present in the name of the inhabitants of the city, and to put the said heart back into the hands of the girl, to whom His Majesty made a gift of it. Which was done immediately, [Saint-Simon] saying to the said girl that she was much obliged to His Majesty and [had] received from him a singular favour, since he was giving her his heart.[79]

One could not dream of a more symbolic act. The city gave its heart to the King; the King gave it back after having made it his own. This slim publication showed little interest in portraying an ideal world in which royal politics, finally made readable, could be discovered in all the intensity of its perfection. Still, it effected a transposition from the real to the ideal (symbolized by the double gift of the heart) in the pact linking the city to its sovereign. Formerly, the King received the keys to the city and gave them back, confirming the city's privileges. The times were no longer right for such a transparent exhibition of the pact. Royal absolutism no longer lent itself to allusions to ancient contracts, even tacit ones. The heart offered and returned in Troyes echoed that practice, but translated it into the language of love. No longer a question of a contract between the institution of the monarchy and the institution of the municipality, from now on it would be a pact of love between the person of the King and the city, represented by a little girl. The exchange remained, but it was totally innocuous politically and both parties came out winners.

The central preoccupations of this *Joyeuse entrée du Roy dans sa ville de Troyes* on the occasion of a celebration of a victory over rebel heretics were to provide a roll of honour for city notables eager for social recognition, and to note the King's reactions to the fidelity of his 'good city'. That the text was reprinted in toto in the *Mercure françois* shows the degree to which this was pleasing to the powers that be.

Four years after the surrender, the royal victory found some sort of consecration in La Rochelle itself. On 20 November 1632, the Queen, Anne of Austria, made a solemn entry into the city. Once again echoes of the ceremonies resounded in Paris. Jean Guillemot immediately published the *Réception royale faite à l'entrée de la Reyne dans la ville de La Rochelle*.[80] At the same time a 90-page booklet on the entry was published in La Rochelle itself.[81] Thus we return to La Rochelle and to the field of battle.

The *Relation de ce qui s'est passé à l'entrée de la Reyne dans la ville de La Rochelle* opened with a dedicatory letter to Richelieu. The cardinal, absent from Paris on 23 December 1628, was absent from La Rochelle as well on 20 November 1632. Stricken with sickness, he had remained in Bordeaux on his way back from Languedoc, where Montmorency's revolt had just been put down and the duke executed.[82] He had organized the entry, however, and dictated rigorous orders and minute requirements for it. In spite of his physical absence, the cardinal-duke was intensely present in spirit, both in the entry and in the print material that gave an account of it.[83] The author of the *Relation* was one Daniel Defos.[84] A lawyer and a member of the city council during the siege, he had been among the vanquished in 1628, and he later participated in the negotiations for the surrender, first with the Maréchal de Bassompierre, then with Richelieu. In 1632, he took up his pen to celebrate the entry of the Queen into the submissive and disarmed city without municipal government, citizen militia, or ramparts.

As everywhere else, there were arches of triumph in La Rochelle, along with statues on pedestals, tapestries and hangings, paintings, festoons, and fluttering ribbons. There was also a fireworks display; there were theatrical performances and receptions. All spoke in unison: the Gallic Hercules ('having the size, the majestic presence and the face of the King') that adorned the face of the second arch was a companion piece for the drama that the Jesuits (promptly installed after the defeat of the city) put on in their *collège*. The victories of the king were 'represented under the name of the Gallic Hercules, in conformity with the design for the arches of triumph'. Similarly, the text shows how the devices paralleled the paintings, how the arrangement of the statues fitted in with the inscriptions, even how the poems were part of the spectacles: 'During the ballet stanzas of poetry in sonnets and on rolled scrolls of all sorts were thrown out.' They were later anthologized.

The surprise lay elsewhere. The central theme of this entry into La Rochelle was the defeat of that city four years earlier. The *Relation* did nothing to avoid recall of that event. It described the long procession of black-mantled figures that left the city and advanced to meet the Queen. Rather than municipal troops, these were two thousand burghers without arms, bare-headed, divided into five companies. All knelt, 'that great people lined up in tight order, resembling, bare-headed, a single line made of many points into infinity', and then a city magistrate, the sieur de Lescale, *lieutenant criminel*, began to speak. The two publications reported his discourse with care and in identical terms. He began: 'This city (if indeed one can still call it so), what remains after the scourges of God and the indignation of the King, the skeleton and the phantom of La Rochelle, resuscitates at the arrival of Your Majesty to come and throw itself at your feet.' He went on to speak of the ramparts razed to the ground, the privileges that had been abolished, and the 'prodigious hunger' that had got the better of the besieged population.

When she entered into the city, the Queen found the first triumphal arch. Forty feet high, it would have towered over the arches in Dijon and Troyes, which rose only 27 and 35 feet high. At its summit there was a relief representing the King, dressed in 'his royal mantle'. On either side of the central arch, however, facing the viewers and at the height best exposed to their gaze, were two large depictions of 'our famine, the chasm of the great woes that devoured our families during the obstinate siege', a strange motif for a festive picture. The renewal could only be solemnized with an ostentatious recall of the city's errors and woes. The treatment of this macabre theme was even more surprising. To one side the dying and the dead were shown, 'their eyes haggard and sunken, their noses long and transparent, their faces earthen and deformed, their skulls enlarged, their necks longer than usual, their ears flapping, their hair tufted and sticking up straight ...'. The narration in no way attenuated the miseries exhibited in the picture. The picture on the other side evoked the daily and exhausting search for food in the besieged city. Daniel Defos reported the tragic spectacle in the following terms:

This prodigious carnival was diversified by strange and frightful mummeries. [One person] was carrying a dead dog by one foot, slung over his shoulder, while a child behind [him] was eating its ears raw. [Another] was removing grease from a trap in which a rat was caught, bloody and half split by the blow, which Monsieur was licking off his fingers. [Another] was drawing a lizard out of a hole by the tail, half of which trembled under his teeth ... You could have seen delicate Demoiselles who had painstakingly scraped a pound of goatskins putting them on a bit of coals and, sprinkling them with spices and suet, gulping them down gluttonously in all haste for fear someone would come along.[85]

This was a strange way to evoke the memory of the thousands of fellow citizens who had died of hunger during the siege. Defos of course notes that it was done to move the queen and 'to pour pity into her natural goodness'. But why then should he speak of Carnival? It does not seem to me that analysis and explanation of the different motifs and inscriptions on the three arches can answer the question satisfactorily. Comparison with other festivities of this solemn entry may permit us to formulate a hypothesis, however.

On 22 November, a final amusement was offered to the Queen: nothing less than a naval battle. Not any battle, but La Rochelle's last chance in 1628, when the English attempted one last time to force the blockade to bring relief. The event was mimed with a good many boardings and cannon salvos. A Turk was added, however: when the English had been routed, an Ottoman ship came along and was captured after a few handsome manoeuvres. This was no theatrical representation but a full-sized reconstruction outside the port. The evening before, as if to serve as transition, the great fireworks display (also representing a naval battle) had taken place over the water. Hundreds of rockets and firecrackers had been set off from ships that gradually came closer, drew up next to one another, and finally bumped together. At that moment a host of firecrackers, fire fountains, and blinding lights were set off. The sight was so impressive that the spectators took fright:

But all the people who were looking on innocently from the shores, being surprised by caprice and fright, fell back pell-mell and there was such a rout in the suffocating throng that all the surrounding spaces were emptied. The court alone remained, immobile, thanks to its long acquaintance with the vainglory and pompous noise of those flickering fires.

The roles had been distributed: the Court, impassive, and, adding to the spectacle of the fictive combats against the rebellious citizens of La Rochelle, a genuine panic among the spectators! How well the spectators collaborated! How spontaneous they were in their willingness to play the role of the vanquished, to commemorate their own defeat, and to adhere to the image of themselves that the victors had disseminated everywhere! If we recall the fantastic *Mémoires* commented on at the beginning of this chapter, listing prices for foodstuffs during the siege, we can see the same atmosphere of the ship of fools in the Parisian print pieces and the macabre Carnival posted on the arch of triumph. As with the panic during the fireworks display – or rather, the report of it in Defos's narrative – the people of La Rochelle zealously displayed their acceptance of the victors' image of their defeat. Just as the people who organized the festivities acquiesced in following the cardinal's orders to the letter, so the citizens and Daniel Defos played and replayed their defeat. They took as their own the power structure's discourse concerning them, and to demonstrate their submission, they displayed it before the world and the queen. This publicizing was a political initiative in both the entry and the printed works. It was a demonstration produced in Paris, where the publications were distributed, as well as in La Rochelle. In both cities, what power prescribed was shown as totally accomplished. No cranny was left in which contestation might ferment or, more simply, any silent manifestation of a remnant of personal pride might lodge.

Power in Print

Three *placards* served the royal policies; thousands of pages published after the siege of La Rochelle joined in the hymn of praise to power glorified. Print pieces related as they exalted; they explained as they celebrated. The Sorbonne, the Jesuits, the Parlement, the convents and monasteries all fed the presses. This unanimity, which lasted scarcely longer than the period of celebration and reconciliation, had an important political dimension. Everything seems to have concentrated on the person of the King: the letters of the King, the words of the King, the facial expressions of the King, the gestures and actions of the King, the virtues of the King, now had the mission of manifesting the perennity of the values of the French monarchy. The hundred or so slim works published at the end of 1628 and the beginning of 1629 helped to create the image of transfers like the one so prettily expressed by the spring-action heart given by the citizens of Troyes.

Power's imagery did not pass inertly into the pages of printed texts, however. Although texts did indeed reproduce the motifs, the maxims, the pictures posted on walls or on arches of triumph, the discourses, and the hymns of joy, they also constructed what they described or related. They raised veils that covered meanings; they divulged contents and they specified rules for decipherment (and for composition). They transpose elements and they allow us to see correspondences, causes, and essential connections. Rhetorical images were produced in order to explain painted or drawn images. The art – the technique – changed from one image to another, but not the ways of signifying. As it changed its medium, the demonstration was transformed in its expression and drew new strength from the reiteration of its effects. Inversely, images were engraved and printed that made long discourses visible to the eyes of those who knew how to see. In the one case as in the other, dissemination of information and prescription were inseparable, for one could not understand without giving adherence: the road to be followed led over lands that belonged to those who had traced the roads, imposing their points of view, distributing the steep places and the gentle slopes. Thus, when a text pre-inscribed in itself the effects of the action that it described, it participated actively in the diffusion of a representation of the efficacy of power, which means that it aided that efficacy.

The official book of the entry poses the insistent question of the relationship between narrative (or commentary) and the acts themselves. As both a public version of the written programme that preceded and gave order to the entry, and the narration of a festivity that had already taken place, it occupied an ambiguous position (from which, what is more, it drew its efficacy). Thus on its own account it could reflect, page after page, the mirage of completeness that inspired the entry. It went beyond and perfected all the readings that spectators had made of it or other print pieces had offered – readings condemned to lose their way struggling with uncertain decipherments. Its own reading was fictional, however, since it only appeared to decipher. It none the less became the measure and the norm of all readings. From that point on, a dual legitimizing process set in. In managing to offer all solutions, in presenting the complete significance of decorations, in justifying factitious decipherments by the results obtained in reconstructing political intent, the book legitimized the order established by the entry. In return, the power of the public ceremony was taken over by the book, which reflected its order and claimed to be its final monument.

In this framework, cognition was almost always recognition. One lifted a veil to show better what was already there. The texts and images that were posted up presented a broad and complex range of levels of recognition, from the implicit references in devices conceived as cultural assemblages to the traces of festivities past and puzzle emblems 'to amuse the commonality'.[86] The discrepancy between the official book and the pamphlets (in particular the *Traduction française des inscriptions et devises faites pour l'entrée du Roy*) points to an essential fact: no reading of the entry could be completely successful without the book. At one moment or another, the spectator would fail to comprehend. And even if he or she did not fail, something would always be left to decipher in a wealth of unattainable meanings. Total failure was just as improbable as total success, however. This leads to the hypothesis that there would be only intermediary decipherments, in part successful and in part not. There are a great many indications of such partial successes present in both the motifs themselves and the external conditions of their reception. There were multiple layers of decipherment proposed by the emblems posted on the arch to clemency; there was a plethora of possible stages in the communication of meaning. The many forms of printed matter that circulated on the day of the King's triumph were among them. Moreover, we have seen incomplete or erroneous decipherments that in the end produced correct and legitimate political interpretations. All this seems sufficient to put aside the hypothesis, often defended, that only a small minority of *literati* were able to understand the decorations and the organization of a celebration such as the one on 23 December.

Failure and success should not be presented as two antagonistic poles, working to divide the citizenry into two clearly distinct groups as active participants or passive bystanders excluded from all comprehension. To deny this too summary, too brutal division does not of course mean to contest inequality of competences and levels of reading. It is, on the contrary, because there was a certain amount of sharing that level differentiation could operate fully, distributing roles and marking social distances according to the complex rules of the hierarchies and social relations in the city under the *ancien régime*. It was also because of shared experience that making something understood was of importance to the power structure that presented for decipherment, in the street and in print, what it wanted people to believe.

Notes

1   The best summaries are still Gabriel Hanotaux, *Histoire du Cardinal de Richelieu* (6 vols, Société de l'Histoire nationale/Librairie Plon, Paris, 1895), vol. 3, pp. 111–90; François de Vaux de Foletier, *Le Siège de La Rochelle (1627–1628)* (Firmin-Didot, Paris, 1931; Rupella, La Rochelle, 1978). Léopold Delayant, *Bibliographie rochelaise* (Impr. A. Siret, La Rochelle, 1882) can also be consulted profitably for older bibliography.

2   This important clause has been emphasized abundantly in the historical literature; neither the victors nor the vanquished paid great attention to it in their writings.

3   *Lettre du roi à MM. Les prévôts et échevins de la ville de Paris, sur la réduction de la ville de La Rochelle, apportée par M. de S. Simon* (30 October) (Paris, P. Rocolet, 1628) BN Lb36 2662.

4   *Lettre du roi, écrite de sa propre main à monseigneur l'archevêque de Paris, avec remercîment des prières extraordinaires qui ont été faites par son clergé* (31 October) (Paris, Impr. de R. Estienne, 1628) BN Lb36 2665.

5   *Articles accordés par le roi à ses sujets de la ville de La Rochelle* (Paris, A. Estienne, P. Mettayer and C. Prévost, 1628) (two printings); *Articles de la grâce accordée par le roi à ses sujets de la ville de La Rochelle, sur le pardon par eux demandé à Sa Majesté* (Paris, A. Vitray, 1628) (at least two printings and perhaps five), B. N. Lb36 2659 and 2660.

6   *Relation véritable et journalière de tout ce qui s'est fait et passé en la reduction de la Ville de la Rochelle à l'obéissance du Roy* (Paris, Impr. de J. Barbote, 1628) BN Lb36 2661; *Relation véritable de tout ce qui s'est passé, dans la Rochelle, tant devant qu'après que le Roy y a fait son entrée le jour de la Toussaincts. La Harangue et les submissions des Maire et*

Habitants de ladite ville, avec la response que leur fit le Roy. L'Ordre qui fut gardé pour les conduire à sa Majesté, et autres particularitéz (Paris, A. Vitray, 1628) BN Lb36 2667; *Les Remarques particulières de tout ce qui s'est passé en la réduction de La Rochelle et depuis l'entrée du roi en icelle; ensemble les cérémonies observées au rétablissement de la religion catholique, apostolique et romaine, avec la conversion de plusieurs habitants de ladite ville* (Paris, N. Rousset, n.d.) BN Lb36 2668 (16pp., 15pp., and 16pp.).

7   *Relation véritable et journalière*, p. 8.

8   *Les Remarques particulières*, p. 6; *Relation véritable de tout ce qui s'est passé*, p. 9.

9   *Harangue faite au roi par les députés de La Rochelle; avec la réponse de Sa Majesté* (La Rochelle, P. Froment, 1628) BN Lb36 2666 (with another edition in Aix-en-Provence).

10   *Relation véritable de tout ce qui s'est passé*, p. 9.

11   Pierre Mervault, *Journal des choses plus mémorables qui se sont passées au dernier siège de La Rochelle* (n.p., 1648), BN Lb36 2679. The edition consulted was *Histoire du dernier siège de La Rochelle ou se voient plusieurs choses remarquables qui se sont passées en iceluy* (A Rouen chez Jean Bertholin et Iacques Cailloué dans la court du Palais, 1648), p. 318.

12   *Les Remarques particulières*, pp. 2–3.

13   On the *occasionnels*, see Roger Chartier, 'Stratégies éditoriales et lectures populaires', in Henri-Jean Martin and Roger Chartier (gen. eds), *L'Histoire de l'édition française* (Promodis, Paris, 1982–), vol. 1, *Le Livre conquérant. Du Moyen Age au milieu du XVIIe siècle*, pp. 596–8 ['Publishing Strategies and What the People Read', in Roger Chartier, *The Cultural Uses of Print in Early Modern France*, tr. Lydia G. Cochrane (Princeton University Press, Princeton, 1987), pp. 145–82]; Jean-Pierre Seguin, *L'Information en France avant le périodique, 517 canards imprimés entre 1529 et 1631* (G. P. Maisonneuve et Larose, Paris, 1964). See also Jean-Pierre Seguin, 'Les occasionnels au XVIIe siècle et en particulier après l'apparition de *La Gazette*. Une source d'information pour l'histoire des mentalités et de la littérature "populaires"', in *L'Informazione in Francia nel Seicento* (Bari and Nizet, Paris, 1983), pp. 34–60.

14   *Mémoire très-particulier de la despense qui a esté faicte dans la ville de la Rochelle, Avec le prix et qualité des viandes qui ont esté excessivement venduës en ladite Ville, Depuis le commencement du mois d'Octobre, jusque à sa Reduction* (A Paris, chez Charles Hulpeau, sur le Pont S. Michel, à l'Ancre double, et en sa Boutique dans la grand'Salle du Palais, 1628) 7 pp., BN Lb36 2669 ,p. 3.

15   *Mémoire véritable du prix excessif des vivres de la Rochelle pendant le Siège. Envoyé à la Royne Mère* (A Paris, par Nicolas Callemont, demeurant ruë Quiquetonne, M.DC.XXVIII) 6 pp., BN Lb36 2670, p. 2.

16   See the presentation by Jean Hébrard in Roger Chartier's seminar at the Ecole des Hautes Etudes en Sciences Sociales, March 1985, publication forthcoming.

17   In periods of high prices and scarcity, the price of wheat fluctuated between 15 and 30 *livres* the *boisseau*, a figure that is already considerable, since the current price was under 10 *livres* the *boisseau* (1 *boisseau* = about 13 litres).

18   The first volume of the *Mercure françois* appeared in 1611, published by Jean Richer, who presented it as the sequel to the chronological summaries of Palma-Cayet. From 1624 to 1638 this annual periodical was directed by the famous Father Joseph.

19 A Paris, chez Jean Petit-Pas, ruë Sainct Iacques, à l'escu de Venise, près les Mathurins, M.DC.XXVIII; 21 pp., BN Lb36 2682.

20 *Ludovici XIII Franciae et Navarrae Regis christianissimi triumphus de Rupella capta, ab alumnis Claromontani colegii societatis Jesu vario carminum genere celebratus* (Parisiis, apud Sebastianum Cramoisy, via Iacobaea, sub ciconiis. M.DC.XXVIII), BN Lb36 2687.

21 Nicolas Caussin, *Le triomphe de la piété à la gloire des armes du Roy, et l'amiable réduction des asmes errantes* (A Paris, chez Sébastien Chappelet, M. DCXXIX) 334 pp., BN Lb36 3560.

22 P. Le Comte, *Le génie de la France au Roy. Sur l'heureux retour de Sa Majesté en Sa Ville de Paris* (A Paris, chez François Iacquin, rue des Massons, et Iulian Iacquin, au Palais, au bas des degrés de la Ste Chappelle (n.d.), BN Ye 25885.

23 *Sortable* in French, a lovely understatement, since an entry on the scale of that of 23 December represented a considerable expenditure that would have ruined some smaller cities.

24 *Au Roy sur la prise de La Rochelle, et triomphe de Paris* (A Paris, chez Louys Boulanger, ruë S. Iacques à l'image S. Louys, près S. Yves. M.DCXXVIII), 17 pp., BN Ye 19696.

25 A Paris, chez Pierre Rocolet, Imprimeur et libraire de la Maison de Ville, au Palais, en la gallerie des Libraires. M.DCXXVIII, 8 pp., BN Ye 22917.

26 Bernard Guénée and Françoise Lehoux, *Les Entrées royales françaises de 1328 à 1515* (Eds du CNRS, Paris, 1968).

27 Jean Boutier, Alain Dewerpe, Daniel Nordman, *Un Tour de France royal. Le voyage de Charles IX (1564-1566)* (Aubier, Paris, 1984), p. 295.

28 Guénée and Lehoux, *Les Entrées royales*.

29 Boutier, Dewerpe, and Dordman, *Un Tour de France royal*, p. 295.

30 Françoise Bardon, *Le Portrait mythologique à la cour de France sous Henry IV et Louis XIII. Mythologie et politique* (A. and J. Picard, Paris, 1974), pp. 20-2.

31 A Paris, chez Pierre Rocolet, Impr. et Libraire ordinaire de la Maison de Ville, en sa boutique au Palais, en la gallerie des prisonniers, M.DC.XXIX, 182 pp., BN Lb36 2711.

32 W. M. Allister-Johnson, 'Essai de critique interne des livres d'entrée français au XVIe siècle', in *Fêtes de la Renaissance* (3 vols, Eds du CNRS, Paris, 1975), vol. 3, pp. 187-200.

33 De Vaux de Foletier, *Siège de La Rochelle*, p. 300.

34 *Eloges et discours*, 'Au Roy'.

35 ibid., p. 1: 'To the Reader. You have greatly desired that the explanation of the paintings made for the Reception of His Majesty appeared immediately after the day it took place [so that] each piece, being presented in all lights simultaneously and being instantly explained, would bring continuing pleasure. One would have spared you the annoying wait that these delays bring with them, and you would have received as a boon what the delays might now persuade you to be a debt; For according to Chrysostom of the Pagans and in the common sense of the world, pleasure expected from someone takes on the nature of a debt when it is long awaited.'

36 ibid., p. 2.

37 Frances A. Yates, *The Art of Memory* (University of Chicago Press, Chicago, 1966).

38 This engraving was also sold separately. It is signed Abraham Bosse.

39 'Toutes les figures contenues en ce Livre ont esté faittes et se vendent à Paris par Melchior Tavernier Graveur et Imprimeur du Roy pour les Tailles douces, demeurant en Lisle du Palais sur le Quay à l'Epic d'Or. Pierre Firens Graveur en Tailles douces demeurant rue St. Iacques à l'enseigne de l'imprimerie en Tailles douces.'

40 Vaux de Foletier, *Le Siège de La Rochelle*, p. 240.

41 ibid., p. 15.

42 Gaston Brière, Maurice Dumolin, Paul Jarry, *Les Tableaux de l'Hôtel de Ville de Paris* (Société d'iconographie parisienne, Paris, 1937): 'At their head is the *prévôt* Christophe Sanguin, seigneur de Livry, the presiding officer of the Parlement, elected in August 1628, having near him the *procureur* Gabriel Payen, named in 1627. Behind them are the four *échevins*, Augustin Leroux, *conseiller* at the Châtelet, and Nicolas Delaistre, elected in August 1627, Etienne Heurlot and Léonard Regnard, king's *procureur* to the Treasury, elected in August 1628. In the background is the court clerk Guillaume Clément and the *receveur* Charles Le Bert, appointed in 1617.'

43 Denis Crouzet, 'La Ligue et le tyrannicide de 1589: une expérience mystique?' (forthcoming); Pietro Redondi, *Galileo Heretic*, tr. Raymond Rosenthal (Princeton University Press, Princeton, 1987), pp. 4–7.

44 Yates, *Art of Memory*, pp. 249–50, speaking of the thirty seals of Giordano Bruno.

45 *Eloges et discours*, p. 32.

46 See pp. 248–9.

47 Bardon, *Portrait mythologique*, p. 113.

48 *Eloges et discours*, p. 22: 'His Majesty is in the chariot itself, open in the modern style; the painter having judged it wise that our eyes could not have supported the orb and the closure in which were the Captains who triumphed in Rome. He is not alone seated in this seat of Honour: all the virtues are there as well, who in this picture do not want to appear in any other visage than his. And although there is combat among the others for the places and ranks that they want to have, as they do on the face of Apollo in Philostratus: none the less Clemency is in possession of the eyes, where she reigns, and takes from her sisters the better part of the Glory.'

49 See ch. 7, n. 15.

50 The notions of right and left are used here only to indicate that the motifs fall in two series, since a great ambiguity reigns in their use in older descriptions. At times they speak of the viewer's right and left (as we do spontaneously today) and at other times they follow the heraldic tradition of speaking of right and left from the viewpoint of the person bearing the colours.

51 *Eloges et discours*, p. 29, 'but all authors agree that when Hercules put down this club near a statue of Mercury, which was at Troezen, it took root immediately and threw out branches, out of which they made crowns for victors'.

52 The relation between what is symbolized and what symbolizes it is not *necessary* since each of them is gifted with an autonomy of significance. It needs to be *motivated*. The word is used here in the sense that linguists give it and bears no psychological connotation.

53 The expression *y estoit représenté*, along with what is known from other sources concerning the construction of triumphal arches, makes one think that paintings

or figurines had been placed on the arch. Still, the ambiguity of the word *représenté* is reinforced by the 'realistic' effects produced by the engraving (in the poses of the figures and the falling hat).

54 *Eloges et discours*, p. 56.

55 BN Lb36 2712.

56 *Camaïeu* should undoubtedly be understood here as a 'drawing made by a painter in which he uses only one colour and he observes the highlights and shadows that are usually represented by bas-reliefs' (Furetière).

57 For example, in the *rencontre huitième* (eighth arch) of the 23 December entry, on the decorations of the façade of the Châtelet, on the rue Saint-Jacques side (see *Eloges et discours*, 87–96).

58 ibid., pp. 24–5. 'Aethera dum triplici mitis Dea sustinet arcu, / Securos aliqua vivere parte jubet. / Tu medios, LODOICE, arcus feliciter imples, / Et cunctos domito pellis ab Orbe metus' became 'Cet heureux Arc-en-ciel que soustient la Clémence, / Semble nous asseurer, que si l'on doit jamais / Voir de tous nos malheurs cesser la violence, / Ce doit estre à l'aspect de ce signe de paix. / Mais cet arc n'estant point parfait en sa figure, / D'un repos accomply ne seroit pas l'augure, / Si ton bon-heur, GRAND ROY, n'achevoit sa rondeur, / Pour nous faire juger que le rond de la terre / Par tes armes conquis, va voir mourir la guerre / Aux pieds de ta Grandeur.'

59 These were the markers around which the racers turned.

60 *Eloges et discours*, p. 169.

61 The French version reads: 'A LOUIS TREZIESME / ROY TRES-CHRESTIEN DE FRANCE ET DE NAVARRE / INVINCIBLE IUSTE DEBONNAIRE / TRES-PUISSANT PAR SA VERTU / TRES-CLEMENT PAR SA PIETE / APRES AVOIR REDUIT LA ROCHELLE EN SON / OBEISSANCE PAR LE SIEGE D'UN AN ET PAR LE / TRAVAIL ADMIRABLE DE LA DIGUE VAINCU TROIS / FOIS LES ESTRANGERS SUR TERRE ET SUR MER / SURPASSE LA GLOIRE ET LA FELICITE DE TOUS / LES PRINCES QUI FURENT JAMAIS / EN SON RETOUR VICTORIEUX ET TRIOMPHANT / LA VILLE DE PARIS.' ('To Louis, thirteenth most Christian king of France and Navarre, invincible, just, debonnaire, most powerful, and, by his virtue, most clement through his piety, [who,] having reduced La Rochelle to his obedience by the siege of one year, and by the admirable labour of the dike, thrice vanquished the foreigners, by land and by sea, surpasses the glory and felicity of all princes who ever were, on his return, victorious and triumphant, the City of Paris.')

62 A Paris chez François Pomeray, au carrefour Ste Geneviève, à la Pomme d'Or. Et au Palais en la Gallerie des Merciers, devant le grand escalier. M.DC.XXIX, BN Lb36 2676A (the Latin text is by 'le Sieur de Sainte-Marthe l'aisné').

63 *Traduction française du Panégyrique du Roy Louis le Juste, sur le sujet de la victoire que Dieu luy a donnée sur les Anglois, en la journée de l'Isle de Ré: fait et prononcé par le Sieur de Merigon, par la libéralité du Roy Professeur et Orateur en langue grecque, le 11 et 26 du mois de novembre 1628 au collège de Harcourt* (A Paris, chez Laurens Saulnier, ruë S. Iacques, à l'enseigne du Soleil d'Or. M.DC.XXIX), 68 pp., BN LB36 2704.

64 A Paris, chez Sébastien Cramoisy, ruë S. Iacques aux Cigognes, M.DC.XXIX, 34pp., BN Lb36 2689.

65 *Sujet du feu d'artifice, sur la prise de La Rochelle que Morel doit faire pour l'arrivée du Roy*

sur la Seine, devant le Louvre. Au Roy (A Paris, chez C. Son et P. Bail, ruë Sainct Iacques à l'Escu de Basle, M.DC.XXVIII), 12 pp., BN Lb36 2713.

66   *Subject du feu d'artifice fait à l'entrée du Roy dans sa ville de Paris. Ensemble le ballet représenté sur la rivière de Seine, devant le Louvre. Au Roy* (A Nantes, par Hilaire Mauclerc, Imprimeur et Libraire, jouxte la coppie imprimée à Paris. M.DC.XXIX), 7 pp., BN Lb36 2714.

67   See above, pp. 246–9.

68   A Paris, chez Jean Guillemot, demeurant ruë Sainct Jean de Beauvais, à l'enseigne de l'Eschiquier, M.DC.XXVIII, 15 pages, BN Lb36 2715.

69   *Prédictions tirées de l'octonaire ou nombre de huit* (Paris, 1629), BN Lb36 3561. The text follows that of the *Mystères de l'octonaire ou conjectures tirées tant de l'Ecriture sainte que des mathématiques, et appuyées sur des raisons naturelles, qui montre évidemment qu'en cette année 1628, pleine de bonheur, le mystère d'iniquité sera exilé, les rebelles rochelais domptés, et les autres hérétiques factieux subjuguées par les armes de notre grand Alcide Louis le Juste*, BN Lb36 2617.

70   Claude François Ménestrier, *La Philosophie des images énigmatiques, ou il est traité des énigmes, hiéroglyphiques, oracles, prophéties, sorts, divinations, loteries, talismans, songes, Centuries de Nostradamus, de la baguette* (H. Baritel, Lyons, 1694).

71   *Dédiés et présentés à Sa Majesté* (Paris, M.DC.XXIX), 12 pp. BN Lb36 2734.

72   *Au Roy* (n.p., n.d.), BN Lb 3556.

73   A Paris, chez Nicolas Touzart, ruë S. Iacques, au Trois Faucilles, M.DC.XXIX, 13pp., BN Lb36 2708.

74   *Traduite de l'Italien en François par le Sieur de Marcilly Dijonnois* (A Paris, chez Jean Martin, au bout du Pont Sainct-Michel, près le chasteau S. Ange [sic], M.DC.XXIX), 14pp., BN Lb36 2673.

75   There is also a *Relatione di quanto è seguito nella resa della Rocella, tanto avanti, che dopo, che il Rè vi habbia fatta la sua entrata il giorno di tutti i Santi . . .* (In Roma, et in Firenze, appresso Pietro Cecconcelli, M.DC.XXVIII), 8pp., BN Lb36 3553.

76   *Ensemble la description des Tableaux et magnificences dressés pour icelle. Par I.S.T.* (A Paris, Chez Iacques Dugast, ruë de la Harpe, à l'enseigne de la Limace, près la Roze rouge, M.DC.XXIX), 14pp., BN Lb36 2723. The same printer published *Les arcs triomphaux érigez à l'honneur du Roy dans sa ville de Dijon. Où ont esté représentées la Deffaite des Anglois, et la réduction de La Rochelle*, 16pp., BN Lb36 2725.

77   *Mercure françois*, vol. 15 (1628–9), pp. 32–110.

78   *La Joyeuse entrée du Roy en sa ville de Troyes, capitale de la Province de Champange. Le Jeudy vingt-cinquiesme jour de janvier 1629* (A Troyes, de l'imprimerie de Jean Jacquard, ruë de la Cordelerie, près le Jeu de Paume, M.DC.XXIX), 50pp., BN Lb36 2722 (reprinted in the *Mercure françois*, vol. 15, pp. 32–60), p. 47 in the Troyes edition.

79   ibid., p. 44.

80   *Avec la harangue à elle faite par le Sieur de Lescale, Lieutenant criminel et Juge de la Police de ladite Ville*, 12pp., BN Lb36 2910.

81   *Relation de ce qui s'est passé à l'Entrée de la Reyne en la Ville de La Rochelle. Au mois de november 1632* (A La Rochelle, pour Mathurin Charruyer, Marchand-Libraire, 1632), 92p., BN Lb36 2911. According to Louis-Etienne Arcère, *Histoire de la ville de La Rochelle et du pays d'Aulnis* (2 vols, La Rochelle and Paris, 1756; reprint edition, Lafitte, Marseilles, distr. H. Champion, Paris, 1975), vol. 2, p. 376, this text also had a Paris edition.

82 See (among others) Georges Mongrédien, *La Journée des Dupes, 10 novembre 1630* (Gallimard, Paris, 1961), pp. 129–48.

83 The *Relation de ce qui s'est passé* opens with a dedicatory letter to Richelieu and contains several allusions to the cardinal's generosity, such as the one on p. 89: 'For whatever is to be found in it [that is] great, superb and magnificent belongs entirely to My Lord the Cardinal, who accepted dressing this ancient amphitheatre at his own expense to make it to some extent worthy of the activities of Her Majesty.'

84 Arcère, *Histoire de la ville de La Rochelle*, p. 376; Delayant, *Bibliographie rochelaise*, p. 293.

85 *Relation de ce qui s'est passé*, p. 22. The third arch was also decorated with macabre drawings, such as a 'skeleton wrapped in its burial clothes, which were falling from the top down in broken and rotten tatters along his arms to his hands' (p. 55). This particular arch was dedicated to the queen and 'her adorable beauties'.

86 Claude François Ménestrier, *L'Art des emblèmes où s'enseigne la morale par les figures de la fable, de l'histoire et de la nature. Ouvrage rempli de près de cins cents figures* (Lyons, 1662; Paris, 1674), p. 157.

Index

Bohemia 191–229; aristocracy in 194, 195, 198; courts in 198–9, 200, 207, 212, 221; Eastern 206, 207; government in 193–4, 198; Patent of Toleration in 195, 196, 198, 216; population movement in 206, 207, 209; Protestantism in 193, 194, 195; schools in 213

Boileau, Nicolas 96, 108

Boissard, Jean-Jacques 278

Bon, F., father 317–18

Bonaventure, frère 145

Bonaventure, St 15–16, 37

Bonfans, widow of Jean 85, 87, 88

Bonhomme, Macé 274

Boniface VIII, pope 31

book, the: attitude towards 17–19, 156; iconography of 16; as object 139, 220; size of 16, 18; uses of 15, 17–18, 215–16

books 2, 3; for Catholic acculturation 199, 201–4, 208; circulation and distribution of 207, 208–10, 211, 212; editions of, studied 4–5, 97–8; 'heretical' 198, 199–203, 204, 208, 209; and heterodoxy 191–229; of hours 2, 141–73; prices for 209–10; prohibited 3, 39

Bordeaux 322

Bornitz, Jakob 266

Boscard, Jacques 59, 61; widow of 59, 61, 73

Bosnia 33

Bosse, Abraham 303

Bouhours, Dominique, father 280–1

Boulle, André Charles 285

Bourbon dynasty 242, 246, 306

Bourges 178, 264

Bourut, Claude 53

Boutard, Blaise 87

Boutier, Jean 297

Bozon, Nicole 267

Brabaneč, Josef 212, 221

Braccio da Montone (Fortebracci da Montone, Andrea) 30, 36, 37

Brémond, Claude 102, 131

Brépols 122

Bresse 70

Brest 122

Brethren *see* Czech Brethren

breviary, the 18–19, 21, 60, 147, 155

Briden, Claude 87

Brignoles 31

Brittany 127

broadsheets/broadsides 6, 121, 132, 176, 208, 218, 235, 250, 290; *see also* placards

brochures 290; *see also* pamphlets

'Brothers of Paradise', the 206

Bruck, Jacobus à 266

Brun, François 31

Burgundy 13, 23

Cabinet des fées 101, 102, 103, 105, 106

Cacace, Giovanni Battista 265

Čačák, Pavel 210, 214, 218, 219–20, 221

Cacchi, Bernardino 29

Cacchio (or Cacchi), Giuseppe 24–31, 34–8, 45

Cadiou, Jean-Baptiste, father 42–4, 45, 46, 47, 48, 49, 53

Caen 106, 116, 122

Caillot 106

Cain 185

Caldora, Jacopo 35

calendar, liturgical 21, 35–6, 40; *see also* saints' days

Calvin, John 145, 151; Calvinism 194, 196, 203, 205–6

Camerarius, Joachim 265

Campanella, Tommaso 270

Camponeschi family 30, 36, 37; Pietro Lalle Camponeschi, count of Montorio 37

canard 5, 59, 88; *see also* occasionnel

Cantelmo family: Béranger, Giovanni, Jacques, Restaino 37

cantor 191, 197, 210, 212, 213

Canzone alla Siciliana… 30

Capetian dynasty 32, 245

Capponi, Orazio, cardinal 25

cards, pedagogical 277

Carnival themes 4, 255, 256, 323, 324

Carpentras 106, 125

object, transitional 266
occasionnel 4, 5, 8, 13, 59–70, 71, 74, 82,
    83–9, 293; *see also*
    broadsheets/broadsides; *canard*;
    pamphlets
Ockham, William of 267
Olibrius 40
Olivi, Peter John 32
Olivier, frère *see* Maillard, Olivier
oral culture/tradition 7–8, 81, 82–3, 95,
    96, 97, 126–32, 219–20; *see also* folk
    tradition; tales, folk tradition in
oratione/orazione 24–5, 26; *see also* prayer
Oratione devotissime del glorioso santo Alvise
    24–7, 28, 30, 31, 34, 37, 39, 45
Orléans 105, 106, 113
Orléans, Gaston-Jean-Baptiste, duc d'
    303, 304
orphans 177, 180
Orta, Johannes de 38
Osnabrück 47, 48, 51
ottava rima 26, 30
Oudot family 115; Jacques 109; Jean
    104, 105, 110, 114; widow of Jean
    105, 110; Nicolas 42, 44, 49; widow of
    Nicolas 49
Oursel, Jean 114
Ovid 146, 248

pagina, sacra 16, 17
paintings 16, 175, 303, 306, 307, 308,
    309, 310
Palatin, Martine 174, 175
pamphlets 51, 218, 190, 291–2, 309; *see
    also occasionnel*
Paolino of Venice 39
Parallèle 96
Pareto, Vilfredo 267
Paris 31, 38, 42, 66–7, 84, 285, 303, 309,
    312, 318; archbishop of 47, 291;
    celebrations in, for fall of La Rochelle
    290, 292–3, 295; entry of Louis XIII
    into 298, 299–318, 320; Notre-Dame
    de 66, 68, 80, 84; Parlement of 237,
    291, 295, 325; *placards* posted in 235,
    237, 250, 256; as place of publication

54, 60, 61, 70, 85, 87, 94, 95, 104, 106,
    117, 264, 274, 276, 277, 278, 318, 322,
    325
Parlement of Dijon 46, 52; of Grenoble
    46; of Paris 237, 291, 295, 325; of
    Rennes 82
Parthenay, Catherine de 269
Pasquino, pasquinade 235
Passion of Perpetua and Felicitas 20
passions, theory of the 279
Pásztor, Edith 32
Patent of Toleration 195, 196, 198,
    216
Paul IV, pope 37
Paul V, pope 49
Paul, St 174, 183, 184, 186, 187, 189,
    214
Paulinus of Nola 312
Paulmy, Antoine-René de Voyer
    d'Argenson, marquis de 101, 130
pedlars 41, 54, 62, 126–7, 198, 200, 204,
    209, 236, 294
Pekář, Jan 218, 221
Pellerin 107, 117, 121, 124
Penne, Gaudibert 125
Périsse brothers 119
Perpetua, St 20
Perrault, Charles, *Contes* attribution of
    92–3, 94, 95, 97, 98, 102, 103, 104,
    106; in *Bibliothèque bleue* 100, 101,
    103, 104–7; classical taste in 100,
    111–12; didactic purpose of 100–1,
    102, 103, 104, 112, 118–21, 125, 130;
    editions of 94–6, 98, 104–14, 116, 121,
    124; *Epitre à Mademoiselle* 94, 113,
    124, 131; *Lettre à Monsieur . . .* 108;
    illustrations for 95, 97, 100–1, 102,
    106, 110, 111, 112, 113, 114, 115, 117,
    120, 121, 122, 124; narrative
    technique in, 98 99, 102, 115–17,
    130–1; oral or folk origins of 95–6,
    103, 104, 119–21, 122–3, 125, 130;
    oral reading of 97, 103, 111, 128, 129,
    130; pirated editions of 95, 97, 98,
    105; *Préface* 94, 96; public for 92,
    93–4, 96–7, 99, 100, 101–2, 103, 129,
    130; typographic procedures in 97,

Index by Lydia Cochrane

This collective illustrated work offers a highly original account of the cultural transformation brought about by the discovery and development of printing in Europe. After Gutenberg, all European culture was a *culture of print* in which the printed word penetrated the entire web of social relations, touching people's deepest selves as well as claiming its place in the public sphere.

In their study of this cultural form, the authors have been guided by three concerns. Firstly, they have focused primarily on printed matter other than books, such as broadsheets, flysheets and posters. Secondly, they have adopted a case-study approach, examining particular texts or printed objects concerning specific events. Thirdly, they have tried to understand the use of these materials by placing them within the local, specific contexts which gave them meaning.

The authors emphasize the multiplicity of ways in which printed materials were used in early modern Europe. Festive, ritual, cultic, civic and pedagogic uses were social activities and involved deciphering texts in a collective way, with those who knew how to read leading those who did not. Only gradually did these collective forms of appropriation give way to a practice of reading – privately, silently, using the eyes alone – which has become common today.

This wide-ranging work opens up new historical and methodological perspectives on one of the most important transformations in western culture. It is a collective work by a group of leading historians, including Roger Chartier, Alain Boureau, Marie-Elisabeth Ducreux, Christian Jouhaud, Paul Saenger and Catherine Velay-Vallantin, and it will become a focal point of debate for historians and sociologists interested in the cultural transformations which accompanied the rise of modern societies.